Chemistry of Peptides and Proteins
Volume 3

Chemistry of Peptides and Proteins

Volume 3

Proceedings of the
Fifth USSR-FRG Symposium on
Chemistry of Peptides and Proteins
Odessa, USSR, May 16–20, 1985

Editors
Wolfgang Voelter · Ernst Bayer
Yuri A. Ovchinnikov · Vadim T. Ivanov

W
DE
G

Walter de Gruyter · Berlin · New York 1986

Editors
Wolfgang Voelter, Professor, Dr. rer. nat.
Head
Department of Physical Biochemistry
Institute for Physiological Chemistry
Hoppe-Seyler-Straße 1
D-7400 Tübingen

Ernst Bayer, Professor, Dr. rer. nat.
Director
Institute for Organic Chemistry
Auf der Morgenstelle 18
D-7400 Tübingen

Yuri A. Ovchinnikov, Professor, D. Sc.
Director
Shemyakin Institute of Bioorganic Chemistry
USSR Academy of Sciences
Moscow, USSR

Vadim T. Ivanov, Professor, D. Sc.
Shemyakin Institute of Bioorganic Chemistry
USSR Academy of Sciences
Moscow, USSR

CIP-Kurztitelaufnahme der Deutschen Bibliothek

Chemistry of peptides and proteins : proceedings of the ... USSR FRG Symposium on Chemistry of Peptides and Proteins. – Berlin ; New York : de Gruyter
In d. Vorlage auch: ... USSR FRG Symposium on the Chemistry of Peptides and Proteins
ISSN 0723-6271

NE: USSR FRG Symposium on the Chemistry of Peptides and Proteins

Vol. 3 = 5. Odessa, USSR, May 16 – 20, 1985. – 1986

ISBN 3-11-010613-2 (Berlin)
ISBN 0-89925-193-5 (New York)

 Printing: Gerike GmbH, Berlin. – Binding: Lüderitz & Bauer GmbH, Berlin. – Printed in Germany.

PREFACE

The fifth USSR-FRG Symposium on the Chemistry of Peptides and Proteins took place in Odessa in May 1985. This bilateral series of symposia began in 1976 in Dushanbe and since then has not lost any actuality. On the contrary, Peptide Chemistry and Biochemistry have grown considerably and entered many new fields such as for example Immunology and Biotechnology. This growing interest in peptides requires a solid methodological basis in Peptide Chemistry. Therefore analytical methods and peptide syntheses have played a fundamental role in previous symposia and also at the Odessa meeting. However, increasingly, parts of the symposia have been devoted to biological problems, and, for the first time, immunological topics were discussed in Odessa.

Again, very fruitful discussions and exchange of knowledge long before publication contributed to the success of the Odessa symposium. The excellent local organisation of the symposium in Odessa has to be gratefully acknowledged as well as the support of the USSR Academy of Sciences and the Deutsche Forschungsgemeinschaft.

For the Editors

E. Bayer

CONTENTS

PEPTIDE SYNTHESIS AND ANALYTICAL PROBLEMS OF SYNTHETIC PEPTIDES

STRUCTURAL FEATURES OF PEPTIDES AND PROTEINS

Peptide Synthesis and Analytical Problems of Synthetic Peptides

NEW POLYMER SUPPORTS FOR SOLID-LIQUID-PHASE PEPTIDE SYNTHESIS

E. Bayer, W. Rapp
Institute for Organic Chemistry, University of Tübingen,
Auf der Morgenstelle 18, 7400 Tübingen / FRG

The increasing importance of peptide synthesis in biochemistry requires reliable peptide synthesis and often a large number of different peptides have to be prepared in a reasonable short time. In principle the Liquid Phase Peptide Synthesis (1) on polyoxyethylene in homogeneous solution fulfills these requirements of a uniform reaction with high reaction rates. However, the Liquid Phase method is difficult to standardize and to be used in automatized synthesizers. Therefore the Solid Phase Method (2) is mainly utilized because it can be easier standardized. However, besides the general problem of failure sequences (3,4) the polystyrene beads show other disadvantages which require improvement. Due to the high compressibility of the polymer beads relatively large diameters (60-200 μ) are used. However it is known that the mass transfer of liquids in microporous beads is dependant upon the bead diameter, and assumes a Gaussian type distribution. Therefore a reduction of particle size should result in a considerable improvement of the rate limiting mass transfer, and consequently in a decrease of the time required for all synthesis steps, like coupling, washing procedures and cleavage.

One approach to unify the advantages of Solid and Liquid Phase Peptide Synthesis would be to immobilze polyoxyethylene (POE) to a polymer matrix, e.g. cross-linked polystyrene. It could be expected that starting from a certain length of the POE chain solvation of such polystyrene-polyoxyethylene (PS-POE) craft copolymers will be dominated by the POE, and mobility of the crafted POE chain will be comparable to soluble POE.

Chemistry of Peptides and Proteins, Vol. 3

Relatively small POE chains have been immobilized to chloromethylated polystyrene by ether synthesis (5-8). However such relatively short POE chains up to 500 dalton do not yet render the polymer the properties of POE. Our attempts to synthesize PS-POE with POE molecular size of 500-10.000 dalton were not very successfull because with increasing chain length of POE the yields of ether synthesis dramatically decrease, and a considerable part of the POE reacts with both terminal OH-groups forming cyclic ethers (5,9). Therefore direct polymerization of ethylene oxide to suitably functionalized polystyrene beads was developped Following scheme 1, tetraethylene glycol was crafted to chloromethylated cross-linked polystyrene beads (1) by ether synthesis. After preparation of the potassium salt of the polymer (2) ethylene oxide is polymerized to (3).

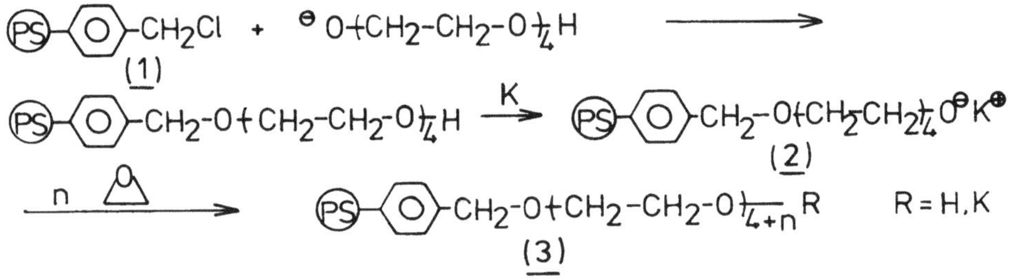

Scheme 1

Since a relatively uniform molecular weight distribution of the crafted POE is desirable, anionic polymerisation was chosen because of formation of living polymers, which yield narrow molecular weight distribution. Depending upon polymerization conditions like monomer concentration, temperature and reaction time PS-POE polymer beads with different average chain lengths of POE between 500 to 10.000 dalton were obtained.

Solvation and mobility of the POE crafted to the cross-linked polystyrene matrix depend on the molecular weight, and can be investigated by ^{13}C-NMR-relaxation time T_1 measurements. In general higher T_1-values indicate an increased mobility. The T_1 values of soluble not immobilized POE decreases with increasing molecular weight. However the T_1 values of immobilized POE on cross-linked polystyrene shows a completely different behavior (Fig. 1). The T_1 values first increase until they reach a maximum at about 2.000 - 3.000 dalton. The slight decrease at molecular weights above 4.000 dalton follows the characteristics of soluble POE. These results indicate that there is a strong influence of the cross-linked polystyrene matrix on the mobility of the POE at lower molecular weights up to 1.500 to 2.000 dalton. The maximum of T_1 depends also on the degree of cross-linking, the degree of POE functionalization and the solvent. In solvents, in which not immobilized POE is insoluble, like ether, small T_1 values are observed, whereas in good solvents for POE high T_1 values are measured. The excellent solvation of the POE part can also be seen from swelling properties. Polystyrene beads do not swell in water, and show a very wide variation of swelling volumes in the different solvents used during Solid Phase Synthesis. This is one reason that peptide synthesis using the conventional Merrifield resins cannot easily be performed in column operation. The PS-POE polymer beads are swelling in different solvents relatively equal. Swelling volume ratios in solvents for POE, like methanol, methylene chloride, DMF are almost equal around 3. On the other hand they do not swell very much in solvents in which POE is insoluble, like ether and hexane. The swelling of PS-POE is exclusively dominated by the POE chains, and no more by the polystyrene matrix. The swollen craft polymers can be used in columns for peptide synthesis, because they are surprisingly pressure stable. Such columns can be operated even at pressures up to 5.000 psi with appreciable flow rates. This is unique for microporous polymer beads, and allows a reduction of bead size.

Mobility of the long POE spacer together with their hydrophilic character and excellent physical properties make these new polymers especially suitable as support for peptide synthesis. Investigations of the interstitial space volume in the polymer beads reveal that the crafted POE spacer molecules occupy only 10-30 % of the available volume, depending upon cross-linking, POE length and solvent. This gives plenty of available volume for growing peptide chains.

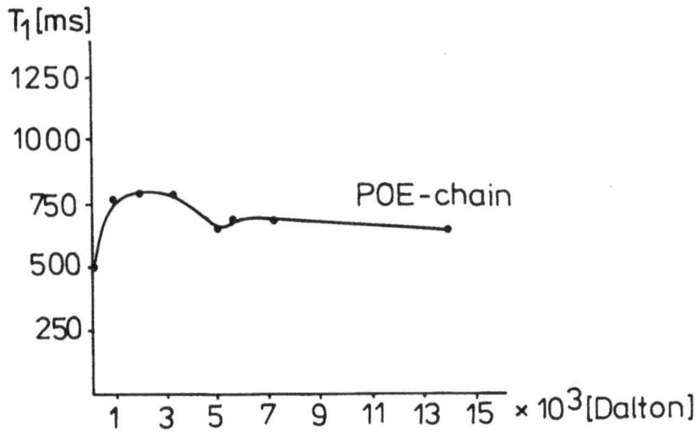

Figure 1.
^{13}C-Spin-lattice relaxation time T_1 of the POE-chain in dependance from molecular weight.

For peptide synthesis a PS-POE craft copolymer prepared according to scheme 1 from polystyrene, cross-linked with 2 % divinylbenzene and an average chain length of POE of 3.300 is recommended. Synthesis cycles correspond to the usual procedures in the Solid Phase method.

Of great interest is the reaction rate of peptide coupling, especially also in comparison to these in Liquid Phase, using POE in homogeneous solution. For these investigations glycine was coupled to the terminal OH-group of PS-POE, and the coupling rate of BOC-Gly-ONP to this polymer followed by UV-spectroscopy. In Table 1 the reaction rates of peptide coupling on the PS-immobilized POE are compared with these on POE in solution. The rate constants on PS-POE polymers are in the region of 0.36-0.32 mol^{-1} $sec^{-1}x10^{-1}$, approximately twice as fast as in the case of soluble POE. In both cases there is no pronounced decrease with increasing molecular weight of the POE above 2.000 dalton. The higher reaction rates on immobilized POE indicate a less pronounced "Knäuel" conformation. These results confirm that the terminal groups on long immobilized POE chains are at least equally accessible for chemical reactions as on POE in solution, and that these new materials combine the advantages of Liquid and Solid Phase Peptide Synthesis.

Table 1.
Comparison of Reaction Rates of Active Ester Peptide Coupling on Immobilized Polyoxyethylene (PS-POE) and Soluble Polyoxyethylene (POE). Methods as described in reference (10).

Polymer (molecular weight of POE)	Rate constant ($mol^{-1}sec^{-1}x10^{-1}$)	Glycine functionalization of polymers (mmol Gly/g polymer)
PS-POE-Gly·HCl (2.000 dalton)	0.37	0.20
PS-POE-Gly·HCl (3.300 dalton)	0.37	0.22
PS-POE-Gly·HCl (5.700 dalton)	0.32	0.12
POE-Gly·HCl (2.000 dalton)	0.16	0.49
POE-Gly·HCl (4.000 dalton)	0.17	0.38
POE-Gly·HCl (6.000 dalton)	0.13	0.27
Gly-OEt·HCl	0.12	-

References

1. Bayer, E., M. Mutter. 1972. Nature (London) 237, 512.
2. Merrifield, R.B. 1963. J. Amer. Chem. Soc. 85, 2149.
3. Bayer, E., H. Eckstein, K. Hägele, W. König, W. Brüning, H. Hagenmaier, W. Parr. 1970. J. Amer. Chem. Soc. 92, 1736.
4. Bayer, E., P. Hunziker, M. Mutter, R. Sievers, R. Uhmann. 1972. J. Amer. Chem. Soc. 94, 265.

5. Doll, W. 1981. Dissertation, University of Tübingen.

6. Warshawsky, A., R. Kaliv, A. Doshe, H. Berkovitz, A. Patchornik. 1979. J. Amer. Chem. Soc. 101, 4249.

7. Beinert, G., P. Chaumont, P. Rempp, J. Herz. 1984. J. Eur. Polym. J. 20, 837.

8. Becker, H., H.-W. Lucas, J. Maul, V.N.R. Pillai, H. Anzinger, M. Mutter. 1982. Makromol. Chem. Rapid Commun. 3, 217.

9. Burgfels, G. 1985. Dissertation, University of Tübingen.

10. Bayer, E., M. Mutter, R. Uhmann, J. Polster, H. Mauser. 1974. J. Amer. Chem. Soc. 96, 7333.

NEW SUPPORTS FOR SOLID-PHASE PEPTIDE SYNTHESIS

R.P. Evstigneeva, E.I. Filippovich, E.N. Zvonkova
Lomonosov Institute of Fine Chemical Technology, Moscow, USSR

Introduction

Modern peptide chemistry possesses two principal methods of higher peptide synthesis: these are the classical method and the solid-phase technique. It is not an easy matter to give preference to either of the methods, as both have their merits and demerits. Yet it should be noted that the majority of successful syntheses of biologically active peptides had been achieved by solid-phase technique. The methodology of solid-phase technique is now fast progressing, and the method has acquired plenty of approaches and know-hows, aimed at circumventing many of its pitfalls.
However in spite of this obvious progress, the synthesis of every new peptide is always a problem to be treated individually; in its course a number of questions does arise, both of general and of specific nature.
The success of the method clearly depends on the nature of the polymer support, tactics of peptide chain building and the procedures employed for truncated peptide removal.
In our laboratory studies were carried out of new polymer supports for peptide synthesis. Two kinds of these were prepared, one based on Teflon-polystyrene graft copolymer and the other - on industrial rubber. On the Teflon support the peptide fragment 57-100 of human hemoglobin β-chain was synthesized.

Results

One of the new supports which is proposed for solid phase synthesis is a macroporous Teflon-polystyrene graft copolymer prepared by γ-irradiation technique. This support markedly surpasses the matrices based on cross-linked polystyrene and is applicable for

Chemistry of Peptides and Proteins, Vol. 3

laboratory syntheses as well as for large-scale peptide production (1). The Teflon matrix that we prepared had polystyrene graft chains containing bromomethyl anchoring groups which are very active in attaching small peptide fragments. Bromomethylated polymer was obtained from the chloromethylated resin by substitution of acetoxy group for chlorine atom using 6-fold excess of cesium acetate. The ester bond was subsequently cleaved with hydrogen bromide in acetic acid.

The second new carrier we used for peptide synthesis was chloromethylated rubber - a copolymer of butadiene and methylstyrene (2). Now chloromethylation is quite a conventional method of solid phase modification, and consequently it is elaborately investigated; synthetic rubbers are however an exception, and there are no data on their chloromethylation.

Therefore we studied chloromethylation of methylstyrene-butadiene polymers which are widely used industrial rubbers.

Chloromethylation was carried out using chlorodimethyl ether in the presence of Lewis catalysts such as $SnCl_4$. Under these conditions the polymers could also undergo Prince addition at the double bond as well as cyclization and cross-linking (Fig. 1).

Since modified rubbers containing 5-6% clorine are still soluble in chlorinated hydrocarbons, we used solutions of our copolymers for elucidation of their structure by ^{1}H-NMR spectroscopy (3). Comparison of ^{1}H-NMR spectra of model compounds - polystyrene and polybutadiene - with the spectrum of copolymer I (n=30; m=70) and then with the spectra of chloromethylated products showed that all the three reactions proceeded simultaneously.

In the spectra of modified polystyrene and modified rubber there appear proton signals with chemical shifts of 4.40 and 4.46 ppm, which should apparently be assigned to $ArCH_2Cl$; in the spectra of modified polybutadiene and modified rubber there are signals with shifts of 3.25 and 3.24 ppm which must be due to $-CH_2-O-CH_3$ and shoulders of 2.05 and 1.99 ppm indicative of cyclization and cross-linking.

Comparative studies on model compounds made it possible to make tentative evaluation of the contribution of each reaction to the structure of the modified polymer.

By variation of modification conditions and chlorine content the polymer can be prepared in soluble form or alternately as a strong-

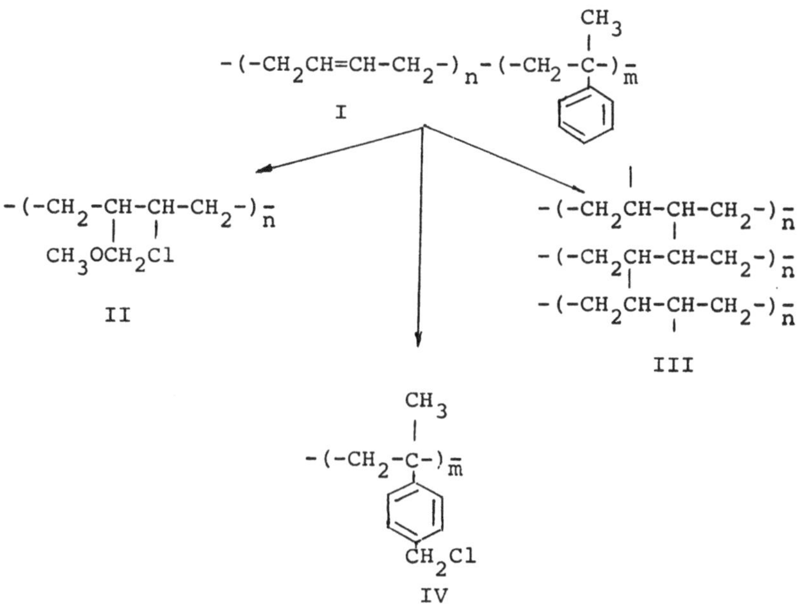

Fig. 1. Chloromethylation of methylstyrene rubbers .

ly swelling gel and as hard, practically unswelling beads. Comparing the conditions of amino acid anchoring for Merrifield resin and for our rubbers we have found that our rubbers exhibited the same relationship as did the Merrifield support. Thus an excess of reagents led to higher yields of polymer-supported amino acids. What regarded the salts of N-substituted amino acids - the best results were obtained with cesium salts (4).

While working out the conditions for anchoring the first amino acid we were confronted with the fact that N-substituted amino acid salts were quite soluble in polar solvents such as DMF, whereas solvated chloromethylated rubbers collapsed in DMF. However by addition of 8-10% (v/v) of polar solvents to the solutions of the polymer in $CHCl_3$ or CH_2Cl_2 we succeeded in carrying the reaction in solution. When the modified polymers were used as swollen gels or unswelling beads, these characteristics still kept true.

We began constructing our peptide chain on a chloromethylated rubber taken as irregular beads and containing 10-12% of chlorine. This polymer bound 0.10-0.12 mmol/g of the anchoring amino acid introduced as cesium salt; it swelled but poorly in routine Merrifield solvents and was readily filtered off. To estimate the possibilities of this support and to compare it and the Teflon-polystyrene copolymer with the Merrified resin we synthesized a model tripeptide Boc-L-Val-L-Leu-Gly-O-Ⓟ, corresponding to the sequence 67-69 of the human hemoglobin β-chain (5). The comparative studies did not reveal any fundamental differences between these three supports. The differences lay mainly in the amounts of solvents requied for the synthesis, as due to different swelling properties of the polymers.

The results of the syntheses were checked by ninhydrin tests and quantitative amino acid analysis of the hydrolyzates of aminoacyl and peptidyl polymers (138°C, 2 hrs, HCl conc.: CH_3COOH 1:1). The analysis was performed on Biotronik LC 7000 instrument (FRG).

Macroporous graft polymer of Teflon and polystyrene bearing bromomethyl anchoring groups was further used for synthesis of 57-100 fragment of human hemoglobin βchain (Fig. 2).

The peptide chain on the polymer was built by fragment condensation of comparatively short peptides which had been prepared beforehand by conventional methods and added one by one with high yields.

The scheme of the synthesis was like this: the polypeptide 57-100 was divided in 16 short peptides. We chose the fragmentation pattern in such a way as to have at the C-terminus the amino acids less prone to racemization, or ended the peptide with glycine.

To obtain optimal loads of peptide on the support and to consequently prevent side reactions, characteristic of the first stages of the solid-phase synthesis, we anchored to the support our C-terminal tripeptide, instead of anchoring the amino acids one by one.

C-terminal tripeptide was coupled to the resin in the presence of DCA; the peptide load on the polymer amounted to 0.15 mmole per gram of polymer. To avoid side reactions at the unsubstituted imidazole cycle of histidine, the residual halogen in peptidyl-polymer was substituted with 6-fold excess of cesium acetate in DMF.

To increase the yield at the stage of N-deprotection peptidyl-polymers were pre-treated with 1.5 M urea in DMF and with various acidic reagents.

To define optimal conditions for fragment condensation of peptides

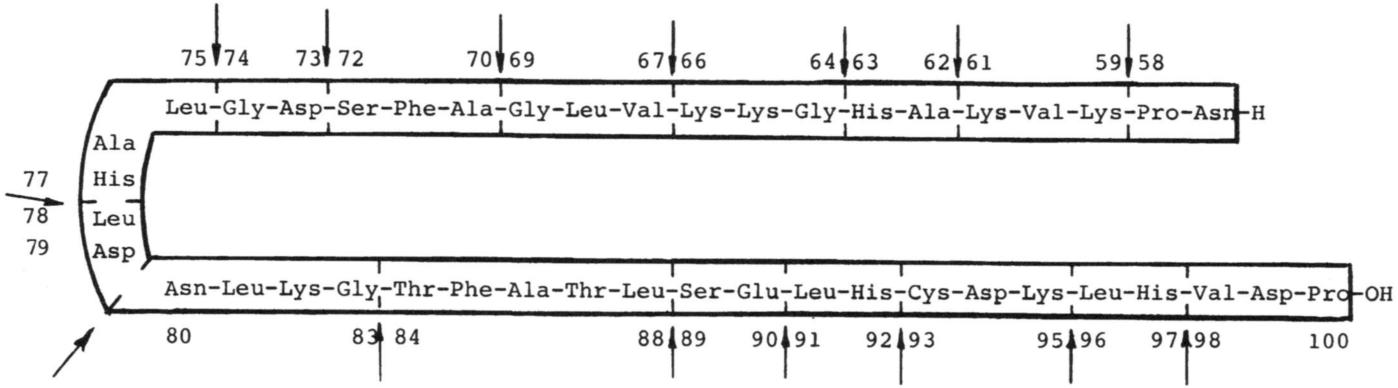

Fig. 2. 57-100 Fragment of human hemoglobin β-chain.

Table I. Comparison of different methods of fragment condensation.

Amino-component	Acylating component	Coupling method	Yields after coupling, %
Boc-Val-Asp(OBzl)-Pro-Ⓟ (98-100); BOC-Cys(Bzl)-...-Pro-Ⓟ (93-100)	BOC-Leu-His-N_2H_3 (96-97),(91-92)	$-N_3$	93-99
BOC-Leu-...-Pro-Ⓟ (96-100)	BOC-Cys(Bzl)-Asp(OBzl)--Lys(Tfa)-OH (93-95)	complex "F" $Ph_3P/(PyS)_2$	99 20-25
BOC-Leu-...-Pro-Ⓟ (96-100)	BOC-Ser(Bzl)-Glu(OBzl)-OPcp	-OPcp	35
	BOC-Ser(Bzl)-Glu(OBzl)-OH (89-90)	complex "F"	99
BOC-Ala-...-Pro-Ⓟ (70-100)	BOC-Val-Leu-Gly-OH	complex "F"	95
	BOC-Val-Leu-Gly-N_2H_3	$-N_3$	70

we have performed a "test synthesis" of our peptidyl-polymer 57-100, that is a pilot experiment using small quantities of peptide with detailed investigation of each reaction step. The yields were determined at each stage, the optical purity of peptides was monitored, different synthetic procedures were assayed (Table I).

The azide procedure was found to be applicable for introduction of His-containing dipeptide and pentapeptide 84-88.

The method of activated pentachlorophenyl esters and oxidative-reductive procedure using triphenylphosphine and dipyridyl disulfide did not produce the results that were hoped for (the yields of the products amounted to 25-30% only). Therefore these methods were not employed any more.

Condensation with the F-complex (DCC complex with pentafluorophenol) every time gave excellent yields (95-99%). GLC showed that condensation with the F-complex practically did not induce racemization.

On the basis of methodological findings made in the course of the "test synthesis" we made a scheme of a preparative synthesis for peptide 57-100 (Fig. 3). The synthesis was performed starting with 15 g of peptidyl-polymer 98-100.

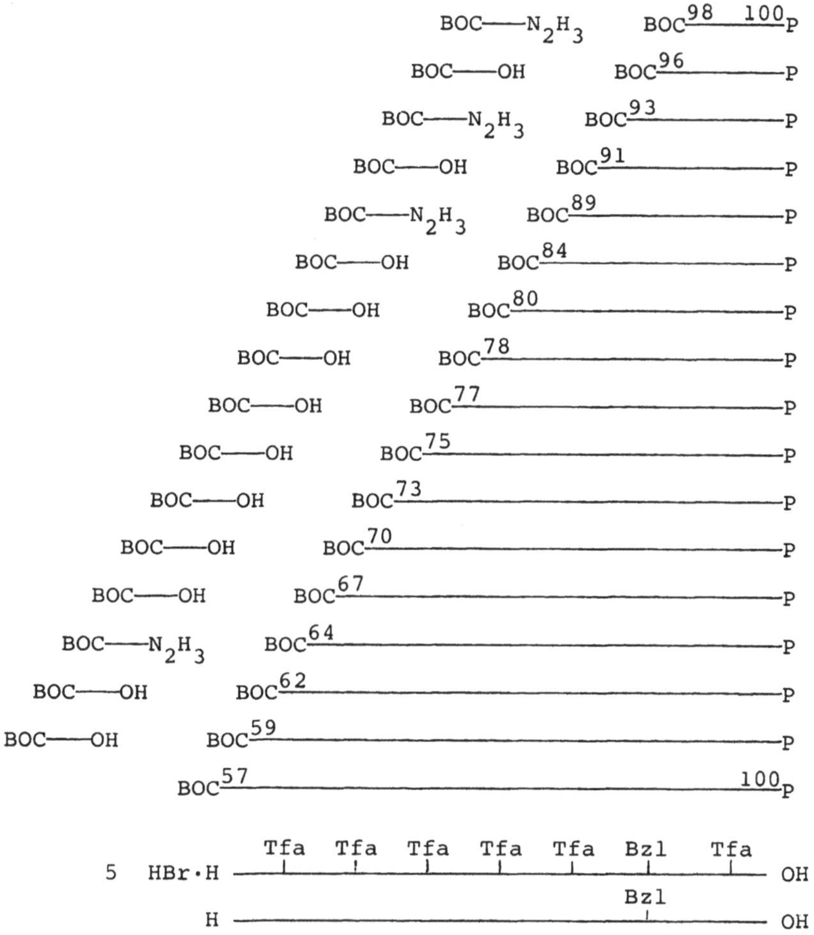

Fig. 3. Synthetic route to polypeptide (57-100).

The preparative synthesis was well as the pilot synthesis was checked by amino acid analysis of peptidyl-polymers and determination of unreacted NH_2-groups using ninhydrin reaction and dansylation.

The yields at the separate stages of the preparative synthesis were practically the same as in the pilot experiment.

With the aim of simplifying the purification of the end-product we introduced a new blocking agent - pentafluorophenyl p-nitrobenzoate

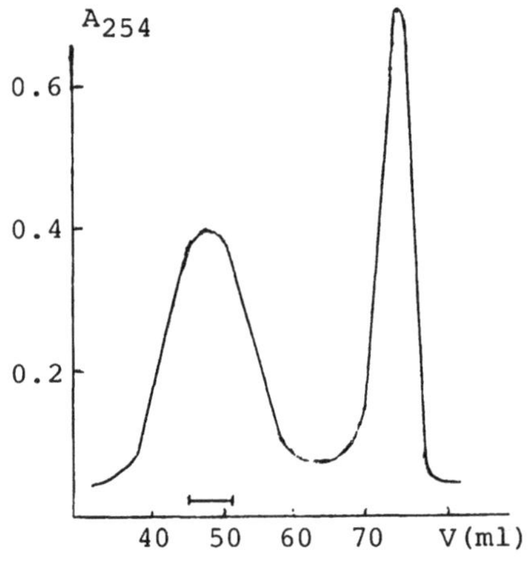

Gel filtration of N-trifluoroacetylated polypeptide (57-100) on Sephadex G-25 (sf) in 3% AcOH.

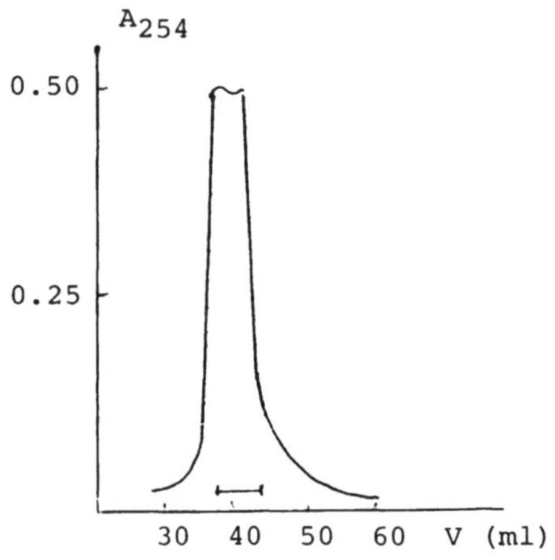

Partition chromatography of N-trifluoroacetylated polypeptide (57-100) on Sephadex G-25 (f) in BAW (4:1:5).

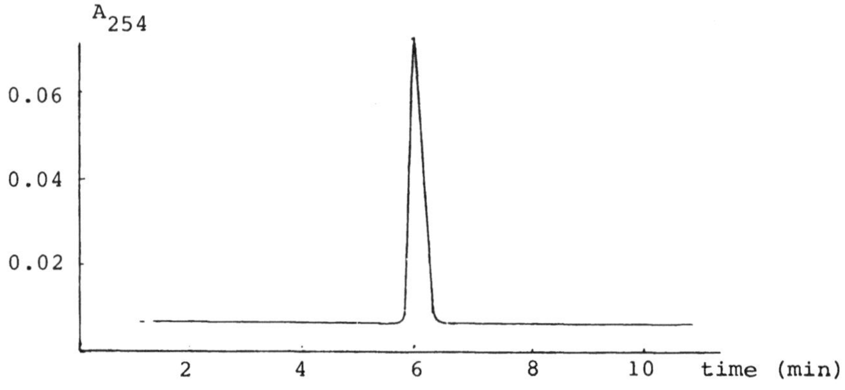

Analytical reverse-phase HPLC of gel-purified polypeptide (57-100). Polypeptide applied onto a LiChrosorb RP-2 column and eluted with CH_3CN gradient; flow rate:1 ml/min.

Fig. 4. Analytical data for peptide (57-100).

(PFP-p-nitrobenzoate), which selectively protected the free amino groups in side chains.

Truncated peptides that were shorter than the title compound, on modification with this bulky reagent became very different physico-chemically from the unmodified title product and could be separated by gel filtration.

The title peptide was cleaved off from the polymer with HBr/TFA in the presence of anisole and chromatographed on Sephadex G-25 (fine) in 3% acetic acid (Fig. 4). On this stage owing to the blocking reagent PFP-p-nitrobenzoate we succeeded in nearly complete separation of our title compound from the p-nitrobenzoylated truncated peptides. Fractions from the 2nd peak were a mixture of colored products well visible under UV-light. According to amino acid analysis data these products were rich in amino acids from the C-terminus of the heme pocket.

Further purification was achieved by partition chromatography on Sephadex G-25 (fine) in BuOH-AcOH-water (4:1:5). The obtained product was additionally analyzed by reverse-phase HPLC in conventional solvent systems.

Removal of residual admixtures could only be effected using tetrabutylammonium bromide.

The title compounds was characterized by high-resolution tlc, HPLC, N-terminal amino acid determination, amino acid analysis data, gel electrophoresis, paper electrophoresis. N-terminal sequence was confirmed by Edman degradation (4 steps).

Thus we can see that the strategy of the synthesis we selected led us to a sufficiently pure product.

N-trifluoroacetyl protection was removed off the lysyl residues with 2 N ammonia for over 24 hrs. The completeness of the reaction was checked by tlc and HPLC.

After deprotection the peptide was chromatographed on Sephadex G-25 (sf). Amino acid composition of the obtained product was as expected.

Completeness of N^{ε}-deblocking was shown by intense Dns-lysine revealed at the N-terminal amino acid determination, marked increase of electrophoretic mobility in acidic buffer and considerable decrease of the retention times on the reverse-phase column. Homogeneity of the obtained product was confirmed by paper chromatography, tlc, paper electophoresis and reverse-phase HPLC.

REFERENCES

1. Sidorova, M.V., G.A. Zheltukhina, E.I. Filippovich, R.P. Evstigneeva. 1983. Soviet-Indian Symposium on the Chemistry of Natural Products. Abstracts. Tbilisi. p. 94.

2. Vasil'ev, V.N., E.N. Zvonkova, E.Ye. Potapov, R.P. Evstigneeva. 1984. Izv. vuzov. Khim., khim. tekhnolog. 27, № 12, 1485.

3. Vasil'ev, V.N., E.N. Zvonkova, V.V. Chupin, R.P. Evstigneeva. 1985. Izv. vuzov. Khim., khim. tekhnolog. (in press).

4. Vasil'ev, V.N., E.N. Zvonkova, R.P. Evstigneeva. 1985. Izv. vuzov. Khim., khim. tekhnolog. (in press).

5. Sidorova, M.V., G.A. Zheltukhina, E.I. Filippovich et al. 1981. Zh. obshch. Khim. 51, № 11, 2605.

NEW STRATEGY FOR PEPTIDE SYNTHESIS USING HIGHLY LIPOPHILIC AND CHROMOPHORIC GROUPS. SYNTHESIS OF OCTAPEPTIDE SEQUENCE [24 - 31] H-LYS-ASN-ALA-TYR-LYS-LYS-GLY-GLU-OH OF HUMAN β-ENDORPHIN

H. Eckert, Y. Kiesel, C. Seidel, C. Kaulberg, H. Brinkmann
Organisch-Chemisches Institut der Technischen Universität München, Lichtenbergstr. 4, D-8046 Garching

Introduction

At the beginning an inductive contemplation shall be done about a peptide, all functional groups of it led back to the most simple scheme one can imagine.

Figure 1 represents the principal structure of any peptide with N- and C-terminus, peptide bonds and side chain functions. These groups are to be classified into 3 classes of its reactivity in order to achieve a directed and skilful planning of synthesis. Groups framed by a square are chemical reactive functional groups, like amino, mercapto, hydroxy and carboxy functions, being able to

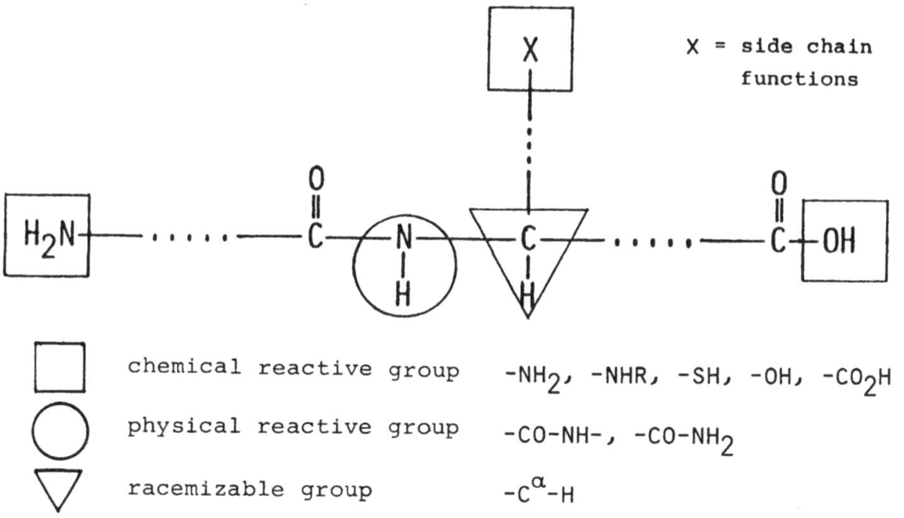

Figure 1. Classification of peptide groups.

Chemistry of Peptides and Proteins, Vol. 3

react during the usual chemical operations at peptides. Its treatment is extensively covered by the conventional protection groups techniques (1,2,3). A circle comprises those groups which are unreactive under customary conditions of peptide chemistry but take a great effect on solubility and formation of stable conformations, like β-turns, of the peptide chain. These shall be named physical reactive groups. All petide bonds belong to it as well as the ω-amido functions. Former works on a regenerative modification of these groups were performed by Weygand (4) respectively by Ugi (5). The problem of racemization at the α-carbon is marked by a triangle. It was tackled as yet mainly by modifying the coupling reagents (1).

Four criteria of peptide synthesis for economically favourable strategies

In order to obtain complete reactions in the chemo-synthesis of peptides, as well as in order to receive products of high purity from intermediates, four criteria of synthesis are set up as a guideline for economically favourable strategies:

1) A maximum of lipophilicity of the intermediate products, including the amide moiety, leads to a good solubility in unpolar organic solvents and therefore to complete reactions.
2) In connection with maximum lipophilicity, chromophores of the intermediate products afford purification by an easy and quick chromatography on low-priced silicagels, needing no expensive equipment.
3) Building up the peptide stepwise beginning at the C-terminus guarantees the growing peptide to be stable against racemization, when coupled with a N-urethane-protected amino acid.
4) Selective and smooth cleavage of an otherwise stable intermediate protection of the α-amino functions in homogeneous solution.

Selecting and designing protection groups

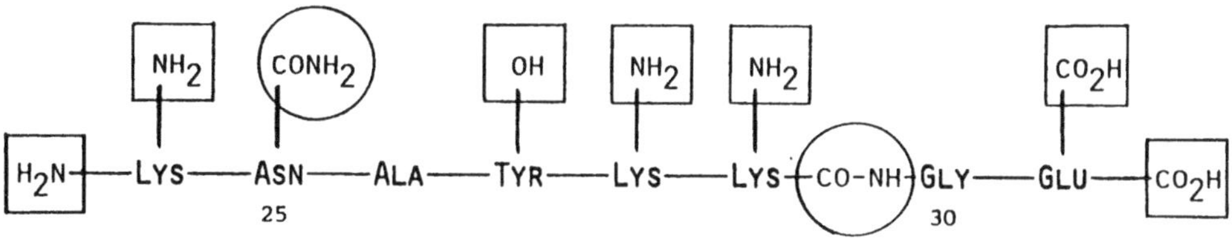

Figure 2. Octapeptide sequence [24 - 31] of human-β-endorphin.

Applying the previous classification of groups to the octapeptide sequence [24 - 31] of human-β-endorphin, as done in figure 2, one will realize seven chemical reactive groups, which will be protected by BOC respectively tBu res. , and at least two physical reactive groups, namely the β-amide moiety of asparagine (pos. 25) and the strong polar peptide bond before glycine (pos. 30).

For both amido groups, which are representative for other ω-amides and peptide bonds, a modification has to be developped, easily to be inserted and cleavable by acids with a comparable rate as N-BOC and OtBu residues.

Orthogonal to the permanent protecting groups cleavable by acids, is the TCBOC res. (2,6), which will be selected as intermediate protection for the α-amino functions.

Ferrocenylmethyl group for β-amide moiety of asparagine

The previous mentioned conditions are accomplished by the ferrocenylmethyl-[FerMe] res. (7). According to figure 3 it can be fixed regiospecifically at the β-carboxylic acid function of TCBOC-Asp(OH)-OH by 2-morpholinoethylisocyanide [MEI] and ferrocenylmethylamine to yield 70 % of the β-ferrocenylmethylcarboxamide besides 23 % of the α-ferrocenylmethylcarboxamide as a byproduct. The two isomers can be separated by chromatography on polyamide in hexane/ethylacetate, considerably facilitated by its colour,arising from the ferrocene chromophore (8). The product TCBOC-Asn(FerMe)-OH is soluble in hexane/ethylacetate (1 : 1), pointing out the noticeable increase of solubility compared with that of the TCBOC-Asn-OH.

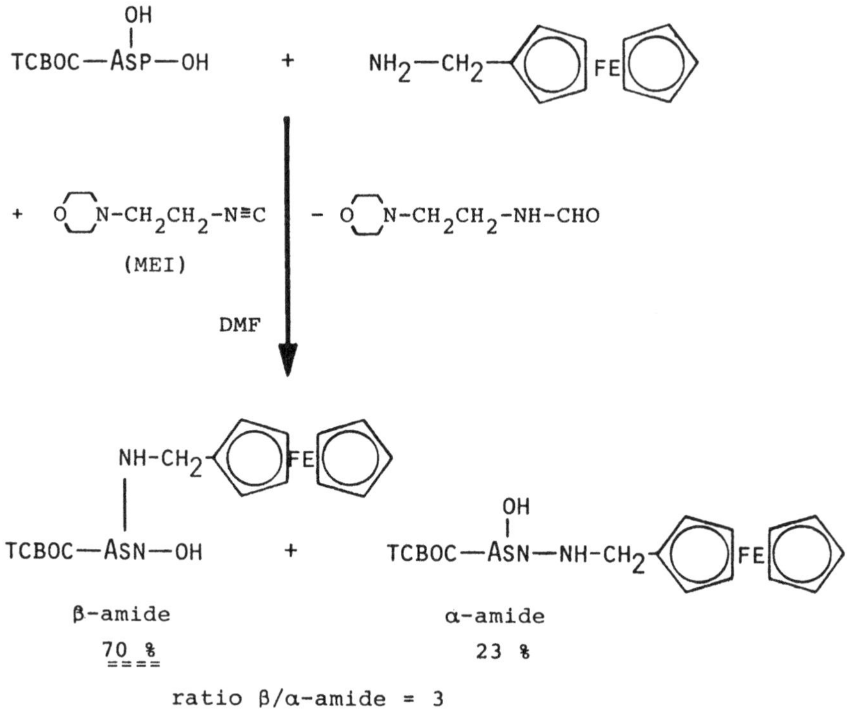

Figure 3. Regiospecific ferrocenylmethylamination of the β-carcoxylic acid function of aspartic acid.

Cleavage of the FerMe res. occurs with triflouroacetic acid [TFA] in dichloromethane within 4 h without any problems.

Ferrocenylmethyl group for peptide bond moiety of glycine

Ferrocenylmethylation of the peptide bond between lysine and glycine ensues from the ferrocenylmethylation of the amino function of glycine by a reductive amination of ferrocenecarboxaldehyde according to figure 4. The latter is an available product. Selective hydrogenation on palladium-phthalocyanine (9) guarantees the arising amine FerMe-glycine derivative to be stable and not reacting to methylferrocene. The extraordinary high stability of the free FerMe-Gly-OMe is impressable. It can be stored and even distilled without any decomposition, b.p. = 235°C/0.1 mm.

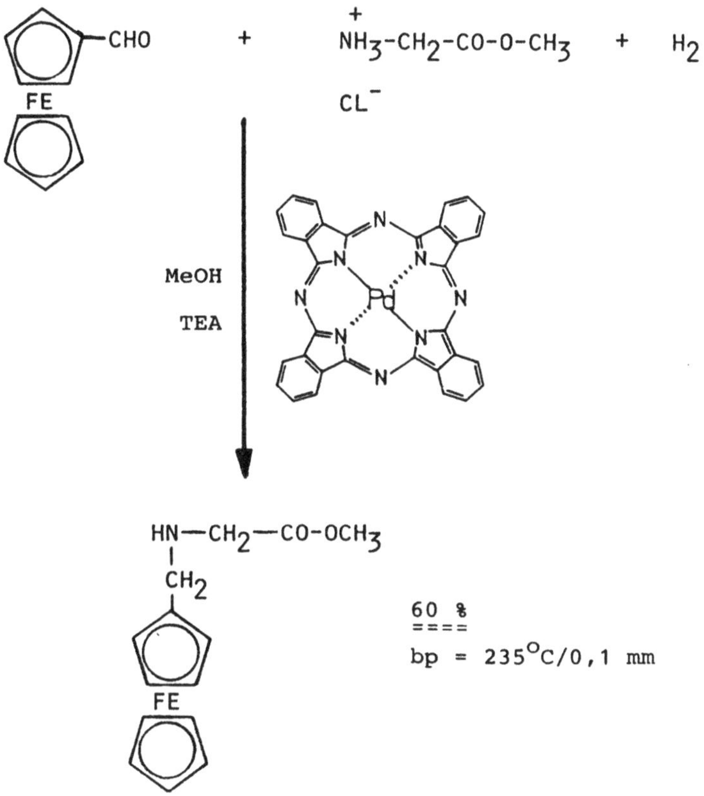

Figure 4. N-Ferrocenylmethylation of glycine.

On the contrary, the free H-Gly-OMe is entirely unstable, because of immediately oligomerizing.

After coupling the FerMe-Gly-OMe with TCBOC-(BOC)Lys-OH the peptide bond moiety can smoothly be deblocked by TFA in dichloromethane within 2 - 4 h.

TCBOC rest as a α-amino group intermediate protection

Each α-amino function at the N-terminus of the growing peptide is blocked after each coupling stage by the TCBOC rest (6). The latter arises from the respective coupled N-TCBOC amino acid and can be inserted by the stable and available reagent TCBOC-Cl. The TCBOC rest is stable under acidic and alkaline conditions, as

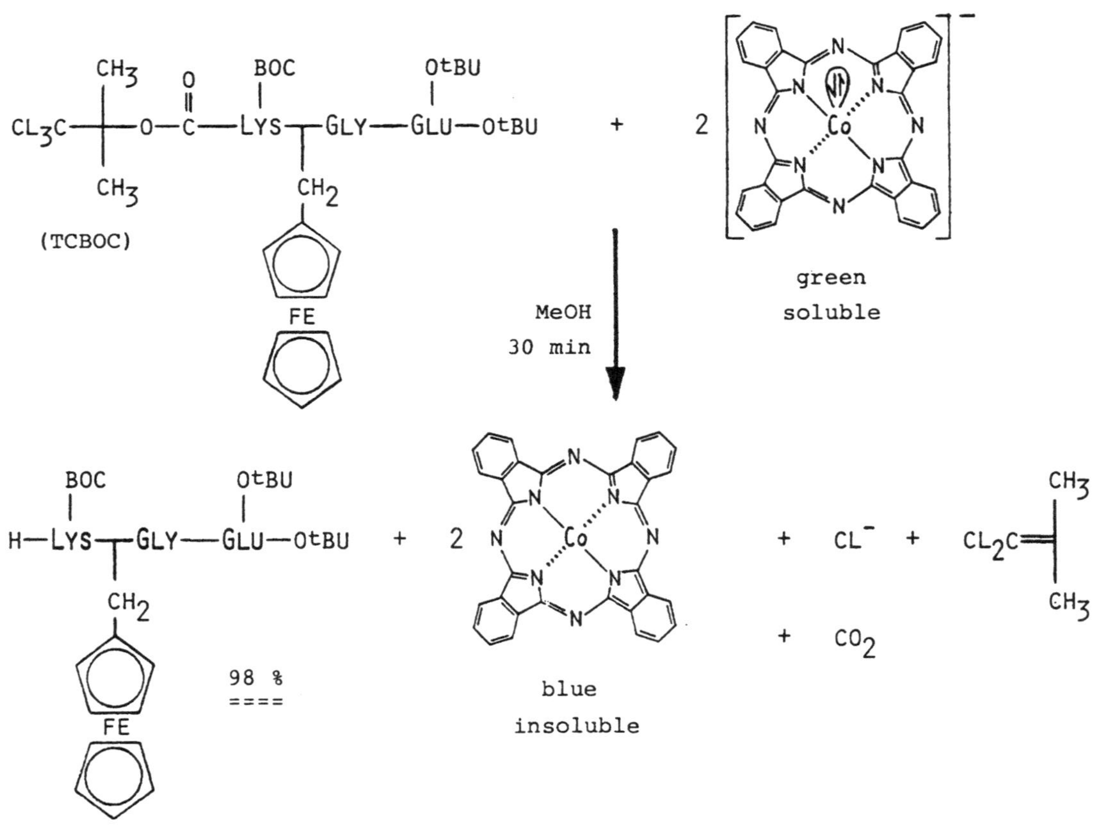

Figure 5. Selective and smooth cleavage of the TCBOC rest with the supernucleophile cobalt(I)phthalocyanine anion, followed by a reductive fragmentation mechanism.

well as during the hydrogenolysis of benzyl esters when using palladium-phthalocyanine as a hydrogenation catalyst (9) (caution: using palladium on coal, reactions difficult to survey will take place !). The TCBOC group is strongly lipophilic and can easily be characterized by the singulett at 1.9 - 2.0 ppm in its ^{1}H-NMR-spectrum.

Its cleavage will be performed entirely selectively and smoothly in homogeneous solution with the supernucleophile, well soluble, green coloured cobalt(I)phthalocyanine anion (10) as a reagent by a reductive fragmentation mechanism (11,12), according to figure 5. After cleavage, this reagent will be obtained as blue coloured cobalt(II)phthalocyanine, entirely unreactive and insoluble, which can easily be separated quantitatively by filtration or centrifugation and regenerated in an external reaction cycle (13). Reaction times for the remove are less than 1 h, yiels of deblocked products are quantitatively.

Synthesis of the octapeptide [24 - 31] H-Lys-Asn-Ala-Tyr-Lys-Lys-Gly-Glu-OH of human-β-endorphin

There is a need in peptide sequences of the C-terminal part of human-β-endorphin for medicinal investigations of metabolic effects recently discovered by Schwandt et al. (14). From a peptide chemist's view one can see the sequence [24 - 31] containing many of the problematic steps occuring in peptide synthesis.

In figure 6 the plan of synthesis for this octapeptide of human-β-endorphin is demonstrated. On principle the peptide chain will be built up stepwise from the C-terminus, causing a coupling (by MEI) without racemization, and in doing so, the advantage of activating always only the respective N-TCBOC protected amino acid and consequently of being careful with the peptide chain.

This strategy was interrupted only when it seemed to be advantageous. Once for the sequence [29 - 30] -Lys-Gly-, allowing TCBOC-(BOC)Lys-OH to react with DCCI and FerMe-Gly-OMe, where the further reaction of the TCBOC-(BOC)Lys-(FerMe)Gly-OH with H-Glu(OtBu)$_2$, to obtain the tripeptide [29 - 31], is not able to cause racemization, and thus some steps of synthesis can be saved.

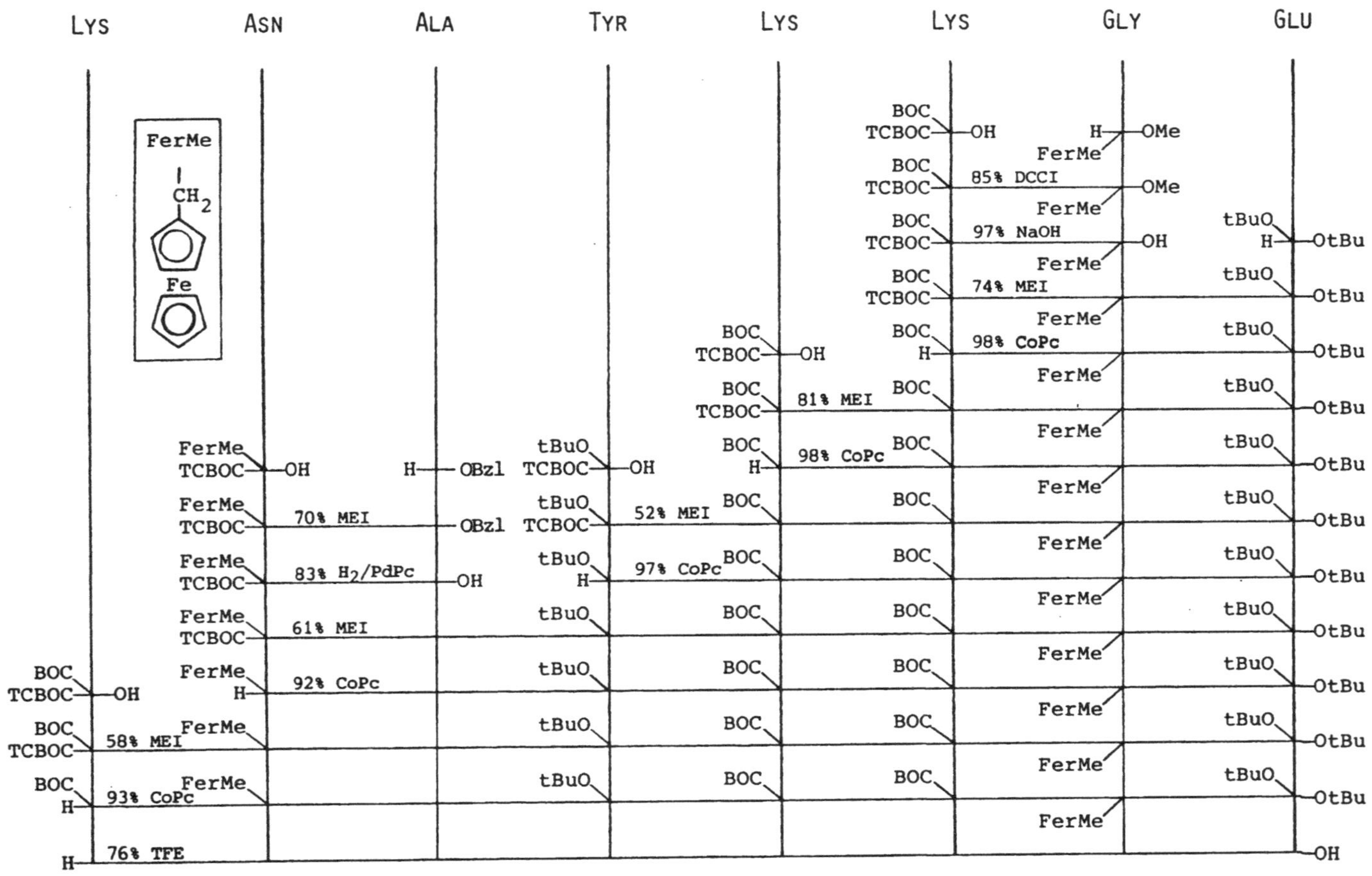

Figure 6. Synthesis of the octapeptide [24 - 31] of human-β-endorphin.

Secondly for the dipeptide sequence [25 - 26] -Asn-Ala-, where separating both isomers, TCBOC-Asn(FerMe)-OH and TCBOC-Asp(OH)-NH-FerMe, formerly obtained, is simplier on the stage of the neutral dipeptide derivative than separating the free acids.

Except coupling of FerMe-Gly-OMe, all coupling reactions are carried out with MEI. Yields vary thereby considerably, the average yield amounts to 66 %.

The TCBOC rest as an intermediate protection for α-amino groups has proved best, cleavable quantitatively on all stages without exception. The average yield amounts to 96 %.

Purification of intermediate products succeeds excellently by simple and quick column chromatography on low-priced silicagels, by reason of the intermediates' colour coming from the ferrocene chromophores, and the application of very unpolar solvents on the basis of hexane, due to intermediates' high lipophilicity. Capacities of about 4 % of the weight of silicagel and a performance on separating products differing by 0.1 R_F-value results in nearly quantitative isolation of the pure product.

Final acidolysis of all BOC, tBu, and FerMe groups from the protected octapeptide derivative with TFA and β-thionaphthol as a scavenger in dichloromethane is achieved in 2 - 4 h, and yields 76 % of the product H-Lys-Asn-Ala-Tyr-Lys-Lys-Gly-Glu-OH.

Acknowledgement

This work was supported by Deutsche Forschungsgemeinschaft.

References

1. Wünsch, E.. 1974. In: Houben-Weyl: Methoden der organischen Chemie, vol. 15. Thieme, Stuttgart.

2. Hubbuch, A.. 1979. Kontakte (Merck). 1979, 14.

3. Büllesbach, E.E.. 1980. Kontakte (Merck). 1980, 23.

4. Weygand, F., W. Steglich, J. Bjarnason, R. Akhtar, N. Chytil. 1968. Chem. Ber. 101, 3623.

5. Marquarding, D., P. Hoffmann, H. Heitzer, I. Ugi. 1970. J. Am. Chem. Soc. 92, 1969.

6. Eckert, H., M. Listl, I. Ugi. 1978. Angew. Chem. 90, 388. Angew. Chem. Int. Ed. Engl. 17, 361.

7. FerMe will be suggested as the abbreviation for ferrocenylmethyl.

8. Eckert, H., C. Seidel, D. Marquarding. Publication in preparation.

9. Eckert, H., G. Fabry, Y. Kiesel, G. Raudaschl, C. Seidel. 1983. Angew. Chem. 95, 894. Angew. Chem. Int. Ed. Engl. 22, 881. Angew. Chem. Suppl. 1983, 1291.

10. Eckert, H., A. Schier. 1979. Angew. Chem. 91, 841. Angew. Chem. Int. Ed. Engl. 18, 794.

11. Eckert, H., I. Ugi. 1976. Angew. Chem. 88, 717. Angew. Chem. Int. Ed. Engl. 15, 681.

12. Eckert, H., I. Ugi. 1976. J. Organomet. Chem. 118, C59.

13. Eckert, H., I. Ugi. 1979. Liebigs Ann. Chem. 1979, 278.

14. Schwandt, P., W. Richter, J.S. Morley. 1981. Neuropeptides 1, 211.

USE OF THE SULFENYL-HYDRAZIDE METHOD FOR THE SYNTHESIS OF UNSYMMETRICAL CYSTINE-PEPTIDES

S. Romani, L. Moroder and E. Wünsch

Max-Planck-Institut für Biochemie, Abteilung Peptidchemie, 8033 Martinsried, FRG

Introduction

Among the various methods known to produce symmetric and unsymmetric disulfides from thiol compounds, the dialkyl azodicarboxylate procedure (1,2) seemed the most adequate for an adaptation to peptide chemistry particularly in view of a selective cysteine pairing to unsymmetrical cystine-peptides. Our model studies clearly indicated that stable and analytically well defined sulfenyl-hydrazides are readily formed upon reacting cysteine-compounds with azodicarboxylic acid derivatives without any side reaction at the level of other amino acid residues, e.g. methionine, tyrosine, histidine and particularly tryptophan (3-5). These sulfenyl-hydrazide derivatives react smoothly with a second cysteine component to generate the desired unsymmetrical cystine-peptides or cystine derivatives in high yields. The usefulness of this new procedure in peptide chemistry is demonstrated on selected examples.

Results

Synthesis of S-tert-butylthio-cysteine

In the last years the S-tert-butylthio-cysteine (6) has found widespread application in the synthesis of cysteine- and cystine-peptides because of its high stability under the usual conditions of peptide synthesis and its clean and selective removal via reduction with phosphines (7). For the preparation of this useful cysteine-derivative we have elaborated a new method (8) to avoid the use of larger excesses of tert-butylmercaptan as needed in

Chemistry of Peptides and Proteins, Vol. 3

the previous procedure (6,9). The new method proceeds via the sulfenyl-hydrazide intermediates as shown in fig. 1.

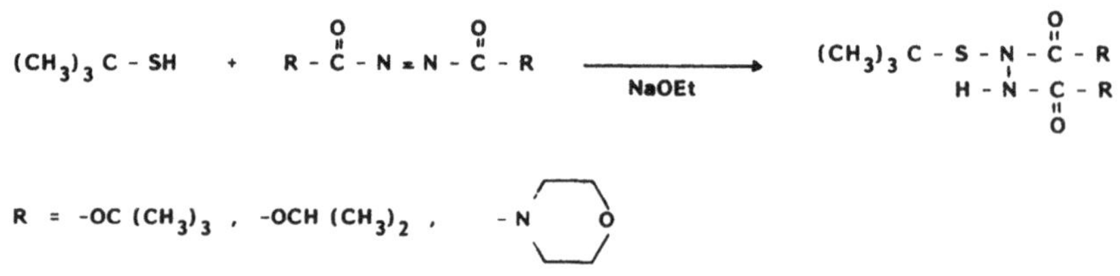

Fig. 1. Synthesis of 1-(tert-butylthio)-1,2-hydrazine dicarboxylic acid derivatives.

The sterically hindered tert-alkylmercaptan was found to react with azodicarboxylic acid derivatives only upon addition of catalytic amounts of sodium alcoholate, to yield quantitatively the corresponding 1-(tert-butylthio)-1,2-hydrazine dicarboxylic acid derivatives as crystalline compounds stable on storage. These tert-butylthio-donators react with cysteine, cysteine derivatives and cysteine-peptides to produce in high yields the related S-tert-butylthio compounds.

Synthesis of the NGF fragment 10-25/75-88

Upon cyanogen bromide cleavage of mouse nerve growth factor (NGF) followed by tryptic digestion, Mercanti et al. (10) succeeded to isolate a double-chain cystine-peptide corresponding to the NGF-sequence 10-25/75-88. This fragment was found to be capable of promoting the outgrowth of neurites at a 1/100th the molar concentration of the native molecule. Since attempts to reproduce these results failed so far, the question of whether this fragment represents an active site-like portion of NGF can only be answered with a synthetic replicate.

The NGF-fragment 10-25/75-88 is an unsymmetrical cystine-peptide containing in both peptide chains a tryptophan residue besides a series of other reactive trifunctional amino acids (see fig. 2). These sequence characteristics preclude the use of the known methods for a selective cysteine pairing, e.g. the method of Kamber (11), except for the sulfenyl-hydrazide procedure recently proposed by us (3-5). Since this new procedure can be applied both in the case of protected peptide derivatives with free cysteine thiol functions (in organic solvent) and of fully unprotected cysteine-peptides (in aqueous media), several synthetic routes to the unsymmetrical cystine-peptide 10-25/75-88 are feasible.

10 25
H-Gly-Glu-Phe-Ser-Val-Cys-Asp-Ser-Val-Ser-Val-Trp-Val-Gly-Asp-Lys-OH

75 88
H-His-Trp-Asn-Ser-Tyr-Cys-Thr-Thr-Thr-His-Thr-Phe-Val-Lys-OH

Fig. 2. Amino acid sequence of the NGF-fragment 10-25/75-88.

Preliminary studies on various approaches (12) led us to focuse our main efforts on the synthetic route outlined in scheme I and II.

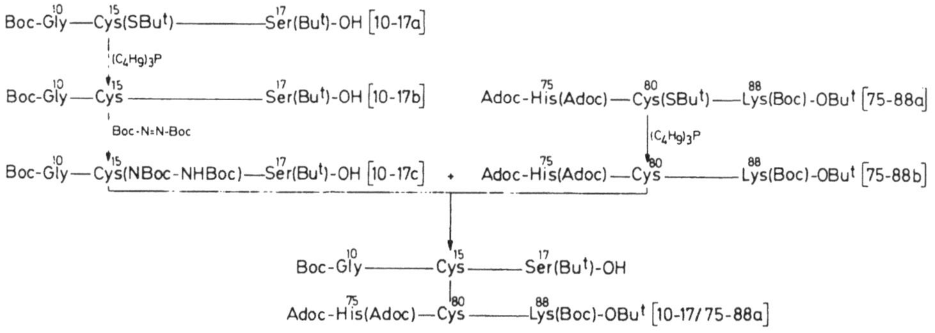

Scheme I. Synthesis of the suitably protected unsymmetrical cystine-peptide 10-17/75-88.

The strategy of synthesis choosen relies mainly on our previous experiences in the preparation of cystine containing peptides (7,13) with a protection scheme based on acid-labile side-chain protecting groups derived from tert-butanol and 1-adamantanol in combination with the S-tert-butylthio group for the cysteine thiol functions. Benzyloxycarbonyl and 2-nitrophenylsulfenyl or tert-butyloxycarbonyl groups served as complementary α-amino protection. To build up the peptide chains, suitably protected fragments were synthetized in stepwise manner via the active ester procedure using preferentially N-hydroxysuccinimide esters in the acylation steps, and catalytic hydrogenolysis for the intermediate N^{α}-debenzyloxycarbonylation or hydrogen halides in presence of excesses of 2-methylindole for the Nps-cleavage.

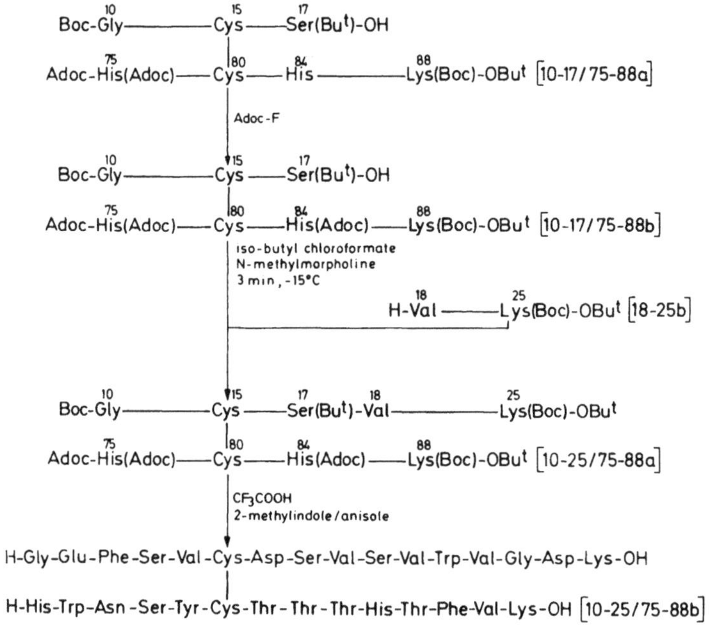

Scheme II. Synthesis of the NGF fragment 10-25/75-88.

The fully protected tetradecapeptide 75-88 was obtained by assembly of three fragments via dicyclohexylcarbodiimide/N-hydroxysuccinimide (14) as shown in scheme III. Conversely, the octapeptide derivative 10-17 with free carboxyl function for further elongation at the C-terminus, was obtained via mixed anhydride coupling of the N^{α}- and side-chain protected pentapeptide 10-14 with the trimethylsilylated tripeptide derivative 15-17 (scheme IV). This new procedure yielded the desired octapeptide derivative in high yields as homogeneous compound without racemization as determined by the gas chromatographic method of Bayer (15).

The two peptide derivatives 75-88 and 10-17 where then selectively deprotected at their cysteine thiol functions by reduction with

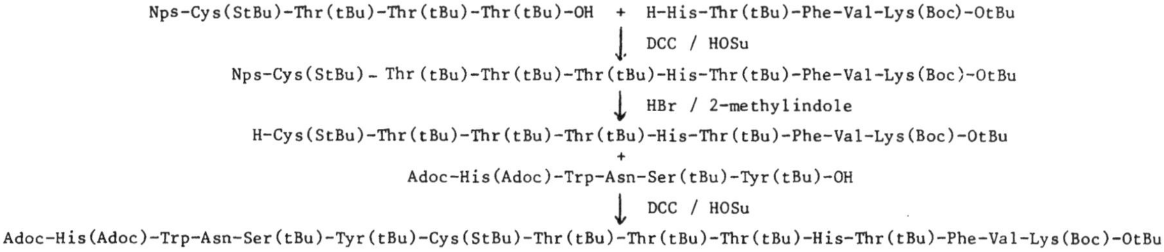

Scheme III. Synthesis of the fully protected tetradecapeptide 75-88.

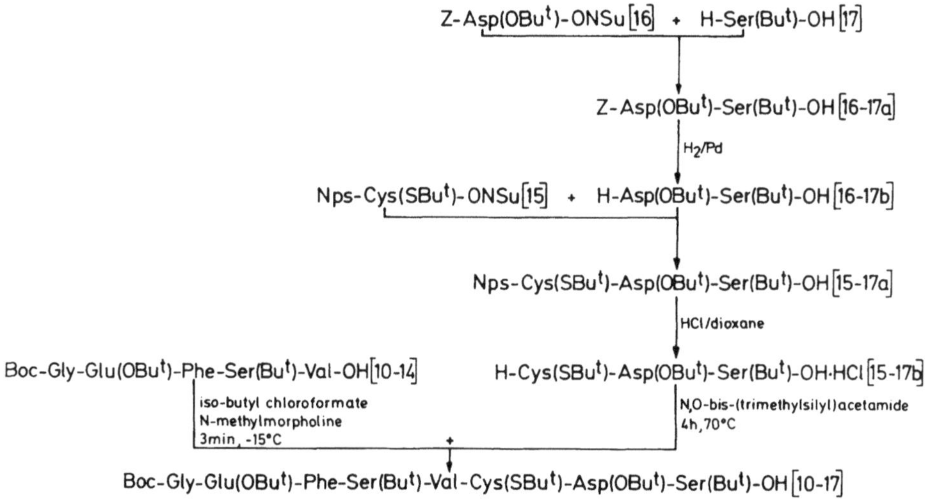

Scheme IV. Synthesis of the suitably protected fragment 10-17.

tributylphosphine in trifluoroethanol containing 1-2 % water. Subsequently the octapeptide derivative 10-17 was activated at its thiol function as sulfenyl-hydrazide derivative and then reacted with the cysteine-peptide derivative 75-88 at a ratio of 1.5 : 1 to yield the unsymmetrical cystine-peptide derivative 10-17/75-88 (scheme I). The excess of S-activated octapeptide derivative was cleanly removed by gel filtration of the reaction mixture on LH-20 using dimethylformamide as eluent.

For the last chain elongation step the imidazole function of histidine-84 was protected a posteriori as adamantyloxycarbonyl derivative (16) and the resulting peptide derivative (see scheme II) was coupled with the N^{α}-deprotected octapeptide 18-25 via the dicyclohexylcarbodiimide/N-hydroxysuccinimide as well as via the mixed anhydride procedure. Gel filtration on LH-20 served again to isolate the fully protected unsymmetrical cystine-peptide at a high degree of purity as judged by thin layer chromatography in various solvent systems, by amino acid analysis of the acid hydrolysate and gaschromatographic chiral analysis.
For the acidolytic deprotection step difficulties were encountered because of the simultaneous presence of the disulfide bridge and the tryptophan residues. Mercaptans normally used for the most efficient cation capture, can not be added; thus, acidolysis by exposure to trifluoroacetic acid containing alternative additives, e.g. 2-methylindole and anisole, led to tert-butylation of the tryptophan residues (17) to larger extents and possibly of the tyrosine residue, too (18). Additionally, the deprotection was not quantitative under standard conditions (2 h at room temperature), particularly at the level of the clusters of hydroxyamino acids. Similar uncomplete deprotection has been reported previously in the case of calcitonin (19). Consequently, the acidolysis was performed with concentrated HCl (0° C for 10 min) and the resulting crude product was purified by partition chromatography on LH-20 using 1 M ammonium acetate buffer (pH 3.7)/2-butanol (10:2) as eluent. By rechromatography of the main peak under identical conditions the NGF-fragment 10-25/75-88 was isolated as homogeneous material as judged by various analytical assays (20).

Selective pairing of the human insulin A- and B-chain

An additional example for the useful application of the sulfenylhydrazide procedure is related to the selective pairing of the human A- and B-chain. Both A- and B-chain of human insulin were prepared by the fragment condensation procedure whereby as much as possible known synthetic routes from the previous synthesis of human insulin (21) were followed. Three selective thiol protecting groups, i.e. the S-tert-butylthio, the S-trityl and the S-acetamidomethyl group, were applied to allow successive correct cysteine pairings. The cysteine residues A7 and B7 were deprotec-

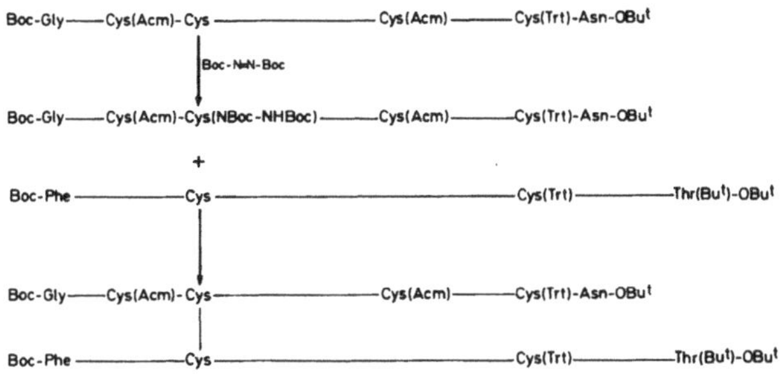

Scheme V. Selective disulfide pairing between human insulin A- and B-chain (A7 - B7).

ted by reduction with tributylphosphine. The thiol function of cysteine A7 was converted into the corresponding sulfenyl-hydrazide derivative which was then reacted with the thiol function of cysteine B7 (scheme V) to produce the first disulfide bridge (A7 - B7). The residual selective disulfide pairings should be feasible following the procedures reported by Sieber et al. (21) in their successful synthesis of human insulin.

Acknowledgments

This study was supported by the Deutsche Forschungsgemeinschaft (Wu 20/12.1 and Mo 377/1.1).

References

1. Mukaiyama, T., K. Takahashi. 1968. Tetrahedron Lett. 56, 5907.
2. Linke, K. H., W. Brandt, H. J. Göhausen. 1973. Chem. Ber. 106, 707.
3. Wünsch, E., S. Romani. 1982. Hoppe-Seyler's Z. Physiol. Chem. 363, 449.
4. Wünsch, E., S. Romani, L. Moroder. 1983. In: Peptides 1982 (K. Bláha, P. Malǒn, eds.) Walter de Gruyter & Co., Berlin-New York, p. 183.

5. Romani, S., W. Göhring, L. Moroder, E. Wünsch. 1984. In: Chemistry of Peptides and Proteins, Vol. 2 (W. Voelter, E. Bayer, Y. A. Ovchinnikov, E. Wünsch eds.) Walter de Gruyter & Co., Berlin-New York, p. 29.

6. Wünsch, E., R. Spangenberg. 1971. In: Peptides 1969 (E. Scoffone, ed.) North-Holland, Amsterdam, p. 30.

7. Moroder, L., M. Gemeiner, W. Göhring, E. Jaeger, P. Thamm, E. Wünsch. 1981. Biopolymers, 20, 17.

8. Wünsch, E., L. Moroder, S. Romani. 1982. Hoppe-Seyler's Z. Physiol. Chem. 363, 1461.

9. Wünsch, E. 1974. In: Houben-Weyl Methoden der Organischen Chemie, Vol. 15, I (E. Wünsch, ed.) Thieme Verlag, Stuttgart, p. 792.

10. Mercanti, D., R. Butler, R. Revoltella. 1977. Biochim. Biophys. Acta, 494, 412.

11. Kamber, B. 1973. Helv. Chim. Acta, 56, 1370.

12. Moroder, L., S. Romani, E. Wünsch, G. Harper, H. Thoenen. 1983. In: Peptides - Structure and Function (V. J. Hruby, D. H. Rich, eds.) Pierce Chemical Company Rockford, Illinois, p. 175.

13. Moroder, L., M. Gemeiner, W. Göhring, E. Jaeger, J. Musiol, R. Scharf, H. Stocker, E. Wünsch. 1981. Hoppe-Seyler's Z. Physiol. Chem. 362, 697.

14. Wünsch, E., F. Drees. 1966. Chem. Ber. 99, 110.

15. Frank, H., G. J. Nicholson, E. Bayer. 1977. J. Chromatogr. Sci. 15, 174.

16. Moroder, L., E. Jaeger, F. Drees, M. Gemeiner, S. Knof, H.-P. Stelzel, P. Thamm, D. Bataille, S. Domschke, W. Schlögl, J. Schulz, E. Wünsch. 1980. Bioorg. Chem. 9, 27.

17. Wünsch, E., E. Jaeger, L. Kisfaludy, M. Löw. 1977. Angew. Chem. 89, 330.

18. Lundt, B. F., N. L. Johansen, J. Markusen. 1979. Int. J. Peptide Protein Res. 14, 344.

19. Riniker, B., B. Brugger, B. Kamber, P. Sieber, W. Rittel. 1969. Helv. Chim. Acta, 52, 1958.

20. Wünsch, E. 1985. Mh. Chem. 116, 505.

21. Sieber, P., B. Kamber, K. Eisler, A. Hartmann, B. Riniker, W. Rittel. 1976. Helv. Chim. Acta, 59, 1489.

APPLICATION OF CROWN ETHERS IN PEPTIDE SYNTHESIS

S.A. Andronati, A.A. Mazurov, T.I. Korotenko
A.V.Bogatsky Physico-Chemical Institute, Academy of Sciences of the Ukrainian SSR, 86 Chernomorskaya doroga, 270080, Odessa, USSR

Crown ethers find a wide application for the development of synthetic methods. In 1976 it was shown that 18-crown-6 catalyzes quantitative formation of esters from potassium salts of tert-butyloxycarbonyl amino acids and chloromethylated resins in DMFA (1).

$$P\text{-}C_6H_4CH_2Cl + BOC\text{-}NHCHRCO_2K \longrightarrow BOC\text{-}NHCHRCO_2CH_2C_6H_4\text{-}P$$

Fluoride ion transported in acetonitrile by a crown ether promotes deprotonation of the trypthophan indole ring thus facilitating the protection of the nitrogen atom by benzyloxycarbonyl or 2,4-dichlorobenzyloxycarbonyl groups (2,3).

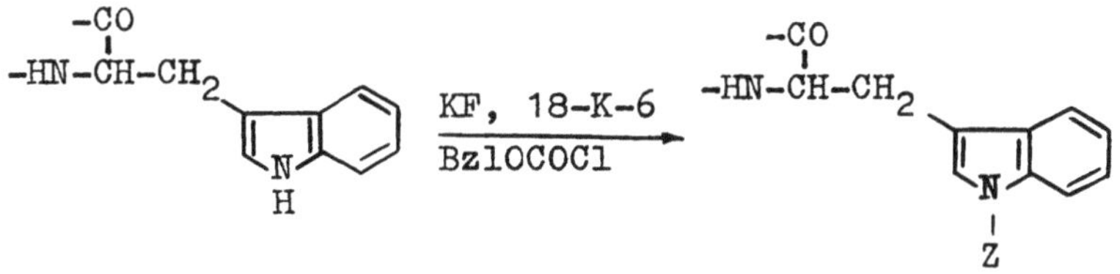

To model enzyme-catalyzed reactions of peptide formation Japanese investigators (4) have carried out synthesis of a peptide by condensation of N- and C-components added to di(mercaptoalkyl)-18-crown-6.
There is a peptide bond formation observed under certain conditions between amino acids bound to a crown ether serving as a peptidase model. After splitting off the dipeptide from the crown ether it is possible to prepare tripeptides according to

Chemistry of Peptides and Proteins, Vol. 3

a similar scheme.

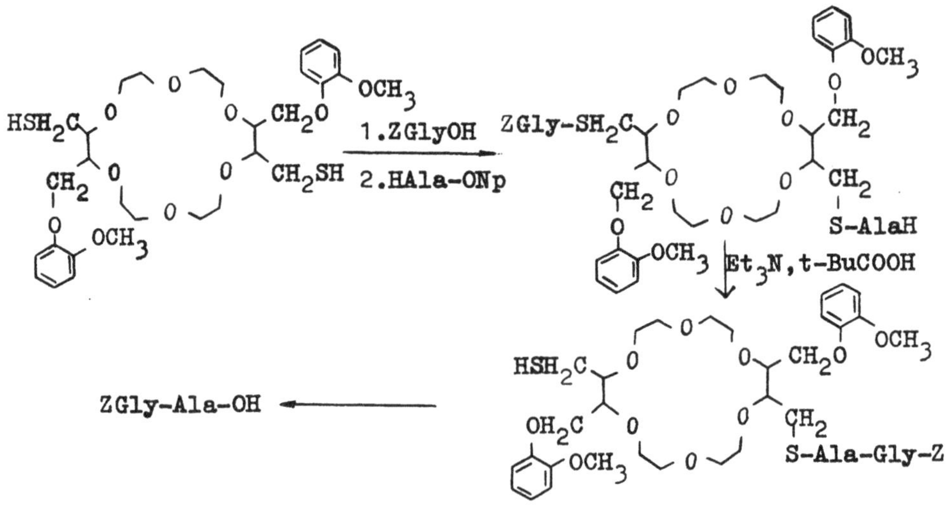

We have proposed a modification of "salt condensation" of amino acid derivatives using crown ethers (15-crown-5, 18-crown-6, dicyclohexyl-18-crown-6, etc.). The most simple procedure for protection of carboxyl function of amino acids consists of their transformation into alkaline metal salts which gives practically quantitative yield. The advantage of this method is the fact that subsequent deblocking of carboxyl groups proceeds readily while the possibility of racemization is minimal, as compared, for example, with alkaline hydrolysis of alkyl esters of amino acids or peptides. The so-called "salt condensation" was carried out by an amino component transfer into the alkaline metal salt followed by treatment of this salt by an activated carboxyl component, the salt itself being not isolated. However, the reaction of peptide formation has to be carried out in an aqueous-organic medium by the methods of mixed anhydrides or activated esters due to the poor solubility of sodium and potassium salts of amino acids in organic solvents. Presence of water in the reaction mixture is known to reduce the yield of peptides, so this method finds rather limited use. We have used the complex of sodium salts of amino acids with crown ethers soluble in organic amino component. The amino component solution is prepared by the addition of dimethyl formamide to the aqueous solution of equimolar quantities of amino acid potassium salt and crown ether, followed by dehydration. Thus, "salt condensation" is realized in organic solvents by the carbodiimide method with good yield.

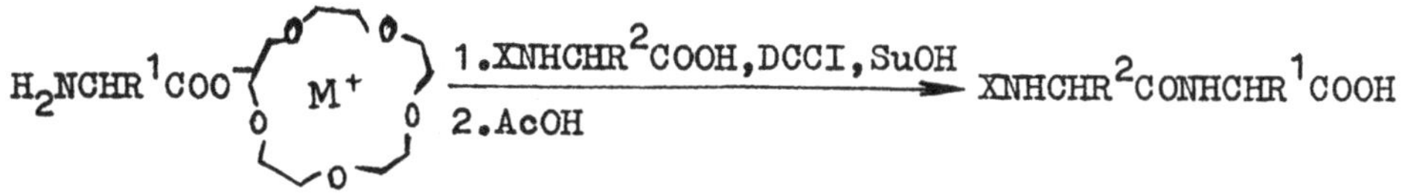

The dimethyl formamide solution of the amino component is treated by a mixture of N-protected amino acid, N,N-dicyclohexylcarbodiimide and N-hydroxysuccinimide. The formed sodium salt of peptide is neutralized by an equimolar quantity of acetic acid, and the peptide obtained is then isolated.
Using as an amino component a complex of amino acid sodium salt with a crown ether "salt condensation" of amino acids and peptides can be carried out in organic solvents by activated esters and mixed anhydrides.
To illustrate the proposed procedure we have prepared some peptides. For instance, thyroliberine is prepared in two stages with good yield. Biologic activity of thyroliberine (the test of motor activity) was identical to that of an authentic sample (Serva, Heidelberg, FRG).

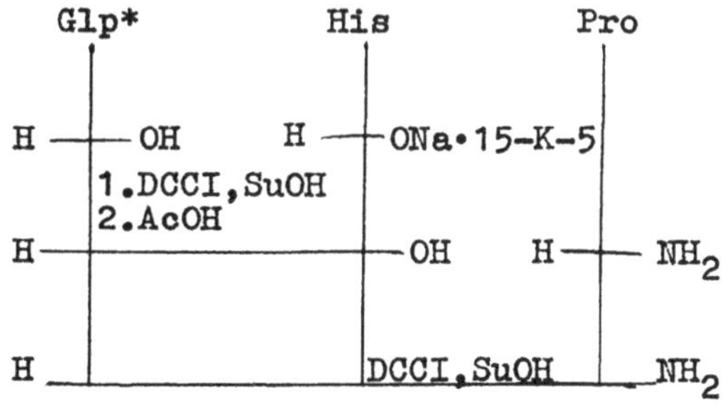

*Pyroglutamic acid

Table. Characteristics of Peptides.

No	Peptide	$(\alpha)^{20}_{578}$ (TFE)	M.p. (°C)	Yield, % Found	Yield, % Lit	Ref	Found / Calcul C	H	N	Formula
1	Glp*-D,L-Phe	-	225	93	65	(5)	61.3 60.9	5.5 6.2	10.0 11.3	$C_{14}H_{16}N_2O_4$
2	Glp-D,L-Phe-Pro	-3.26° C 0.46	85	90	-	-	61.0 61.1	6.6 6.2	11.5 11.3	$C_{19}H_{23}N_3O_5$
3	Z-Gly-Gly	-	175	97	75	(7)	54.3 54.1	5.5 5.3	10.4 10.5	$C_{12}H_{14}N_2O_5$
4	Z-Gly-D,L-Phe	-	161	92	70	(8)	64.5 64.0	5.8 5.6	8.1 7.8	$C_{19}H_{20}N_2O_5$
5	BOC-Gly-Gly	-	125	95	65	(6)	47.0 46.6	7.1 6.9	12.3 12.1	$C_6H_{16}N_2O_5$
6	Glp-Gln	-329.09° C 0.55	87-89	91	-	-	46.4 46.7	6.0 5.8	16.7 16.3	$C_{10}H_{15}N_3O_5$
7	Glp-His	-14.48° C 0.58	217	91	55	(4)	49.9 49.6	5.2 5.3	21.5 21.1	$C_{11}H_{14}N_4O_4$

* Pyroglutamic acid

References

1. Roeske, R.W., P.D.Gesellchen. 1976. Tetrahedron Lett., N 38, 3369.

2. Chorev, M., Y.S. Klausner. 1976. Chem. Commun., N 15, 596.

3. Klausner, Y.S., M. Chorev. 1977. J. Chem. Soc., Perkin I, N 6, 627.

4. Sasaki, S., M. Shinoya, K. Kogak. 1983. Heterocycles, 20, 124.

5. Rips, R., E. Morier. Patent Gr. Br. 1592552. Int. cl. C 07 C 103/52

6. Sievertsson, H., J.-K. Chang, K. Folkers, C.Y. Bowers. 1972. J. Med. Chem., 15, 8.

7. Voelter, W., H. Kalbacher, E. Pietrzik. 1976. Z. Naturforsch., 31b, 1015.

8. Vilkas, E., M. Vilkas, J. Sainton. 1980. Int. J. Peptide and Protein Res., 15, 29.

9. Reig, F., J.M. Garcia Anton, G. Valencia, J.J. Garcia Dominique. 1981. Fette, Seifen, Anstrichmittel, 83, 367.

SYNTHESIS AND BIOLOGICAL ACTIVITY OF THE SAUVAGINE FRAGMENT 18-40

G. Wendlberger, P. Thamm* and E. Wünsch
Max-Planck-Institut für Biochemie, Abteilung Peptidchemie, 8033 Martinsried, FRG

V. Erspamer
Istituto di Farmacologia Medica, Universitá di Roma, Roma, Italy

Introduction

The skin of frogs and toads has been found to represent an inexhaustible source of physiologically important compounds such as biogenic amines and peptide factors (1-2). A straight correlation, both, concerning structure homology and pharmacological activities, was revealed for the skin peptides and their counterparts in mammalian brain and gut, leading to the concept of a skin-gut-brain peptide triangle (2). In this context a new tetracontapeptide has recently been isolated by Montecucchi et al. (3-4); it was named sauvagine according to its source, the Phyllomedusa sauvagei (fig. 1). This peptide factor displays characteristic pharmacological actions as intense mucosal vasodilatation, hypotension, stimulation of ACTH and ß-endorphin release and potent inhibition of prolactin, somatotropin and tyrotropin release. The observed

Pyr-	Gly-	Pro-	Pro-	Ile-	Ser-	Ile-	Asp-	Leu-	Ser-	Leu-	Glu-	Leu-	Leu-	Arg-	Lys-	Met-	Ile-	Glu-	Ile-
1	2	3	4	5	6	7	8	9	10	11	12	13	14	15	16	17	18	19	20
-Glu-	Lys-	Gln-	Glu-	Lys-	Glu-	Lys-	Gln-	Gln-	Ala-	Ala-	Asn-	Asn-	Arg-	Leu-	Leu-	Leu-	Asp-	Thr-	Ile-NH_2
21	22	23	24	25	26	27	28	29	30	31	32	33	34	35	36	37	38	39	40

Fig. 1. Primary structure of sauvagine.

---

* Present address: Diamalt AG, Munich

Chemistry of Peptides and Proteins, Vol. 3

biological activities parallel those of the mammalian counterpart, the corticotropin-releasing factor (CRF) isolated subsequently by Vale et al. (5) from ovine hypothalamus. An alignement of the primary structures of sauvagine and CRF indicates a high degree of homology with almost half of the residues indentical and 12 additional amino acids resulting from single point mutations. The presence of two adjoining basic residues Arg-Lys in positions 15-16 in the sauvagine sequence as possible site for ulterior processing of these tetracontapeptide to shorter peptide hormone(s) prompted us to synthetize for structure-function studies in a first approach the sequence portion 18-40 corresponding to the C-terminal fragment resulting from the cyanogen bromide cleavage.

Results

The synthesis of the tricosapeptide amide was performed following our general strategy (6) successfully applied in previous syntheses of peptide hormones (e.g. Ref. 7,8):

i) a side-chain protection with acid-labile groups derived from tert-butanol and 1-adamantanol in combination with N^{α}-benzyloxycarbonyl or N^{α}-2-nitrophenylsulfenyl derivatives in the intermediate chain-elongation steps. The arginine guanido function is protected by protonation with hydrogen halides to enhance the solubility of both fragments and intermediate seqments via an interchange of hydrophobic and hydrophilic regions. The symmetry of the side-chain protection is additionally disrupted by using various protecting groups (Boc/Adoc) to further enhance the solubility.

ii) assembly of shorter fragments in sequence order via dicyclohexylcarbodiimide in the presence of 1,2-dinucleophiles.

iii) deprotection under relatively mild conditions, i.e. trifluoroacetic acid treatment.

Consequently four suitably protected fragments corresponding to sequences 32-40, 27-31, 22-26 and 18-21 (fig. 2) were prepared by the stepwise active ester procedure using 4-nitrophenyl and N-hydroxysuccinimide esters in the acylation steps, and catalytic

Fragment I : H-Asn-Asn-Arg(HBr)-Leu-Leu-Leu-Asp(OBu^t)-Thr(Bu^t)-Ile-NH$_2$ [32-40]

II : Nps-Lys(Boc)-Gln-Gln-Ala-Ala-OH [27-31]

III : Nps-Lys(Adoc)-Gln-Glu(OBu^t)-Lys(Boc)-Glu(OBu^t)-OH [22-26]

IV : Boc-Ile-Glu(OBu^t)-Ile-Glu(OBu^t)-OH [18-21]

Fig. 2. Protected fragments for the synthesis of sauvagine 18-40.

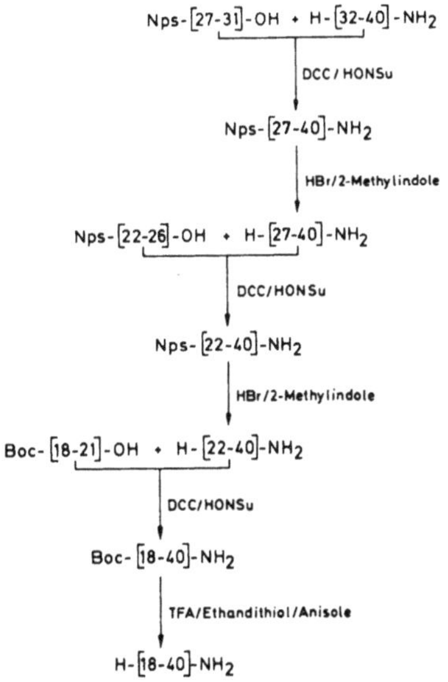

Fig. 3. Synthetic route to sauvagine 18-40.

hydrogenolysis for the intermediate N^{α}-debenzyloxycarbonylation. The fragments were then condensed via the dicyclohexylcarbodi-imide/N-hydroxysuccinimide procedure following the route outlined in fig. 3. The intermediate selective N^{α}-deprotection was achieved by exposure to slight excesses of hydrogen bromide in organic media upon addition of larger excesses of 2-methylindole for the Nps-capture. Under these conditions the proton activity is sensibly reduced thus preventing hydrolysis of particularly acid-labile groups, e.g. N^{ϵ} -tert-butyloxycarbonyl groups.
The fully protected tricosapeptide amide was then exposed to trifluoroacetic acid in the presence of 1,2-ethanedithiol/anisole and the resulting crude sauvagine fragment 18-40 was purified by partition chromatography on Sephadex G-25 (fine) under the conditions reported in fig. 4. Fractions of the main peak were pooled to yield after repeated lyophilization with intermediate millipore filtrations, the desired product at the high degree of purity as judged by different analytical assays, e.g. thin layer chromatography in various solvent systems, hplc unter the conditions reported in fig. 5, amino acid analysis and gas chromatographic racemization test (see table 1); the recovery from enzymatic digestion with aminopeptidase M coincides with the peptide content as determined by quantitative amino analysis.

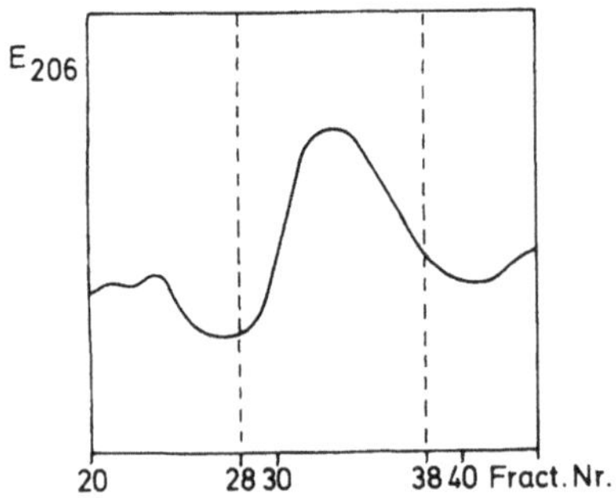

Fig. 4. Partition chromatography of the crude tricosapeptide amide on Sephadex G-25 (fine); elution with the solvent system 2-butanol/0.05 M ammonium acetate buffer (pH 6.8), 1:1.

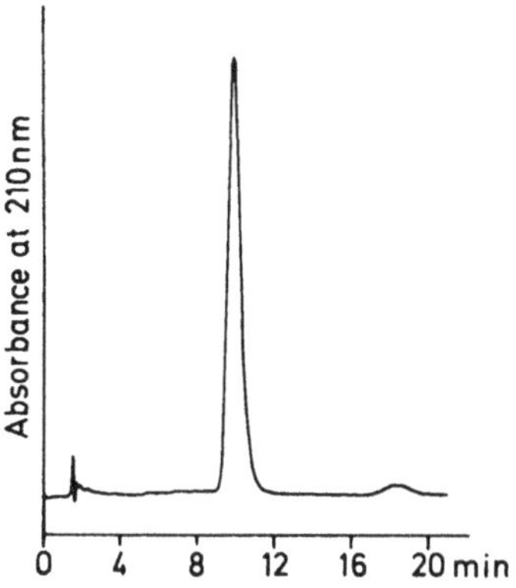

Fig. 5. Hplc of sauvagine 18-40 on μ-bondapak C 18; eluent: acetonitrile/0.02 M ammonium acetate buffer (pH 6.8), 23:77; flow rate: 1.5 ml/min.

Table 1. Amino acid analysis of the acid hydrolysate (6 M HCl; 48 h; 110° C) (I) and gas chromatographic racemization test (II); peptide content: 78.5 %.

	I (amino acid ratios)	II (% D-amino acid)
Asp	3.01 (3)	2.0
Thr	1.01 (1)	<0.5
Glu	6.70 (7)	2.5
Ala	2.00 (2)	0.8
Ile	2.82 (3)	0.2
Leu	3.02 (3)	0.5
Lys	2.95 (3)	0.9
Arg	1.02 (1)	1.2

The synthetic sauvagine fragment 18-40 was found to be practically inactive (< 0.5 % as compared with natural sauvagine), both, in the rat antidiuresis test and in the inhibition of prolactin release in rats. Elongation of the C-terminal fragment by the methionine residue-17 did not restore any sauvagine activity as reported by Santangelo et al (9). These observations are further confirmed by a similar structure function study on CRF which showed that the CRF-fragment 10-41 corresponding to sauvagine fragment 9-40 is less than 0.1 % as potent as the the intact molecule (5,10).

References

1. Erspamer, V., P. Melchiorri. 1980. TIPS, 391.
2. Erspamer, V., P. Melchiorri. 1983. In: Neuroendocrine Perspectives, Vol. 2, (E.E.Müller and R.M.Macleod, eds.) Elsevier Science Publishers B.V., p. 37.
3. Montecucchi, P.C., A. Anastasi, R. de Castiglione, V. Erspamer. 1980. Int. J. Peptide Protein Res. 16, 191.
4. Montecucchi, P.C., A. Henschen. 1981. Int. J. Peptide Protein Res. 18, 113.
5. Vale, W., J. Spiess, C. Rivier, J. Rivier. 1981. Science 213, 1394.
6. Wünsch, E. 1967. Z. Naturforsch. 28b, 1269.
7. Riniker, B., M. Brugger, B. Kamber, P. Sieber, W. Rittel. 1969. Helv. Chim. Acta, 52, 1058.
8. Wünsch, E. 1972. Naturwissenschaften, 59, 239.
9. Santangelo, F., P.C. Montecucchi, L. Gozzini, H. Henschen. 1983. Int. J. Peptide Protein Res. 22, 348.
10. Rivier, J., J. Spiess, C. Rivier, R. Galyean, W. Vale, K. Lederis. 1983. In: Peptides 1982 (K.BláHa, P.Maloň, eds.) Walter de Gruyter & Co., Berlin-New York, p. 597.

Reprint from
Chemistry of Peptides and Proteins, Vol. 3
Editors: W. Voelter, E. Bayer, Y. A. Ovchinnikov, V. T. Ivanov

SYNTHESIS AND PROPERTIES OF A NEW SOMATOSTATIN STRUCTURAL ANALOG

Yu.P. Shvachkin, A.P. Smirnova, N.M. Ermak, V.P. Fedotov, N.V. Sadovnikova, G.N. Pluzhnikova

Institute of Experimental Endocrinology and Hormone Chemistry, Academy of Medical Sciences of the USSR, Moscow, USSR

Introduction

Synthesis of a new structural analog of somatostatin with Ala^5 and Orn^9 instead of Asp^5 and Lys^9 has been performed. It is found that $[Ala^5,Orn^9]$somatostatin preserves the typical for somatostatin ability to suppress somatotropin, insulin and glucagon secretion, but - contrary to the natural somatostatin - does not inhibit prolactin secretion.

Results

The hormone somatostatin is a tetradecapeptide of structure (I) (all asymmetric amino acids have L-configuration) (1).

```
 1              5                   10                  14
H-Ala-Gly-Cys-Lys-X-Phe-Phe-Trp-Y-Thr-Phe-Thr-Ser-Cys-OH
          |_______________________________________|
```

I. X = Asn, Y = Lys; II. X = Ala, Y = Orn;
III. X = Asn, Y = Orn; IV. X = Ala, Y = Lys.

At present this hormone is of interest for peptide chemists as well as for clinicians. Discovered as a hormone inhibiting somatotropin secretion, somatostatin has been found to inhibit secretion of insulin, glucagon, prolactin, thyreotropin hormone and some gut hormones. A number of structural analogs of somatostatin have been synthesized and studied with the aim of investigating general rules of structural-functional organization of the hormone and creating preparations with selective activities. Nowadays it is considered that Lys^9 is rather important for specific activity of the hormone (2,3) because positively charged ε-amino group of the amino acid residue interacts with receptor anionic centre of target cells, the contact being an important condition of successful hormone binding by the receptor (4). On the other hand, the presence of a positively charged group in the side chain of ninth amino acid residue of somatostatin alone is not sufficient, though necessary. The compound (III), containing Orn^9 instead of Lys^9, does not possess the specific somatostatin biological activity (5), though electronic characteristics of Orn^9 δ-amino group do not differ from ones of Lys^9 ε-amino group. The only difference between (I) and (III) is the absence of methylene group in the side chain of the ninth amino acid residue of the compound (III). The result shows the importance of steric factors. Probably, absence of the 1,27Å long methylene bridge makes it impossible for the NH_2-group of Orn^9 to reach the surface and to interact with the receptor. To compensate for the steric hindrance leading to loss of the biological activity, it is necessary to take into consideration the whole conformation of the somatostatin molecule and the peculiarities of intramolecular interactions of its amino acid residues.

The analysis of CD, NMR spectra, of the sedimentation and diffusion of somatostatin in aqueous solutions (3,6) shows that the hormone molecule forms a hairpin structure with disulfide bridge between Cys^3 and Cys^{14} with Trp^8(6) at the U-turn region. In such a conformation Lys^9 is located near the Asn^5 and the substitution of Orn^9 for Lys^9 makes the δ-amino group of Orn^9 sterically blocked by Asn^5 side chain which blocks its participation in hormone - receptor interaction.

This kind of steric relations in somatostatin molecule brought us to the conclusion that the only way to retain the biological activity of the molecule after substitution of Orn^9 for Lys^9 is the simultaneous replacement of Asn^5 by another amino acid residue with a small and nonpolar side chain to prevent sterical hindrance and intramolecular binding of Orn^9 δ-amino group. These conditions are met with alanine. Such a substitution looks quite reasonable if one takes into consideration that Ala^5-somatostatin possesses hormonal activity (7). So, analysis of experimental data and theoretical suppositions concerning the specificity of biologically active conformation of somatostatin molecule in solution lead us to a hypothesis: simultaneous two-point re-

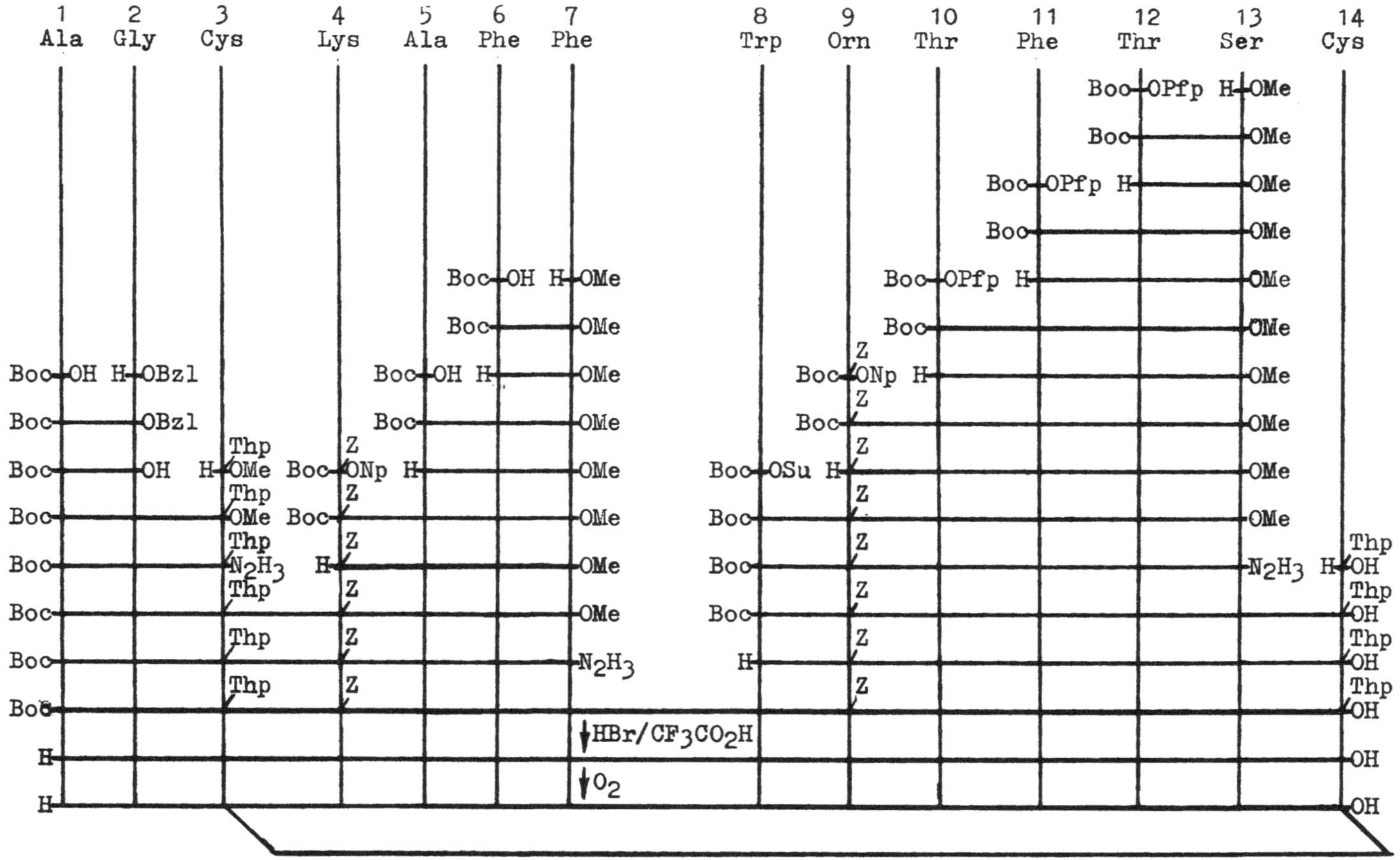

Abbreviations: BOC, tert-butyloxycarbonyl; Z, benzyloxycarbonyl; Thp, tetrahydropyranyl; OSU, N-hydroxysuccinimide ester; ONp, p-nitrophenyl ester; OPfp, pentafluorophenyl ester.

placement of Lys^9 by Orn and of Asn^5 by Ala may result in formation of biologically active somatostatin analog.

To examine the hypothesis, chemical synthesis of previously unknown [Ala^5,Orn^9] somatostatin (II) was undertaken in solution according to (1-7)+(8-14) scheme. The peculiarity of the scheme is utilisation of tetrahydropyranyl group for protection of sulfhydryl functions of Cys^3 and Cys^{14} at intermediate steps. The reason of using 7+7 scheme is better separation of the desired product (protected tetradecapeptide) from the starting compounds. To obtain the heptapeptides, the step-by-step elongation of the peptide chain (fragments 1-3, 4-7, 8-14) or block condensation was employed. The tert-butyloxycarbonyl group was used for protection of α-amino functions; the benzyloxycarbonyl - for ω-amino groups of Lys^4 and Orn^9; esterification or salt formation - for carboxylic groups; the hydroxyl groups of Thr^{10},Thr^{12} and Ser^{13} were left unprotected. The mixed anhydrides or activated esters methods were applied for producing the short peptides with Gly residue on the C-end; the longer fragments were joined by azide method.

The tetradecapeptide synthesized was deprotected with HBr in trifluoracetic acid. Formation of the disulfide bridge between Cys^3 and Cys^{14} residues was achieved at the final stage of the synthesis. Analytically pure compound (II) was isolated by gel-chromatography.

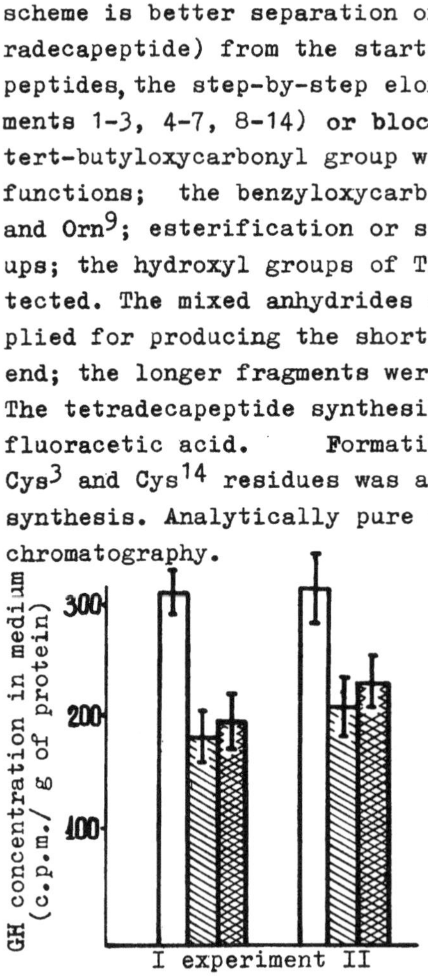

Fig.1. Influence of somatostatin and its analog (2 μg/ml) on secretion of T-labelled GH by adenohypophysis cells.
□-control ▩-somatostatin
▧-[Ala^5, Orn^9] somatostatin

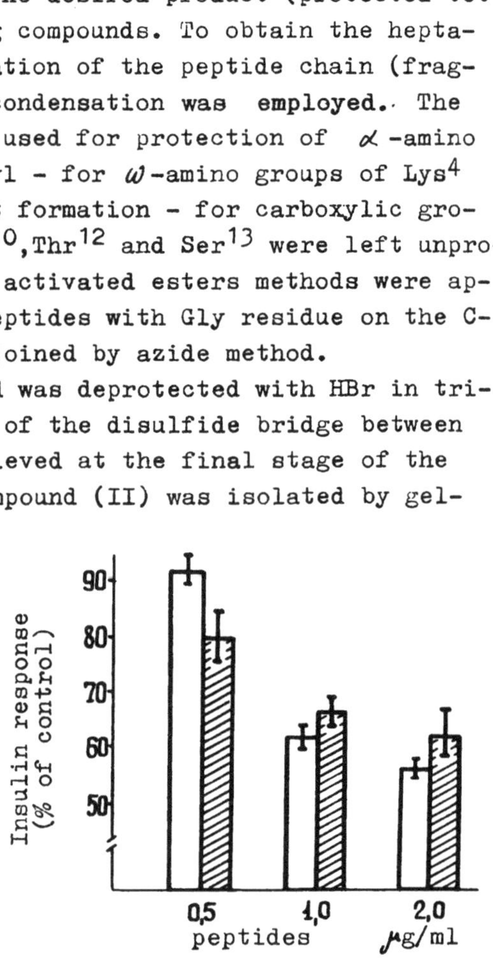

Fig.2. Inhibition of insulin secretion by somatostatin and its analog in islet cells culture of new-born rats. □-somatostatin
▨-[Ala^5, Orn^9] somatostatin

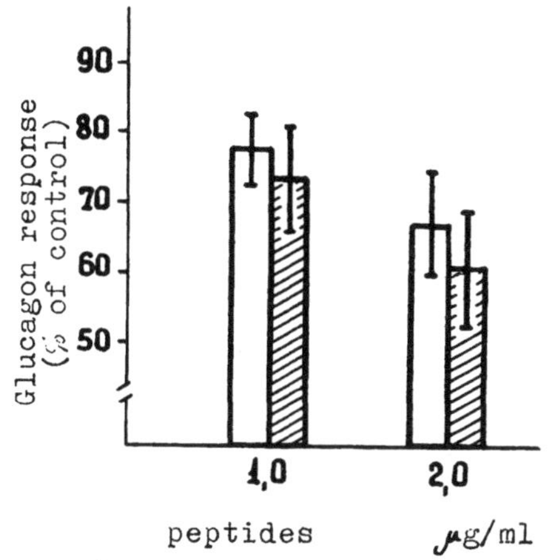

Fig.3. Inhibition of glucagon secretion by somatostatin and its analog in islet cells culture of new-born rats.

□ -somatostatin
▨ -[Ala^5, Orn^9] somatostatin

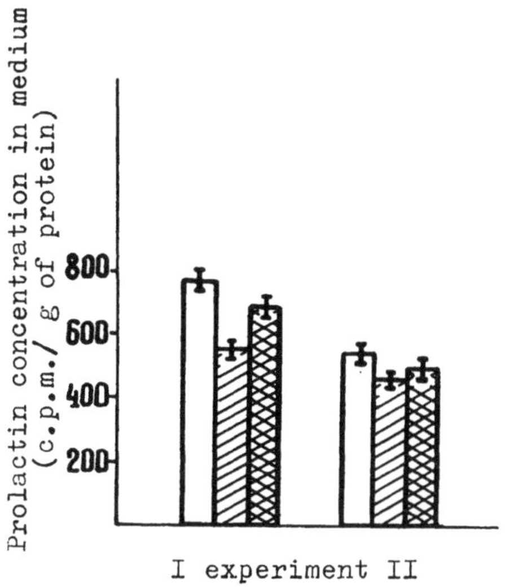

Fig.4. Influence of somatostatin and its analog (2μg/ml) on secretion of T -labelled prolactin by adenohypophysis cells.

□ -control ▨ -somatostatin
▩ -[Ala^5, Orn^9] somatostatin

Investigation of the biological activity of compound (II) was performed in primary culture of isolated adenohypophysis cells and rat islets cells. The specific hormonal activity of [Ala^5,Orn^9] somatostatin was estimated by measuring the inhibition of somatotropin, prolactin, insulin and glucagon secretion with the use of somatostatin as a control standard (8). The data obtained show that [Ala^5,Orn^9] somatostatin preserves characteristic ability to suppress somatotropin, insulin and glucagon secretion, the effect being dose-dependent and statistically reliable($P<0,01$)(Fig.1,2,3). The result confirms the suggestion of constructing a new biologically active analog by simultaneous replacement of two amino acid residues in somatostatin molecule.

Isolated rat adenohypophysis cells were used to investigate the influence of the analog on prolactin secretion. The data obtained show, that, **contrary to** somatostatin, its [Ala^5,Orn^9] analog does not inhibit this function (Fig.4). So, it was found that somatostatin and its new analog have statistically reliable different effects concerning somatotropin and prolactin.

References

1. Brazeau, P., W. Vale, R. Burgus, N. Ling, M. Butcher, J. Rivier, R. Guillemin. 1973. Science 179, 77-79.

2. Vale, W., J. Rivier, N. Ling, M. Brown. 1978. Metabolism 27, 1391-1401.

3. Nutt, R.F., D.F. Veber, P.E. Curley, R. Saperstein, R. Hirschmann. 1983. Int.J.Peptide Protein Res. 21, 66-73.

4. Hirst, B.H., D.H. Coy. 1984. Regul.Pept. 8, 267-271.

5. Rivier, J., M. Brown, C. Rivier, N. Ling, W. Vale. 1976. In: Peptides (ed. A. Loffet). Bruxelles, p. 453-465.

6. Holladay, L.A., D. Puett. 1976. Proc.Natl.Acad.Sci.USA 73, 1199-1202.

7. Brown, M., J. Rivier, W. Vale. 1976. Endocrinology 98, 336-343.

8. Shvachkin, Yu.P., A.P. Smirnova, A.A. Shishkina, N.M. Ermak, V.P. Fedotov, I.S. Komolov, L.G. Morozova, G.N. Pluzhnikova, N.V. Sadovnikova. 1979. Bioorgan.Chemistry 5, 169-180.

IMMUNOLOGICALLY ACTIVE PEPTIDES OF THE THYMUS

Wolfgang Voelter, Hartmut Echner, and Hubert Kalbacher
Abteilung für Physikalische Biochemie des Physiologisch-chemischen Instituts der Universität Tübingen, Hoppe-Seyler-Straße 1, D-7400 Tübingen

Introduction

The endocrine thymus plays on important role in immune regulation (1). A lymphopoietic factor, isolated from rat and mouse thymus,was first reported from the laboratory of A.L. Goldstein et al. (2,3) in1965 and 1966. Since then several laboratories isolated pure peptides mainly from calf thymus fractions and determined their primary structure.

Thymus polypeptide β_1, showing no "thymic hormone" activity, is homologous with ubiquitin and the N-terminal amino acids of the nonhistone chromosomal protein A-24 (4,5). Since then it could be demonstrated that ubiquitin plays an important role in protein modification and breakdown (6).

```
1               5                  10
MET-GLN-ILE-PHE-VAL-LYS-THR-LEU-THR-GLY-
                15                 20
LYS-THR-ILE-THR-LEU-GLU-VAL-GLU-PRO-SER-
                25                 30
ASP-THR-ILE-GLU-ASN-VAL-LYS-ALA-LYS-ILE-
                35                 40
GLN-ASP-LYS-GLU-GLY-ILE-PRO-PRO-ASP-GLN-
                45                 50
GLN-ARG-LEU-ILE-PHE-ALA-GLY-LYS-GLN-LEU-
                55                 60
GLU-ASP-GLY-ARG-THR-LEU-SER-ASP-TYR-ASN-
                65                 70
ILE-GLN-LYS-GLU-SER-THR-LEU-HIS-LEU-VAL-
LEU-ARG-LEU-ARG-COOH
```

Figure 1. Primary structure of thymus polypeptide β_1.

For the peptides, isolated from calf thymus extracts, the following nomenclature is applied according to their isoelectric points (pI): α thymosins (pI < 5.0), β thymosins (pI 5.0-7.0) and γ thymosins (pI > 7.0. Thymosin α_1 was the first immunologically active polypeptide isolated from bovine thymosin fraction 5 (1). It has been shown that thymosin α_1 amplifies T cell immunity and

Chemistry of Peptides and Proteins, Vol. 3

induces terminal deoxynucleotidyl transferase in prothymocytes (1).

```
 O    1             5
 ‖
H3C-C-N-SER-ASP-ALA-ALA-VAL-ASP-THR-SER-
    10                  15
SER-GLU-ILE-THR-THR-LYS-ASP-LEU-LYS-GLU-
    20                  25
LYS-LYS-GLU-VAL-VAL-GLU-GLU-ALA-GLU-ASN-
COOH
```

Figure 2. Primary structure of thymosin α_1.

Most recently the primary structure of rat prothymosin α was determined (7) and it has been shown that thymosin α_1 corresponds to des-(113-29)-prothymosin α; therefore it seems that thymosin α_1 is produced in the thymus from a larger peptide.

```
ACSER-ASP-ALA-ALA-VAL-ASP-THR-SER-SER-GLU(10)-
  ILE-THR-THR-LYS-ASP-LEU-LYS-GLU-LYS-LYS(20)-
  GLU-VAL-VAL-GLU-GLU-ALA-GLU-ASN-GLY-ARG(30)-
  ASP-ALA-PRO-ALA-ASN-GLY-ASN-ALA-GLN-ASN(40)-
  GLU-GLU-ASN-GLY-GLU-GLN-GLU-ALA-ASP-ASN(50)-
  GLU-VAL-ASP-GLU-GLU-GLU-GLU-GLU-GLY-GLY(60)-
  GLY-GLU-GLU(ASX,GLY,GLY,GLX,GLX,GLX,GLX(70),
  GLX,GLX,GLX,GLX,GLX,GLX)ASN-GLY-ASP-GLU(80)-
  ASP-GLU-GLU-ALA-GLU-ALA-PRO-THR-GLY-LYS(90)-
  ARG-VAL-ALA-GLU-ASP-ASP-GLU-ASP-ASP-ASP(100)-
  VAL-GLU-THR-LYS-LYS-GLN-LYS-LYS-THR-ASP(110)-
  GLU-ASP-ASPOH
```

Figure 3. Primary structure of rat prothymosin α.

From the group of G.Goldstein three closely related peptides inducing selectively T cell differentiation were isolated from bovine thymus (thymopoietin I and II) and bovine spleen (thymopoietin III) (8). It has been shown that thymopoietin 32-36 has almost identical biological activity in a number of assays as the parent peptide.

```
     1                  5
H2N-PRO-GLU-PHE-LEU-GLU-ASP-PRO-SER-VAL-
10                  15
LEU-THR-LYS-GLU-LYS-LEU-LYS-SER-GLU-LEU-
20                  25
VAL-ALA-ASN-ASN-VAL-THR-LEU-PRO-ALA-GLY-
30                  35
GLU-GLN-ARG-LYS-ASP-VAL-TYR-VAL-GLU-LEU-
40                  45              49
TYR-LEU-GLN-SER-LEU-THR-ALA-LEU-LYS-ARG-COOH
```

Figure 4. Primary structure of thymopoietin II.

The first biologically active and completely characterized polypeptide from the ß thymosin series, isolated from thymosin fraction 5, was thymosin $ß_4$ (Ac-Ser-Asp-Lys-Pro-Asp-Met-Ala-Glu-Ile-Glu-Lys-Phe-Asp-Lys-Ser-Lys-Leu-Lys-Lys-Thr-Glu-Thr-Gln-Glu-Lys-Asn-Pro-Leu-Pro-Ser-Lys-Glu-Thr-Ile-Glu-Gln-Glu-Lys-Gln-Ala-Gly-Glu-Ser-OH) (9). In the meantime homologous peptides(thymosin $ß_8$, $ß_9$, $ß_{10}$, $ß_{11}$), were also structurally determined. Through thymosin $ß_4$ induces the terminal deoxynucleotidyl transferase, which is specific for prothymocytes, it was recognized that this peptide occurs almost ubiquitously in many tissues (spleen, brain, lung, liver, heart) and one expects to discover further biological activities caused by this compound (10).

Results

For testing immunological activities in vitro as well as in vivo we developed solution syntheses of thymopoietin 32-36, the phagocytosis-stimulating peptide tuftsin and fragments of thymosin $ß_9$.

The strategy for our thymopoietin 32-36 synthesis is outlined in Figure 5. For amino protection the Adpoc, developed in our laboratory (11,12), and the Z group are used. The Adpoc group can be cleaved off under very mild acidolytic conditions (3 % TFA/ CH_2Cl_2) without the attack of the Boc and the t-butyl residues used for the side chain protection. Coupling with DCCI/HOSu or DCCI/HOBt results in similar yields.

In the course of our synthesis of thymopoietin 32-36 (Figure 5) we found, that the p-methoxybenzenesulfonyl group (13) is not fully cleaved under the chosen conditions from the guanidino residue of arginine. Therefore we introduced for the first time the 2,4,6-triisopropylbenzenesulfonyl (TIP) residue for the protection of arginine which can be cleaved off in methanesulfonic acid to 1oo %. The first application of the TIP group for the synthesis of arginine-containing peptides is seen from Figure 6 for the preparation of tuftsin.

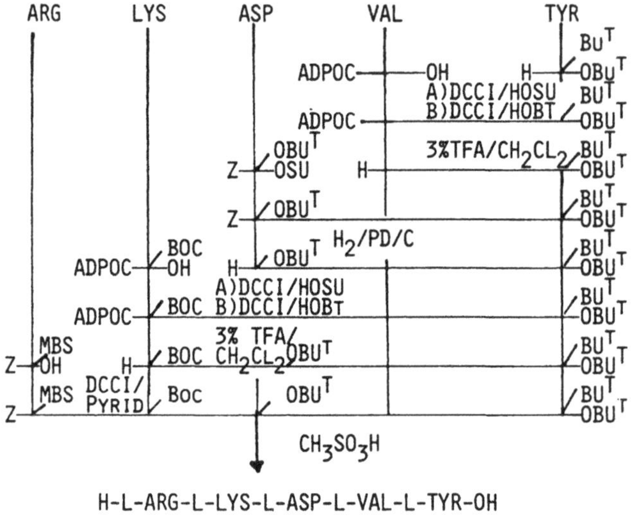

Figure 5. Strategy for the synthesis of thymopoietin 32-36.

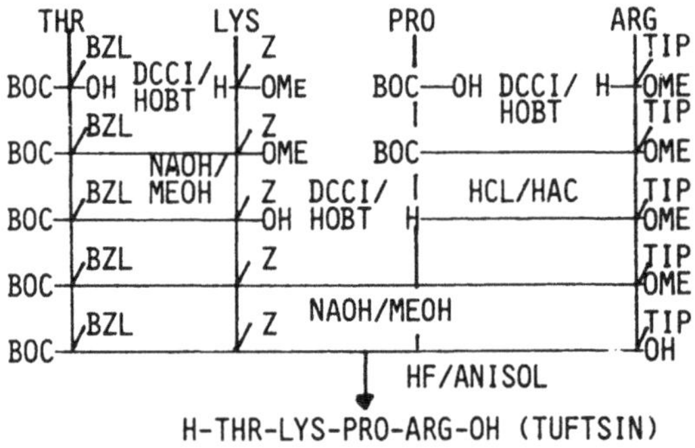

Figure 6. Strategy for the synthesis of tuftsin.

In connection with the total synthesis of thymosin β_9 the fragments, collected in Figure 7 were synthesized. The side chains of Asp, Glu, Ser, Thr were protected by t-butyl those of Lys by Boc residues.

Detailed information about the applied coupling and cleavage methods for the syntheses of the thymosin β_9 fragments is seen from Figure 8 which, as an example, demonstrates the strategy for the synthesis of the thymosin β_9 1-7 fragment.

1
AC-L-ALA-L-ASP(OBU^T)-L-LYS(BOC)-L-PRO-L-
7
ASP(OBU^T)-L-LEU—GLY-OH
8
Z-L-GLU(OBU^T)-L-ILE-L-ASN-L-SER(BU^T)-L-PHE-
14
L-ASP(OBU^T)-L-LYS(BOC)-OH
15 17
Z-L-ALA-L-LYS(BOC)-L-LEU-OH
18 21
Z-L-LYS(BOC)-L-LYS(BOC)-L-THR(BU^T)-L-GLU-
(OBU^T)-OH
22 24
Z-L-THR(BU^T)-L-GLN-L-GLU(OBU^T)-OH
25
Z-L-LYS(BOC)-L-ASN-L-THR(BU^T)-L-LEU-
29
L-PRO-OH
30 32
Z-L-THR(BU^T)-L-LYS(BOC)-L-GLU(OBU^T)-OH
33
Z-L-THR(BU^T)-L-ILE-L-GLU(OBU^T)-L-GLN-L-
37
GLU(OBU^T)-OH
38 41
H-L-LYS(BOC)-L-GLN-L-ALA-L-LYS(BOC)-OBU^T

Figure 7. Protected peptide fragments to the total synthesis of thymosin β_9.

The synthetic peptides as well as a peptide mixture, isolated from bovine thymus, were tested for their phagocytosis enhancing as well as their carbon clearance effect. From the peptides tested thymopoietin pentapeptide shows relatively strong activities (Tables 1,2) (14).

Table 1. Phagocytosis enhancing effect (%) of peptides for human blood granulocytes (5 x 10^6 g/ml) subjected to yeast particles (3 - 5 x 1o^7 p/ml).

Conc. (mg/ml)	1	1o^{-1}	1o^{-2}
Thymopoietin pentapeptide	12	33	18
Thymus peptide mixture (MW > 6000)	15	22	11
Tuftsin	36	24	9

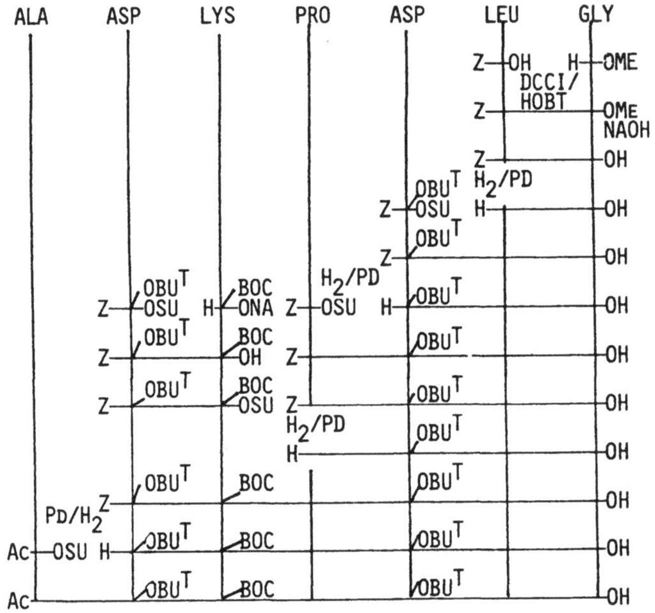

Figure 8. Strategy for the synthesis of thymosin β_9 fragment 1-7.

Table 2. Effect of peptides in carbon clearance in female NMRI mice (injection of 0.3 ml carbon particle (200-250 Å) suspension (16 mg/ml)).

Peptide		RC_{TR} / RC_C
Tuftsin	1o mg/kg	1.14
	30 mg/kg	1.56
Thymopietin pentapeptide		
	1o mg/kg	1.62
	30 mg/kg	2.10

The effect of thymopoietin 32-36 on the growth of a fibrosarcoma induced by methylcholanthrene (Meth.A.asc.) is tested in female CFB_1 (Balb/c ♀ C57 o) mice. Although the tumor growth is decisively reduced during the first 14 days, none of the animals survives longer than 28 days. Thymosin β_9 1-7 and the whole thymus extract show similar effects (cf. Table 3).

Table 3. Influence of thymopoietin 32- 36, thymosin β_9 1-7, and whole thymus extract on the growth of the fibrosarcoma induced by methylcholanthrene.

	Tumor volume in %			
	7	14	21	28 days after transplantation
Control	100	100	100	100
Thymopoietin 32-36	96	67	93	93
Thymosin β_9 1-7	110	43	55	73
Thymus extract	97	55	50	71

References

1. Goldstein, A.L., Low, T.L.K., Thurman, G.B., Zatz, M.M., Hall, N., Chen, J., Hu, S.-K., Naylor, P.B., McClure, J.E.: Rec. Progr. Hormone Res. 37, 369 (1981).
2. Klein, J.J., Goldstein, A.L., White, A.: Proc.Natl.Acad.Sci. U.S.A. 53, 812 (1965).
3. Klein, K.J., Goldstein, A.L., White, A.: Ann. N.Y.Acad.Sci. 135, 485 (1966).
4. Low, T.L.K., Goldstein, A.L.: J.Biol. Chem. 254, 987 (1979).
5. Hunt, L.T., Dayhoff, M.O.: Biochem. Biophys. Res. Commun. 74, 650 (1977).
6. Hershko, A.: Cell 34, 11 (1983).
7. Haritos, A.A., Blacher, R., Stein, S., Caldarella, J., Horecker, B.L.: Proc.Natl. Acad. Sci. U.S.A. 82, 343 (1985).
8. Audhya, T., Schlesinger, D.H., Goldstein, G.: Biochemistry 20, 6195 (1981).
9. Low, T.L.K., Hu, S.-K., Goldstein, A.L.: Proc.Natl. Acad.Sci. U.S.A. 78, 1166 (1981).
10. Hannappel, E.: Habilitationsschrift,Erlangen-Nürnberg 1983.
11. Kalbacher, H., Voelter, W.: Angew. Chem. 90, 998 (1978); Angew.Chem.Int.Ed.Engl. 17, 944 (1978).
12. Voelter, W.: Advances in and Comparison of Protective Groups for Peptide Synthesis, in: Polypeptide Hormones, p.135, Beers, R.F.Jr., Bassett, E.G. (eds), Raven Press, New York 1980.
13. Nishimura, O., Fujino, M.: Chem.Pharm.Bull. 24, 1568 (1976).
14. Voelter,W., Kalbacher, H., Echner, H., Wagner, H.: Liebigs Ann.Chem., im Druck.

SYNTHESIS AND BIOLOGICAL ACTIVITY OF IMMUNOGLOBULIN G PEPTIDE FRAGMENTS

O.A.Kaurov, N.I.Kolodkin, G.P.Kazakov, B.D.Halaypin
Institute of Pure Biochemicals of the Main Board of Microbiological Industry, Pudozhskaya 7, Leningrad, 197110, USSR

Introduction

Activation of the classical complement cascade by immunoglobulin G (IgG) involves the binding of the first complement component (C1) to a multivalent antigen-IgG complex. The binding occurs by the interaction of a site on the surface of the C_H2 domain of IgG with a globular head of the $C1_q$ subcomponent (1,2).

Thomas J.Lukas et al.(2) estimated the ability of the synthetic peptides from the 281-292 region of the C_H2 domain of IgG to inhibit the C1-mediated immune hemolysis. This region contains some positively charged residues (His-285, Lys-288, Lys-290, Arg-292) which may be involved in the $C1_q$-binding site of human IgG.

In order to estimate the role of the positive charges in IgG-$C1_q$ interaction it was of interest to synthesize the 285-292 region peptides having different charges of the amino acid residues in position 285.

The 282-292 peptide (Val-11 peptide) and tuftsin were synthesized in order to compare our results concerning C1 inhibition assay with those of Lukas.

Results

All peptides were prepared by the solid-phase method using Merrifield resin (1% divinylbenzene). The protective groups and the activation methods are shown in Fig.1. The first amino acid was esterified by the resin by the method of Gisin (3). The amount of Boc-Arg(NO_2)-OH per gram resin, which was measured by the

Chemistry of Peptides and Proteins, Vol. 3

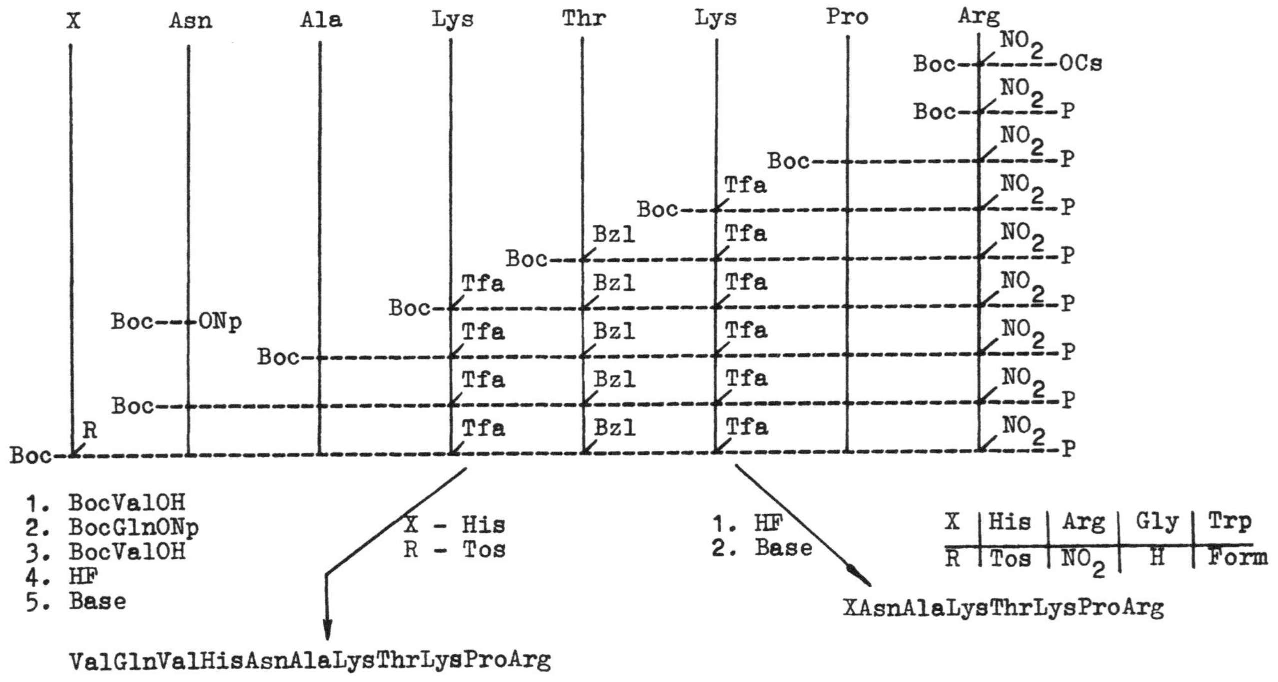

Fig. 1. Synthesis of IgG 282-292 peptide fragment and its analogues.

picrate monitoring method of Gisin (4) was 0.4 mmol/g. A detailed protocol for one synthetic cycle is given in Table 1.

Table 1. Protocol for One Synthetic Cycle .

Step	Reagent or Solvent	Time(min)
A, deprotection	CH_2Cl_2	1
	50% TFA in CH_2Cl_2	2
	- " -	30
	CH_2Cl_2, 6 times	1
B, neutralization	10% Et_3N in CH_2Cl_2, 3 times	2
	CH_2Cl_2, 6 times	1
C, coupling	3 eq Boc-AA	2
	3 eq DCC	120
	CH_2Cl_2, 3 times	1
	i-PrOH, 3 times	1
	CH_2Cl_2, 3 times	1
D, ninhidrin test		

The coupling reaction was monitored with ninhidrin assay. If the result was unacceptable the coupling cycle was repeated. Asparagine and glutamine were coupled as nitrophenyl esters in the presence of 1-hydroxybenzotriazole. The time of the reaction was 5 hours. Peptides were cleaved from the resin by treatment with anhydrous HF (1 hour, 0°C).

Analytical reverse-phase HPLC of the crude TFA-protected peptide 289-292 (TFA-tuftsin) is shown in Fig.2A. Some peaks of the by-products were identified using standard peptides specially synthesized. The percentage content of some impurities is given in brackets. It is of interest to note that some quantity of unprotected tuftsin was discovered after anhydrous HF treatment of the peptidyl-resin containing ε-TFA-lysine. The crude TFA-tuftsin was treated with 1N piperidine during 2 hours at 0°C. Analytical HPLC of the final product is given in Fig.2B. Analytical HPLC of the mixture of the diastereomeric tuftsin analogues (5) is given

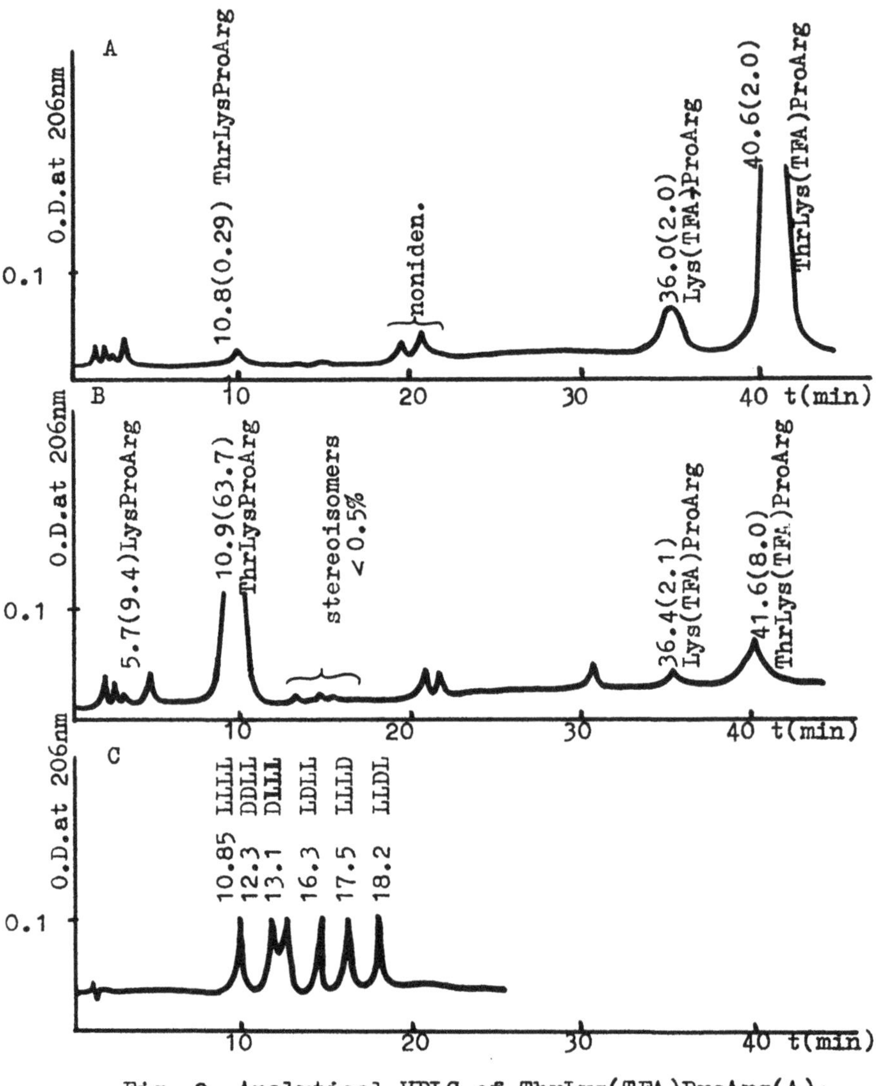

Fig. 2. Analytical HPLC of ThrLys(TFA)ProArg(A); ThrLysProArg(B); tuftsin stereomers(C). Column - Ultrasphere ODS (5μ, 0.4×25cm); Flow rate-1ml/min Eluent - linear gradient CH_3CN from 0 to 20% for 45 min. in 0.5M ammonium phosphate buffer, pH-4.6.

in Fig.2C.

Comparison of the chromatograms B and C suggests that the degree of the racemization in this synthesis was less then 0.5%. It is clear also from the chromatogram B that some ε -TFA-lysine groups were not cleaved under the used conditions. That is why we decided to study several conditions of the TFA-group cleavage using HPLC control. The results are given in Fig.3.

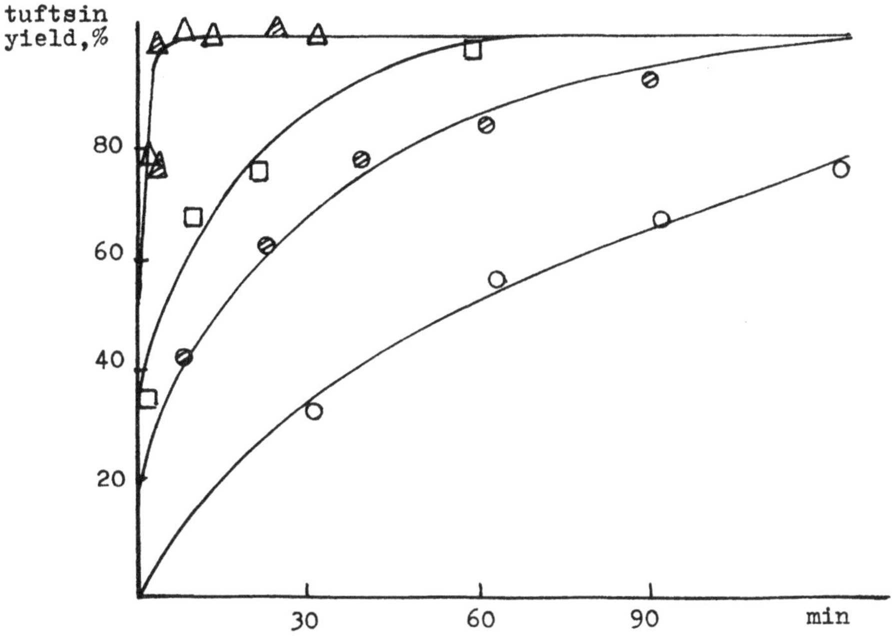

Fig.3. Kinetic curves of TFA-tuftsin hydrolysis with different bases; (▲)-1N piperidine, 25°C; (△)-0.2N NaOH, 25°C; (□)-0.2N TEA, 25°C; (⊘)-1N piperidine, 0°C; (○)-1N NH_4OH, 25°C.

TFA-group was removed with the highest rate by 0.2N NaOH or 1N piperidine at room temperature. It takes much more time to cleave that group with 0.2N piperidine at 0°C. The comparative data of the TFA cleavage for tuftsin and Boc-Lys(TFA)-OH are given in Table 2.

Table 2. TFA-group Removal with Different Bases.

Conditions	Time of TFA-group hydrolysis		
	Experimental data		Lit.data
	BocLys(TFA)OH	ThrLys(TFA)ProArg	
0.2N NaOH, 27°C	6 min	6 min	10 min (6)
1N piperidine, 0°C	3 hours	3 hours	1-2 h (7)
1N piperidine, 37°C	20 min	10 min	3 h (8)
7N NH_4OH, 27°C	20 hours		2 h (9)

One can see that there are some differences in comparison with the data found in literature. To our opinion the treatment of the peptides with 0.2N NaOH during 6-10 minutes at room temperature is the optimal condition for the TFA-group removal. This procedure was used for the preparation of all the final peptides. The peptidyl-resin was treated with anhydrous HF and 0.2N NaOH. After that they were desalted and purified by various methods (e.g. ion-exchange chromatography, gel filtration, etc.). Some peptides were purified by means of reverse-phase HPLC, using semipreparative column with the following desalting. If 0.1% TFA was used as a buffer the following desalting was unnecessary. The homogeneity and the composition of the purified synthetic peptides were confirmed by means of TLC, HPLC and amino acid analysis.

The peptides were tested for their ability to inhibit the classical complement pathway by a modified method(10). The serum containing human complement, sheep erytrocytes bearing rabbit anti-sheep erytrocyte IgG and veronal buffer containing 0.2 mM $MgCl_2$ were used as usual. To measure the complement quantity in the test system a novel kinetic method was elaborated.

Several years ago Boakle and Pruitt(11) showed that erytrocyte concentration is directly proportional to the optical density of the hemolytic system at λ=700-800 nm. We used these data for the kinetic research of the human serum complement lytic action.

The kinetic curves obtained by means of Hitachi model 330 spectrophotometer are given in Fig.4.

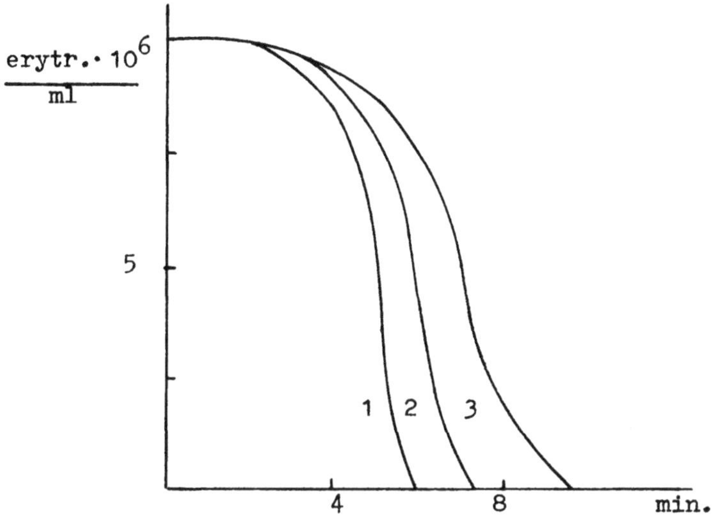

Fig.4. Kinetic curves of human complement lysis. 1-25µl standard serum in test-system; 2-20µl; 3-15µl.

The value of the maximal rate of the target-cell lysis (V_m) is determined as the value of the slope of the line tangent to the kinetic curve in the point of maximal slope. It may be used as a measure of the complement activity. The V_m-values are directly proportional to the complement quantity in this system. The V_m-values as the measure of the complement activity was used in our complement inhibition assay.

The peptides resolved in veronal buffer (50 ml) were incubated with serum (50µl) for 30 minutes at 37°C. Then the solution containing $1.5 \cdot 10^8$ erytrocyte was added and the V_m-values were measured. The percentage of the complement-mediated hemolytic residual activity was calculated by comparison of the V_m-values in the presence and absence of peptides. The concentrations of the peptides at which 50% of the complement-mediated immune hemolysis was inhibited (I_{50} constant) were calculated by means of the linear regression analysis of the experimental points. The I_{50} constants and relative molar activity of the synthesized peptides are given in Table 3.

Table 3. Inhibition of the Classical Complement Pathway by Synthetic Peptides.

PEPTIDE	peptide code	exp. I_{50} (μM)	rel. mol. activity	lit. I_{50} (μM)
ValGlnValHisAsnAlaLysThrLysProArg	Val-11	185	[1.00]	175
ThrLysProArg	tuftsin	670	0.28	720
HisAsnAlaLysThrLysProArg	His-8	820	0.23	-
ArgAsnAlaLysThrLysProArg	Arg-8	450	0.41	-
GlyAsnAlaLysThrLysProArg	Gly-8	*		
TrpAsnAlaLysThrLysProArg	Trp-8	*		

* Peptides were inactive at the concentrations 100–500 μM .

It is obvious from this Table that our simplified method for estimation of peptide ability to inhibit the classical complement pathway gives the same results as those of Lukas. According to Lukas, the Val-11 synthetic peptide containing four positively charged side chains inhibits C1 mediated immune hemolysis as efficiently as monomeric IgG $C1_q$-binding site itself(2). Histidine residues about 10% of which exist in the positively charged imidazolium form at pH 7.4 are considered as positively charged. However the I_{50} value of the His-8 peptide is lower than for tuftsin. Thus these data conflict with the assumption concerning the role of the histidine positive charge in IgG-$C1_q$ interaction. On the other hand the replacement of the imidazole-ring in octapeptide sequence (Gly^8 and Trp^8 analogues) destroys the biological activity. Introduction of a positively charged guanidinium function instead of the imidazole group leads to the increase of the inhibition activity. So it may be supposed that the decrease in the I_{50} value of the His-8 peptide in comparison with that of tuftsin is to be accounted for by conformational reasons. Attachment of ValGlyVal sequence to His-8 peptide is likely to

change the spatial arrangment of the peptide positively charged groups. It becomes similar to that in IgG molecule which gives vise to the increase of inhibition activity.
If it is correct, attachment of the same sequence to Arg-8 peptide has to increase the inhibition activity of the final undecapeptide. Such peptide synthesis is in progress in our laboratory. In order to study the mechanism of IgG-C1 interaction we now are synthesizing some peptides from the complementary site on the peripheral globular heads of the first complement component $C1_q$ subcomponent.

References

1. Prystowsky, M.B., Kehoe, J.M., Erickson, B.W. 1981. Biochemistry 20, 6349.
2. Lukas, T.J., Munoz, H., Erickson, B.W. 1981. J.Immunol. 127, 2555.
3. Gisin, B.F. 1973. Helv.Chim.Acta 56, 1476.
4. Gisin, B.F. 1972. Anal.Chim.Acta 58, 248.
5. Kaurov, O.A., Prusakov, A.N., Pivovarov, A.M., Pasechnik,V.A. 1983. In: Peptides 1982 (K.Blaha and P.Malon, eds.), Walter de Gruyter Berlin New York, p.257.
6. Weigand, F., Csends, E. 1952. Angew.Chem. 66, 136.
7. Anfinsen, C.B., Ontjes, D., Ohno, M., Corley, L., Eastlake,A. 1967. Proc.Natl.Acad.Sci. U.S.A. 58, 1806.
8. Osterman, D., Mora, R., Kezdy, F.Y., Kaiser, E.T., Meredith S.C. 1984. J.Am.Chem.Soc. 106, 6845.
9. Moore, A.T., Rydon, H.N., Smithers, M.J. 1966. J.Chem.Soc.(C) 24, 2349.
10. Haljapin, B.D., Prokopjev, A.A.; (in press).
11. Boackle, R.J., Pruitt, T.M., Mestecky, J. 1974. Immunochemistry 11, 543.

SYNTHESIS OF PEPTIDE FRAGMENTS OF HORSE HEART CYTOCHROME C

R.P. Evstigneeva
Lomonosov Institute of Fine Chemical Technology, Moscow, USSR

Introduction

The general development of peptide synthesis to-day allows to carry out rather sophisticated investigations aimed at understanding the structure-function relationships in naturally occurring proteins. The synthesis of more elaborate peptides in its turn enriches and broadens the methodology of peptide synthesis. In this paper the synthesis is described of peptide fragments of horse heart cytochrome *c* that was undertaken with the purpose of immunochemical study of cytochrome *c* (1,2).

Results

Fragments 1-13; 23-56; 57-81; 82-104; 23-104 of the horse heart cytochrome *c* peptide chain were synthesized. In the syntheses of these peptides the tactics of minimal protection of trifunctional amino acids was generally adhered to, though the lysyl residues were always N -substituted with trifluoroacetyl groupings (Fig. 1). In the long run we had many opportunities to see that this tactical approach was quite advantageous for the synthesis of more complex peptides, because the final purification of the reaction products was considerably facilitated.

I. Synthesis of fragment 1-13. This is the N-terminal fragment of the cytochrome *c*, beginning with an acylated glycine. Consequently we have chosen the following scheme of fragment condensation as a more rational one (Fig. 2) Ac-Gly was introduced at the final stage of the synthesis.

The stepwise chain elongation for shorter peptides was carried out using pivaloyl chloride and isobutyl chloroformate (IBC). The

Chemistry of Peptides and Proteins, Vol. 3

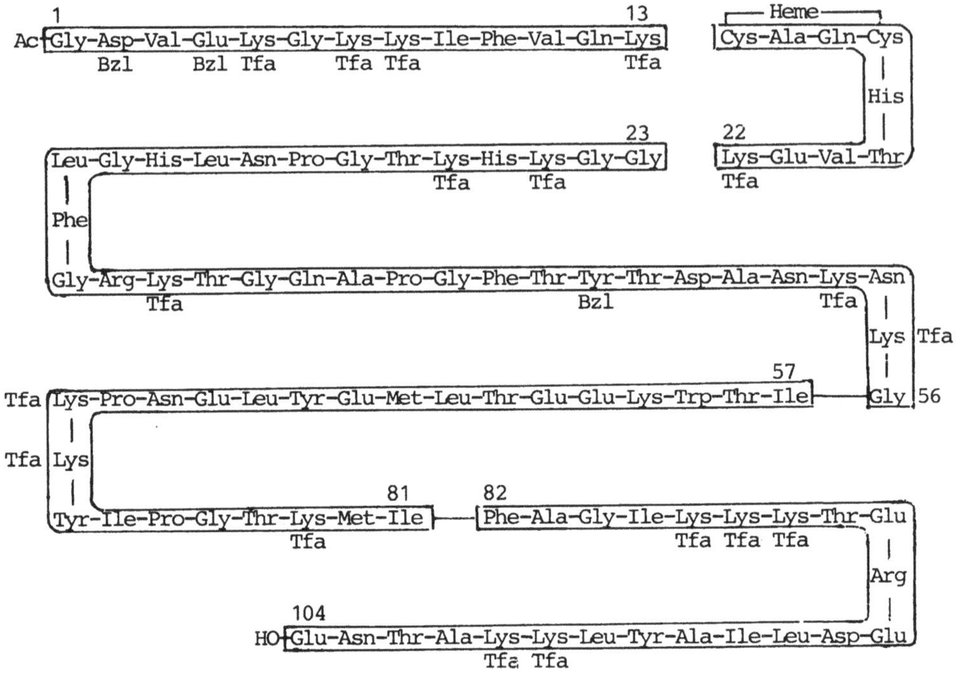

Fig. 1. Peptide fragments of amino acid sequence of horse heart cytochrome *c*.

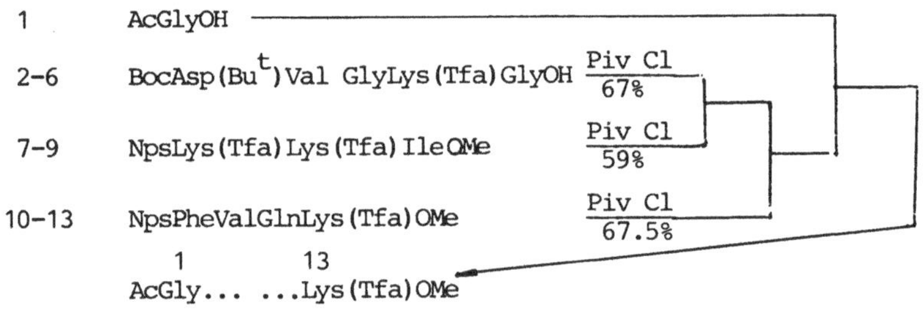

Fig. 2. Synthetic route to N-terminal peptide fragment 1-13.

fragments were coupled using pivaloyl chloride.
On the example of the synthesis of this peptide one can see the wide possibilities of the mixed anhydride approach in synthetic peptide chemistry.

II. Synthesis of fragment 23-56. Synthesis of the polypeptide 23-56 fragments was reported earlier (3). It is noteworthy that condensation of fragments with C-terminal N-trifluoroacetyllysine (27,53,55) and Asn (31,52) was carried out using pivaloyl chloride (Fig. 3). The same reagent was used for tetrapeptide 46-49 synthesis by fragment condensation; in this case Nps-Phe and Boc-Tyr(Bzl) were activated with pivaloyl chloride; unsubstituted Thr was used as the amino component. This was the only example of Tyr (48) hydroxyl group protection in the total synthesis of cytochrome c amino acid sequence (Polypeptide 82-104 also contained tyrosines 67,74,97, which were not protected). Pivaloyl chloride was also used for coupling of fragments corresponding to 39-56 sequence of cytochrome c. Undecapeptide 46-56 was prepared with a yield of 93%, and a 18-membered peptide 39-56 with a yield of 72%. (If fragment 50-56 was used with unsubstituted C-terminal Gly, the condensation proceeded with yields of 68 and 38% respectively). The described scheme of fragment condensation is the optimal one as the 46-49 fragment can be used as carboxylic component only, because after N^{α}-deblocking of Phe-46 the tetrapeptide will be practically insoluble in conventional organic solvents. Subsequent cleavage of the benzyl esters (Tyr-48) was performed by catalytical transfer hydrogenation over 10% Pd/C catalyst. The yield was 96% (Fig. 3).

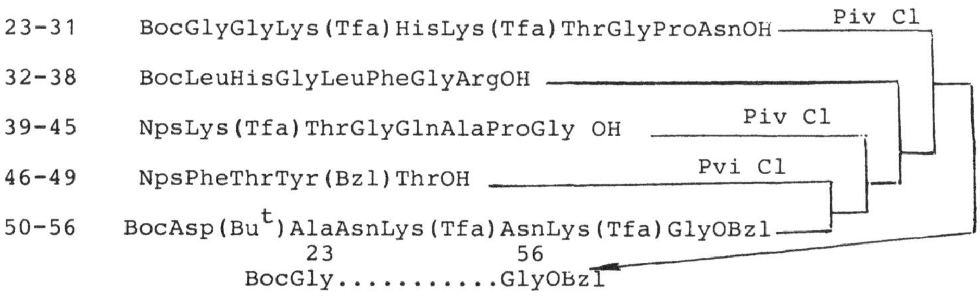

Fig. 3. Synthetic route to peptide fragment 23-56.

For the synthesis of the 34-membered peptide corresponding to sequence 23-56 we have assayed several coupling routes. The initially suggested coupling of the two peptides 23-38 and 39-56 could not be accomplished because of the low yield (12%) of the 16-membered Arg-containing peptide, obtained by condensation of nonapeptide 23-31 and heptapeptide via pivaloyl chloride activation (4). This low yield resulted from extremely unsatisfactory solubility of the amino component in organic solvents, including DMFA and DMSO. More effective was another combination of fragments, where the Arg-containing peptide 32-38 was used as a C-component. Condensation with unprotected 18-membered peptide was carried out using ethyl chloroformate in $CHCl_3$ in the presence of N-methylmorpholine. The reaction mixture was very diluted because of the low solubility of the peptide. The yield of the 25-membered peptide after chromatography on silica gel and recrystallization amounted to 41%. Fragment 32-56 was N-deprotected with 2N methanolic HCl and coupled with nonapeptide 23-31 containing a C-terminal Asn. The coupling was carried out using pivaloyl chloride with addition of 1.5 equiv. of 1-hydroxybenzotriazole in THF for 24 hrs. The end-product was isolated after chromatography with 62% yield (5).
To assess the homogeneity of peptides we extensively used HPLC. Standard mixtures of the title compound and parent substances were analyzed on a LiChrosorb RP-8 column. Practically all the compounds separated well on LiChrosorb, except the Arg-containing peptide (irreversible sorption).
Homogeneity of the synthesized peptides was checked by HPLC data and dansylation of the N-terminal amino acids. For peptides containing more than 20 amino acid residues the C-terminal acids were determined by digestion with carboxypeptidase A and subsequent identification of the cleaved amino acids by dansylation.
III. Synthesis of fragment 57-81. The 25-membered peptide corresponding to the 57-81 sequence of cytochrome *c* contains 4 out of the total 9 glutamyl residues of the cytochrome molecule, and carries also functionally important Trp-59, Tyr-67, and Met-80. Significantly, the amino acid sequence 70-80 is the longest invariant continuity in the peptide chain of cytochromes *c* of different species, what probably indicates its importance for biological function of the protein, or, perhaps, for its structure.

These features could not but strongly influence our choice of the synthetic approach to this peptide. The chosen route involves condensation of four fragments (Fig. 4).

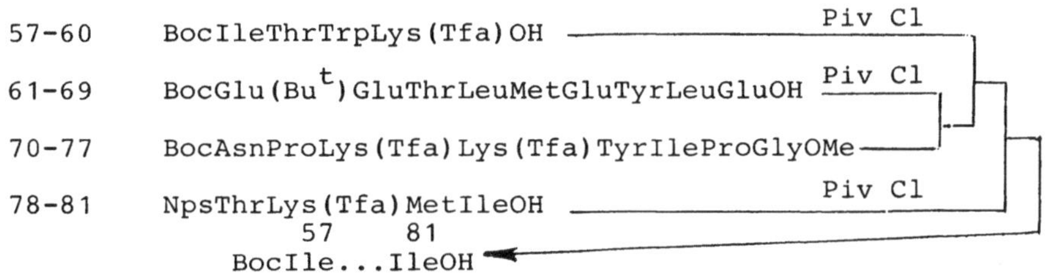

Fig. 4. Synthetic route to peptide fragment 57-81.

Especially interesting from the point of view of the primary structure and the resultant peculiarities of preparation was the peptide corresponding to the 61-69 sequence of cytochrome c and carrying four Glu-residues. Its synthesis involved coupling of 3 short peptides (61-62, 63-66, 67-69) in such a way that every time at the C-terminus there was an α-Me ester of L-glutamic acid. In the synthesis of this fragment we repeatedly resorted to the activated ester method with deprotection by HCl-gas in dioxan (15-20 min, 5 equiv.). For the condensation of Glu-containing peptides with unsubstituted γ-COOH groups we again resorted to pivaloyl chloride activation, that had been employed earlier for chain elongation in hemopeptide complexes with unprotected propionyl groups on the heme. Before coupling C-terminal Me esters were hydrolyzed with 1N methanolic NaOH (1 hr, 50°). Peptides corresponding to sequences 61-69 and 61-77 were obtained by this procedure (1-1.5 hrs, yields of 75 and 81% respectively). Peptide 70-77 was synthesized according to literature recommendations. The peptides in question were also alternately synthesized by oxidation of their hydrazides derived from corresponding Me esters. The yields after two-fold chromatography on Sephadex LH-20 were (%): 71.4 (63-69); 64.8 (61-69); 56 (61-77) (6).
Physicochemical properties, namely: $[\alpha]_D^{20}$, melting points, ORD-curves, HPLC data for compounds obtained by both routes were identical. Peptides obtained via pivaloyl chloride activation were optically

very pure, as was shown by exhaustive hydrolysis with papain and aminopeptidase M.
Nonapeptide 61-69 unlike peptides 61-77 and 57-77 had a positive Cotton effect.
This part of our study connected with the synthesis of Glu-containing peptides with unsubstituted γ-COOH groups apparently indicates, that the use of pivaloyl chloride considerably simplifies the process of peptide synthesis, for it allows to use well available glutamic acid derivatives and reduces the number of reaction steps. Activation with pivaloyl chloride also favourably increases our choice of condensing agents for peptide synthesis via the minimal protection approach.
Tetrapeptide 57-60 was synthesized according to the scheme using activated ester method (tripeptide 57-59) and DCC/1-hydroxybenzotriazole procedure. The yield was 65%. Tetrapeptide (57-60) Me ester was hydrolyzed, and the resulting peptide was coupled with the 61-77 fragment in the presence of pivaloyl chloride and 1.5 equiv. of 1-hydroxybenzotriazole (24 hrs, yield 71%). Tetrapeptide 78-81 was synthesized by fragment condensation using pivaloyl chloride, the constitutive blocks being prepared by DCC coupling (80-81) and activated ester procedure (78-79). By an analogous procedure peptide 57-77 was coupled with tetrapeptide 78-81. The coupling was carried out in THF in the presence of 1.5 equiv. of 1-hydroxybenzotriazole, with a double excess of the amino component. After chromatographic purification the end-product was obtained in 69% yield.
IV. Synthesis of fragment 82-104. The C-terminal fragment of cytochrome c contains an Arg-residue in position 91. Therefore, planning the tactics of our research, we divided this sequence schematically in two parts: a peptide 82-91 and a 92-104 one, so that the Arg-residue was positioned at the C-terminus of the first peptide (Fig. 5). However, while in the earlier synthesis the Arg-containing heptapeptide was very nicely prepared by stepwise chain elongation, the same approach to the synthesis of decapeptide 82-91 proved quite unsuccessful. Stepwise chain elongation gave good results up to tripeptide 89-91 only. Therefore we worked out a scheme based on fragment condensation of shorter peptides 82-85, 86-88, 89-91. In the syntheses of these peptides we mostly

resorted to the activated ester method. Aminolysis was carried out in the presence of the following catalysts: 1-hydroxybenzotriazole, 1,2,4-triazole (for the reactions carried out in THF or a dioxan-water mixture), glacial acetic acid for reactions proceeding in $CHCl_3$. Nps-Phe addition in the synthesis of tetrapeptide 82-85 as well as condensation of fragments was achieved using IBC in a mixture of THF and DMFA.

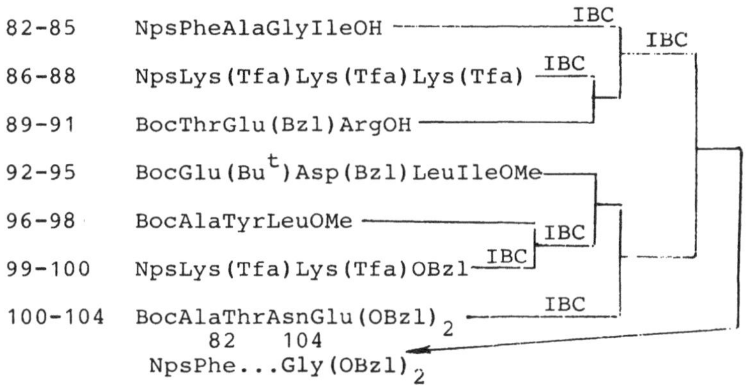

Fig. 5. Synthetic route to peptide fragment 82-104.

In the synthesis of tridecapeptide we mainly recoursed to the procedure of Wolman et al. (7,8). Coupling of fragments 82-91 and 92-104 was performed using IBC in THF. After column chromatography on silica gel L 40/100 and recrystallization from $CHCl_3$ the chromatographically pure peptide was isolated in 43% yield.

V. Synthesis of peptide 23-104 by fragment condensation. Dioctacontapeptide 23-104 was obtained by two routes (Fig. 6).

The first route was a consecutive condensation of N-terminal tetratriacontapeptide 23-56 with fragments 57-81 and 82-104. This route afforded a 59-membered fragment 23-81 that contained several structurally and functionally important amino acid residues, such as Met-80 which serves as sixth ligand of cytochrome c heme iron. The second route involved coupling of a 34-membered peptide 23-56 with a large peptide corresponding to the 57-104 sequence of cytochrome c. The latter was prepared by coupling of fragments 57-81 and 82-104. The couplings were carried out using IBC in THF with a

10% excess of the amino component for over 24-48 hrs. The only exception as what concerned the reaction conditions was the coupling of the 34-membered peptide 23-56 with the 25-membered fragment 57-81, which had been carried out with pivaloyl chloride in the presence of 1.5 equiv. of 1-hydroxybenzotriazole. After chromatographic purification and recrystallization the yields of 82-membered peptide from both synthetic routes were comparable, being 54 and 62% respectively. In both cases we had met no particular difficulties connected with unsufficient solubility of large peptides. On the contrary the peptides of increasing length were better soluble in organic solvents. This is consistent with conclusions earlier made by Wünsch et al. in his synthesis of larger glucagon fragments (9).

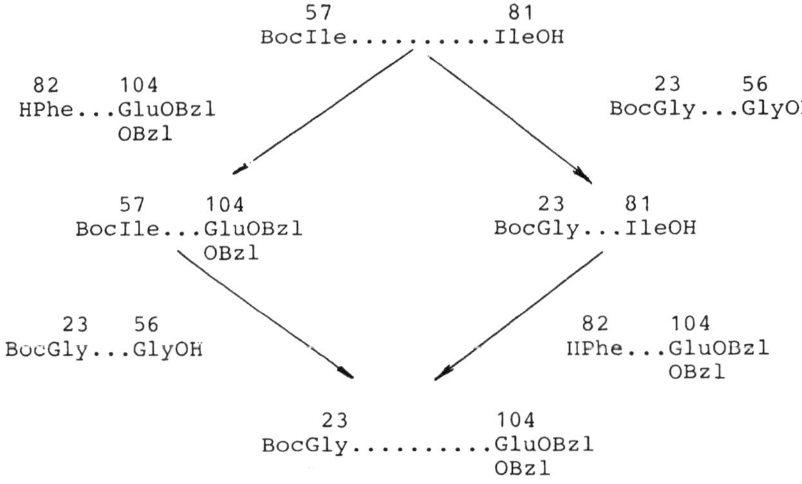

Fig. 6. Synthetic routes to peptide fragment 23-104.

The homogeneity of the substituted 82-membered peptide was assessed by TLC in different solvent systems, HPLC (Fig. 7), determination of the N-terminal amino acid (after N-deprotection with 3N HCl in dioxan) and paper chromatography of α-unsubstituted peptide. After cleavage of the CF_3CO- protection and hydrolysis of benzyl esters (Glu-90,104; Asp-93) by treatment with 0.1N ammonia (pH 10.7, 2 hrs) and purification of 0.2 mg of deprotected peptide on Sephadex G-25 (Superfine), the 82-membered cytochrome c fragment was found to be homogeneous according to paper chromatography on Whatman 3 MM in different solvent systems, TLC (Silufol UV 254), paper electrophoresis, disc electrophoresis on polyacrylamide gel. N-terminal and C-terminal amino acids were determined to be Gly- and Glu-respectively. Molecular mass was 10200-10400 as determined by gel electrophoresis on PAAG according to Weber-Osborne.

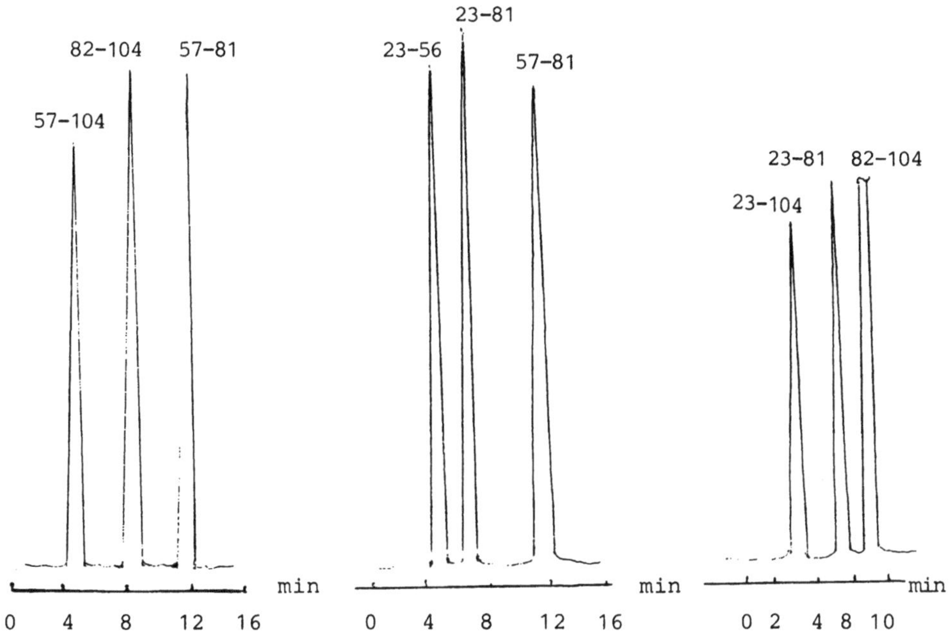

Fig. 7. HPLC of peptide fragments of horse heart cytochrome c. Column: LiChrosorb RP-8 (250x4.6 mm), eluent: methanol - 0.01 M ammonium acetate buffer (70-30), detection: u.v. 260 nm.

Homogeneity of shorter peptides was established by HPLC and TLC on Silufol UV 254 (after column chromatography on silica gel and Sephadex LH-20); these peptides were also routinely characterized by elemental analysis, melting points, $[\alpha]_D^{20}$. For peptides of medium length (18-20 amino acid residues) the N-terminal amino acids were identified by dansylation, and for sequences longer than 20 amino acid residues the C-terminal amino acids were cleaved off by carboxypeptidase A and then also dansylated. Amino acid composition was confirmed by quantitative analysis on Biocal 201 and Durrum instruments after acid hydrolysis with 5.7N HCl at 110° for 24 (48 and 72) hrs or 2 hrs at 150° or with p-toluenesulfonic acid at 150° for 2 hrs. The hydrolysis was carried out under nitrogen-gas.

Optical purity of peptides obtained by mixed anhydride method was established using HPLC. To-day HPLC is regarded as a good method for optical purity determinations, as it allows to reliably discern an admixture of less than 1% of D-isomer. To check this we have taken the tetrapeptide 82-85 carrying an N-terminal Phe, which is prone to racemization. LLL- and DLL-diastereomers of the tetrapeptide Nps-Phe-Ala-Gly-Ile-OMe were synthesized. The synthesis was carried out by the following routes: a) stepwise chain elongation using activated ester method; b) addition of the fourth residue via pivaloyl chloride activation; c) the same addition performed via IBC activation. The second and third routes do not exclude the DLL-isomer formation. Both diastereomers were analyzed by HPLC, and so were the standard mixtures containing 50, 5 and 1% of DLL-isomer respectively (Fig. 8). Comparison of HPLC data indicated that under the reaction conditions (THF, 30 min) the synthesis using IBC activation is not accompanied by more than 1% racemization, and in the synthesis involving pivaloyl chloride activation ($CHCl_3$, 1 hr) racemization is practically unnoticeable. Quite reliable demonstration of the optical purity of peptides prepared by pivaloyl chloride activation can be obtained from the exhaustive enzymatic hydrolysis data with subsequent analysis of the amino acid composition and HPLC of the peptides.

Optical purity of the synthetic 82-membered peptide corresponding to the 23-104 sequence of cytochrome c̲ was also confirmed by enzymatic hydrolyses with papain and aminopeptidase M.

The synthesized peptides were used for determination of antigenic determinants of cytochrome c (10). In toto four antigenic determinants, recognizable by rabbit antibodies, were localized on the cytochrome c peptide chain, the methods used being fluorescence quenching, gel filtration and radioimmunoassay.

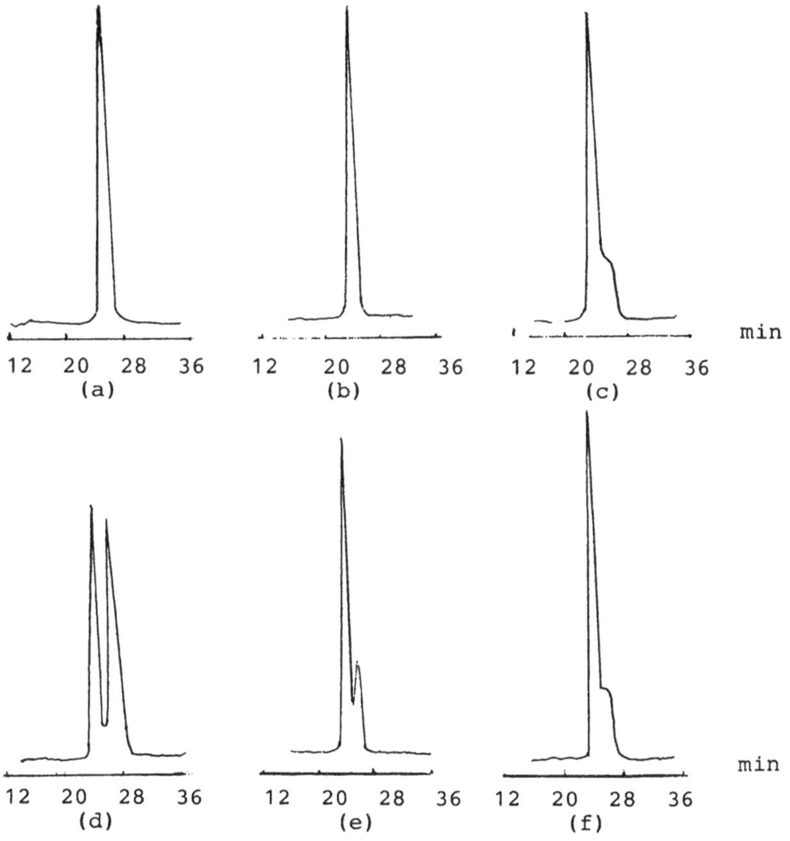

Fig. 8. HPLC of tetrapeptide NpsPheAlaGlyIleOMe.
Column: Spherisorb ODS (250x4.6 mm). Eluent: MeOH - 0.01 M ammonium acetate buffer (60-40).
a - LLL-isomer; b - NpsPhe activated with pivaloyl chloride; c - NpsPhe activated with IBC; d - 50% LLL + 50% DLL; e - 95% LLL + 5% DLL; f - 99% LLL + 1% DLL.

Synthetic peptides 46-56, 61-69, 92-104, 23-104 were immunoactive. It is interesting to note that peptide 23-104 exhibited activity of epitope 46-56 only, and did not have 61-69 and 92-104 activities, what may be connected with the absence of the heme. Fragments 2-13 and 14-22 are not antigenic when taken separately, but introduced together they produce the fourth antigenic determinant of the cytochrome c molecule.
Glutaric aldehyde polymerization, which is generally used for cytochrome immunogenicity increase, changes its antigenic properties and shields the epitope of peptide 61-69.

References

1. Komissarenko S.V., M.V. Skok, G.A. Vasil'eva, R.P. Evstigneeva. 1982. Dokl. Akad. Nauk SSSR 264, N 3, 752.

2. Skok M.V., P.V. Denisyuk, S.V. Komissarenko, G.A. Vasil'eva, R.P. Evstigneeva. 1982. Dokl. Akad. Nauk USSR, Ser. B, N 10, 70.

3. Evstigneeva R.P., G.A. Vasil'eva, N.E. Vol'skaya, A.F. Mironov. 1978. Proceedings of the II FRG-USSR Symposium on the Chemistry of Peptides and Proteins. Grainau/Eibsee, p. 3.

4. Vol'skaya N.E., G.A. Vasil'eva, V.N. Amelina, N.V. Filenkova, R.P. Evstigneeva. 1979. Bioorgan. khim. 5, N 9, 1302.

5. Vol'skaya N.E., G.A. Vasil'eva, M.B. Efros, R.P. Evstigneeva. 1979. Bioorgan. khim. 5, N 9, 1310.

6. Vol'skaya N.E., G.A. Vasil'eva, A.F. Mironov, R.P. Evstigneeva. 1979. Bioorgan. khim. 5, N 2, 181.

7. Wolman J., J.S. Klausner. 1971. Israel J. Chem., 9, N 2, 217.

8. Wolman J., J.S. Klausner. 1971. Israel J. Chem. 9, N 2, 229.

9. Wünsch E. 1978. Proceedings of the II FRG-USSR Symposium on the Chemistry of Peptides and Proteins. Grainau/Eibsee, p. 141.

10. Komissarenko S.V., M.V. Skok. 1984. 16-th Meeting of FEBS, Abstracts, 370.

STRUCTURE OF BLASTOLYSIN. SYNTHESIS AND PROPERTIES OF RELATED GLYCOPEPTIDES

M.V. Bezrukov, O.S. Reshetova, B.V. Rosynov, Ya.D. Gavrilov, S.A. Kozmin, M.V. Astapova, T.I. Barkova, T.M. Andronova, V.T. Ivanov

Shemyakin Institute of Bioorganic Chemistry, USSR Academy of Sciences, Ul. Miklukho-Maklaya, 16/10, 117871 Moscow V-437, USSR

Introduction

Bacterial preparations such as BCG (1), *Corynebacterium parvum* (2, 3), prodygiazans (4), etc, are broadly used as immunomodulators or antitumour agents. To such preparations also belongs a mixture of glycopeptides called blastolysin, isolated from the lysozyme digest of the *Lactobacillus bulgaricus* cell wall by bulgarian group headed by I.Bogdanov and used by him in cancer therapy (5,6). Early attempts to establish the structure of one of the active components of blastolysin resulted in a formula with a unique cyclopeptide moiety (7). However that structure had to be abandoned since total synthesis of respective compound afforded a weakly active product with quite different physical properties relative to the active fraction (8). In order to clarify the resultant uncertainty we continued the structural study of blastolysin.

Results

Successive fractionation of blastolysin on Sephadex G-50 and DE-32 (Cl^-) allowed to separate a number of inactive fractions (B-3 - B-9) (Figs 1,2).
The major fraction, (B-10) slightly surpassing in activity the starting material was isolated with 43% overall yield.
Study of B-3 - B-9 fractions was limited to their amino acid and carbohydrate analysis (Table 1) while fraction B-10 was subjected to further structural analysis. In HPLC analysis and SDS electrophoresis this fraction behaved as an individual product, accordingly it had only one N-terminal amino acid (Asp), as determined by

Chemistry of Peptides and Proteins, Vol. 3

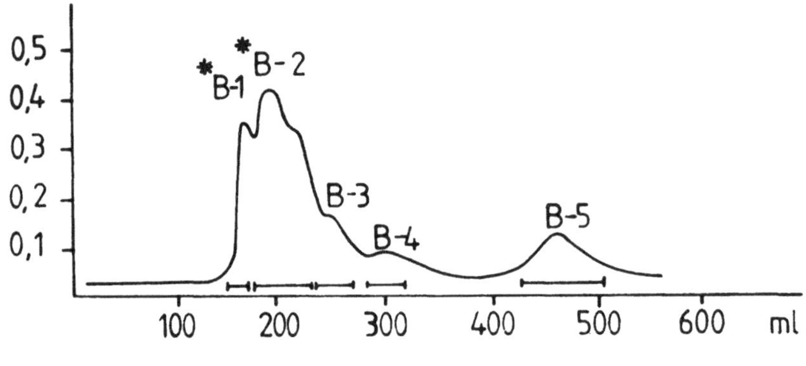

Fig. 1. Chromatography of blastolysin (100 mg) on Sephadex G-50 superfine column (2.6x80 cm) in 0.1 M AcOH. Flow rate 16 ml/min. UV (206 nm) detection. * Fractions displaying antitumour activity.

standard phenylthiohydantoin procedure (9). Judging from size exclusion chromatographic data (calibrated Sephadex G-50 and TSK-SV-3000 columns) molecular mass of B-10 is ca. 10000. The B-10 hydrolyzate contains glucosamine, muramic acid, Ala, Glu, Lys and am-

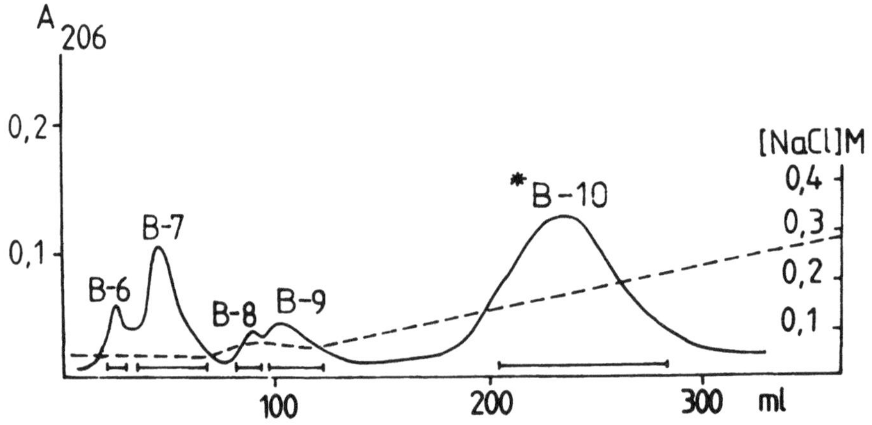

Fig. 2. Chromatography of fraction B-10 (40 mg) on DE-32 (Cl^-) cellulose column (0.9x30 cm). Flow rate 1.67 ml/min, pH 6.0. The NaCl gradient is shown by dotted line. * Fraction displaying antitumour activity.

Table 1. Amino Acid and Carbohydrate Composition of Chromatographic Fractions of Blastolysin.

Component \ Fraction №	1	2	3	4	5	6	7	8	9	10
	Composition, nmol/mg									
Asp[a]	660	645	295		488	444	630	536	679	633
Mur[a]	584	532	700		545	460	656	620	606	620
Glu[a]	745	940	680		942	922	1320	1410	1100	816
Ala[a]	1100	1470	1170		1120	1030	1470	1910	1420	1370
$GlcNH_2$[a,b]	500	505	1060	290	1060	372	532	536	635	706
Lys[a]	580	555	337		296	321	456	426	496	712
Gal[b]	370	400	2080	4260			373		369	418
Glc[b]	800	706	2670	2300	6450	1570	340	392	242	750
$GalNH_2$[a,b]	230	167				171			128	
Admixtures[a]	Arg, Ser, Gly	-	-	-	-	-	-	-	Arg	-

a - Amino acid analysis
b - GLC

monia in the approximate 1:1:2:1:1:1 ratio characteristic for *Lactobacilli* peptidoglycan (10) (excepting ammonia which was partially lost during acidic treatments used for preparation of blastolysin). GLC and elemental analysis also revealed glucose, galactose, glycerol and phosphorus indicating the presence of teichoic acid attached to the glycopeptide core. However its presence is not essential for biological activity since partial splitting of galactose, glucose and glycerophosphate by mild acidic or alkaline treatment resulted in fully active preparation (Table 2). On the contrary, HJO_4 oxidation (which in addition to glucose and galactose also destructs the muramic acid residues) gives rise to a totally inactive product.

The next question to be answered was structure of the polysaccharide chain built of GlcNAc and MurNAc residues. Since the process of blastolysin production includes lysozyme treatment of the bacterial cell wall B-10 is expected to contain $(GlcNAc\text{-}MurNAc)_n$ segments.

The known resistance of respective tetrasaccharide to lysozyme (11) allows to suggest that n=2. This suggestion was confirmed by $NaBH_4$

Table. 2. Antitumour Activity and Composition of Blastolysin, the B-10 Component and its Derivatives .

Sample / Activity and composition	Blastolysin	B-10	B-10 treated with 0.05M NH_4OH (20°C,12hr)	B-10 treated with 5% CCl_3CO_2H (40°C,100hr)	B-10 treated with 0.05% HJO_4 (+4°C,260hr)
	Antitumour activity on S-180 (mice, dose 50 mg/kg)				
Necrosis vs total member of mice, %	71	85	70	74	0
Inhibition of S-180 tumour growth, %	47	50	50	50	0
	Composition, nmol/mg				
Ala[a]	1053	1370	1400	1500	1600
Glu[a]	567	816	850	866	900
Lys[a]	513	712	760	780	800
Asp[a]	540	633	706	720	800
Mur[a]	486	626	686	690	39
$GlcNH_2$[a,b]	486	706	730	755	480
Gal[b]	135	418	355	320	117
Glc[b]	1093	756	407	370	79
Glycerol[b]	1242	1270	556	500	400
Phosphorus[c]	1242	1300	656	640	500
$GalNH_2$[b]	297				
Man[b]	5.4				
Lact[b]	148.5				

a - Amino acid analyser
b - GLC data provide the relative quantities (12); the absolute figures are calculated starting with the results obtained for $GlcNH_2$ by method a
c - Elementary analysis

treatment of B-10 which as expected resulted in reduction of one half of muramic acid (i.e. the terminal residue) to muramitol. ^{13}C-NMR spectra of the above mentioned product obtained by mild alkaline treatment of B-10 and a series of model spectra provided additional evidence favouring such conclusion. As seen from Fig. 3, the C-1 carbons participating in the glycoside link resonate

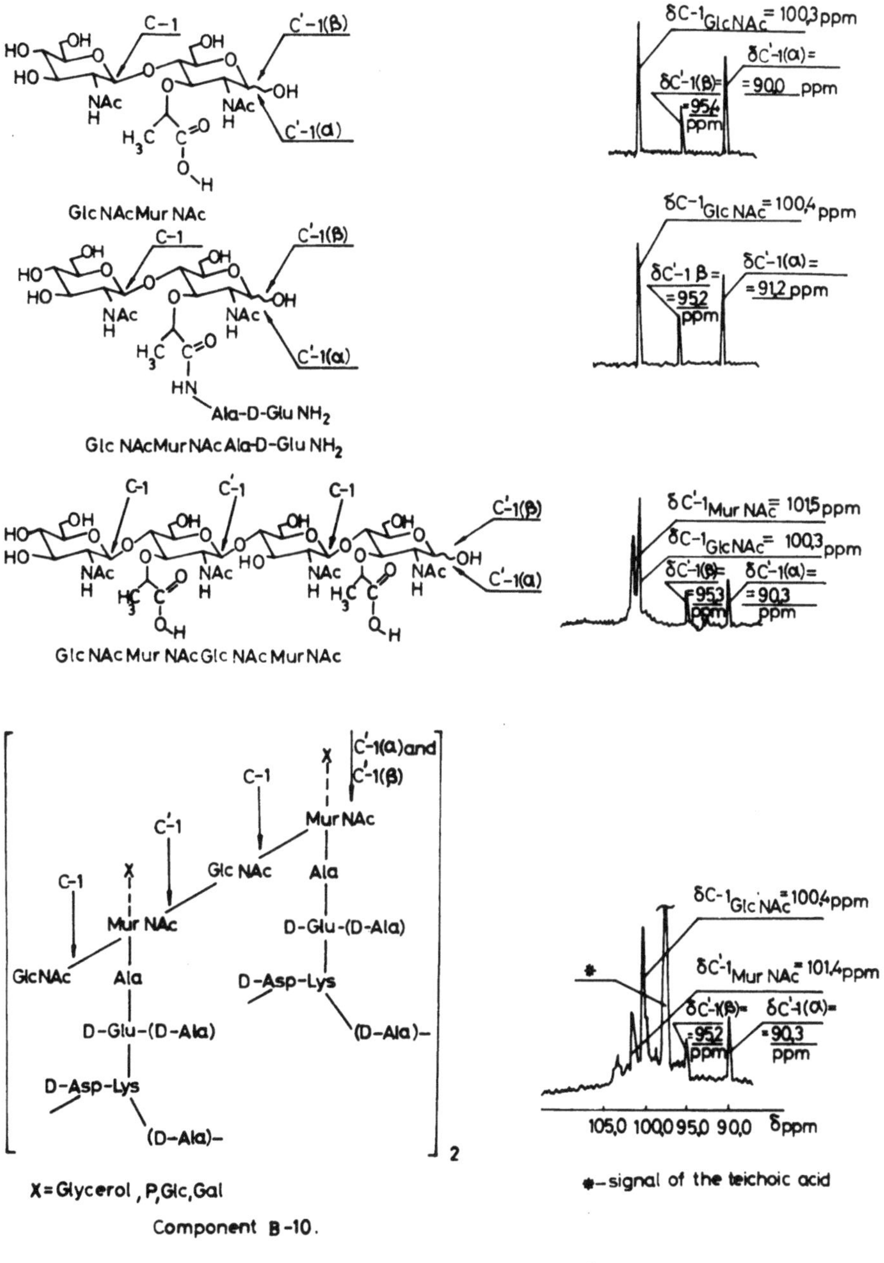

Fig. 3. C-1 signals in the proton decoupled ^{13}C-NMR spectra at 20° in D_2O of the disaccharide, GMDP, tetrasaccharide and modified by mild alkali B-10 fraction. The three upper spectra are taken at 20.0 MHz and the lower spectrum - at 125.7 MHz.

at ca. 100.3 ppm (GlcNAc) and 101.7 ppm (MurNAc). Respective pairs of signals with the characteristic 2:1 intensity ratio are clearly present in the spectra both of the tetrasaccharide and of the B-10 derived product. Signals at 90.0-91.2 ppm and 95.0-95.4 ppm due to the anomeric C-1 (α) and C-1 (β) carbons of the terminal muramic acid as expected are present in all four spectra. Presence of sugar units belonging to the teichoic acid part of the B-10 derived product is reflected in a series of additional signals of which the most intensive one is the α-glucose C-1 signal at 97.9 ppm. All assignments are taken from (13,14,15,16).

The partial hydrolyzate of B-10 (6N HCl, 30 min, 100°C) was subjected to trifluoroacetylation and methylation followed by mass-spectral analysis which afforded important structural information on peptide portion. GLC separation of individual components allowed detection of a series of short peptide sequences shown in Table 3.

Table 3. Oligopeptide Segments Deduced from Chemical Ionization Mass-Spectra of Blastolysin Hydrolysate.

Peptide \ Fragment ion (m/z)	$[MH]^+$	$[MH+NH_3]^+$	$[MH-32]^+$	$[MH+NH_3-32]^+$
Ala-Glu*	343	360		328
Glu-Ala*	343	360		328
Ala-Asp*	329	346		314
Lys-Ala*	424	441		
Asp-Lys*	482	499	450	467
Glu-Lys*		513		
Lys-Ala-Asp* (Asp-Lys-Ala)	553	576	521	538
Ala-Glu-Lys (Glu-Lys-Ala)	567	584		
Glu-Lys-Asp (Asp-Lys-Glu)	625	642	593	610
Ala-Glu-Lys-Asp (Asp-Lys-Glu-Ala)	696	713	664	681
[Ala-Glu-Lys-(Asp)-Ala]	767	784	735	752

* - presence of these peptides is confirmed by GLC - MS

A few longer sequences, shown in the same table, were determined from chemical ionization mass spectra of the partial hydrolyzate (17). The total hydrolyzate of B-10 (6N HCl, 16 hr., 90°C) was trifluoroacetylated, esterified with isopropyl alcohol and subjected to enantiomeric analysis by GLC on lauroyl-L-valyl-tret-buthylamide as described in (18). Glu and Asp residues were shown to belong to D-family, while both, L and D enantiomers of Ala were found in the hydrolyzate.

The data presented above coupled with the available data on the structural principles of the cell wall of *Lactobacilli* lead to formula presented in Fig. 4.

The structure contains 2 tetrasaccharide segments to which teichoic acids and peptide moieties are attached. The segments are linked with D-Ala and D-Asp containing bridges. Some uncertainty remains with respect of the position of D-Ala residues; this is reflected in placing that residue within brackets.

With the above structural information at hand and bearing in mind very high immunomodulatory activity of blastolysin we attempted synthesis of tetrasaccharide containing glycopeptides. A few years ago the first member of that family was obtained in our laboratory by direct coupling of the unprotected tetrasaccharide (isolated from lysozyme treated cell wall of *Micrococcus lysodeikticus*) with Ala-D-Glu(Bzl)-NH_2 and subsequent hydrogenation (19,20). However, such approach which proceeds very successful with GlcNAc-MurNAc-

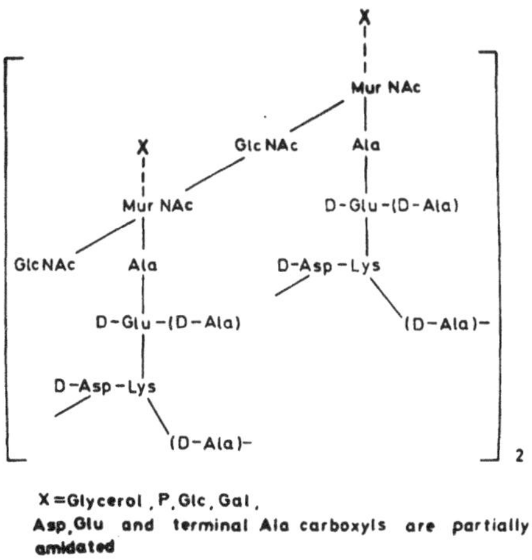

Fig. 4. Structure of the active principle of blastolysin (component B-10).

containing glycopeptides gave very low yields with the tetrasaccharide. To reduce the extent of side reactions which obviously involved the hydroxyls of the sugar moiety we developed an alternative route implying temporary protection of these hydroxyls with 4,4'-dimethoxytrityl residue. This protection is widely used at present in nucleotide synthesis (21). The grouping is easily introduced and removed under mild conditions; at the same time it is sufficiently stable under standard coupling conditions.

The carboxyl and the C_6-hydroxyls of tetrasaccharide were arylated by DMTrCl in pyridine in the presence of triethylamine. The protection was then selectively removed from the carboxyls at pH 3 and the partially protected tetrasaccharide isolated as a N-methylmorpholine salt. Subsequent mixed anhydride (isobuthylchloroformate) or DCCI coupling with benzyl protected dipeptides followed by DMTr removal afforded the relevant glycopeptides with good yields. Removal of benzyl protection from the C-terminus proceeds with quantitative yields on catalytic hydrogenation.

Scheme

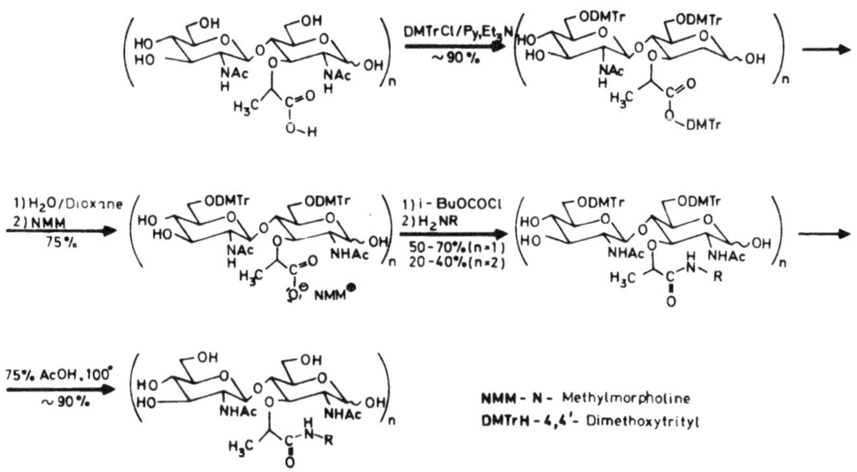

n=1, R=Ala-D-Glu(OBzl)NH_2; n=2, R=AlaOBzl, Ala-D-Glu$(OBzl)_2$, Ala-D-Glu(OBzl)NH_2, D-Ala-D-Glu$(OBzl)_2$, Gly-D-Glu$(OBzl)_2$, Val-D-Glu$(OBzl)_2$, Leu-D-Glu$(OBzl)_2$ or Ser(Bzl)-D-Glu$(OBzl)_2$.

Table 4. Antitumour and Immunostimulating Activity of Synthetic Glycopeptides.

№№	Compound	Antitumour activity a)			Immunostimulating activity			
		Dose µg/kg	Necrosis %	Tumour growth inhibition	Antibody titers against bovine serum albumin (Exp/Contr.) b)		Stimulation of leukocyte migration in agar c) %	
					Dose µg/mouse	Effect	Dose µg/mouse	Effect
1.	Blastolysin	50	70	47	100	19.0	30	22
		5	40	20	1	2.0	5	5
2.	GlcNAc-MurNAc-Ala-D-Glu-NH_2 (GMDP)	50	51	40	100	10.2	30	52
		5	56	36	10	5.3	5	54
		0.5	53	29	1	2.2		
3.	MurNAc-Ala-D-Glu-NH_2 (MDP)	50	45	38	100	2.0	30	21
		5	45	29	10	5.2	5	4
		0.5	30	13	1	-1.0		
4.	GlcNAc - MurNAc - GlcNAc - MurNAc Ala-D-Glu-NH_2 Ala-D-Glu-NH_2	50	63	46			30	55
		5	40	12	-	-	5	59
		0.5	20	12				
5.	GlcNAc - MurNAc - GlcNAc - MurNAc Ala-D-Glu(OBzl)NH_2 Ala-D-Glu(OBzl)NH_2	50	50	44			30	25
		5	50	48	-	-	5	25
		0.5	50	28				
6.	GlcNAc-MurNAc-GlcNAc-MurNAc Ala-D-Glu Ala-D-Glu	50	50	59	100	47.0	30	50
		5	45	58	10	16.6	5	35
					1	1.0		

Table 4. Antitumour and Immunostimulating Activity of Synthetic Glycopeptides (Continued).

№№	Compound	Antitumour activity a)			Immunostimulating activity			
		Dose μg/kg	Necrosis %	Tumour growth inhibition	Antibody titers against bovine serum albumin (Exp/Contr.) b)		Stimulation of leukocyte migration in agar c) %	
					Dose μg/mouse	Effect	Dose μg/mouse	Effect
7.	GlcNAc - MurNAc - GlcNAc - MurNAc Ala-D-Glu(OBzl)$_2$ Ala-D-Glu(OBzl)$_2$	50 5	50 40	54 50	100 10 1	2.2 1.5 1.0	30 5	53 55
8.	GlcNAc - MurNAc - GlcNAc - MurNAc D-Ala-D-Glu D-Ala-D-Glu	-	-	-	-	-	30 5	2 4
9.	GlcNAc - MurNAc - GlcNAc - MurNAc D-Ala-D-Glu(OBzl)$_2$ D-Ala-D-Glu(OBzl)$_2$	50 5	16 15	46 0	100 10 1	4.0 1.0 1.0	30 5	-3 -1
10.	GlcNAc - MurNAc - GlcNAc - MurNAc Val-D-Glu Val-D-Glu	50 5	0 0	27 0	-	-	-	-
11.	GlcNAc - MurNAc - GlcNAc - MurNAc Val-D-Glu(OBzl)$_2$ Val-D-Glu(OBzl)$_2$	50 5	0 0	22 22	100 10 1	30.0 4.1 -1.0	30 5	18 6
12.	GlcNAc - MurNAc - GlcNAc - MurNAc Gly-D-Glu(OBzl)$_2$ Gly-D-Glu(OBzl)$_2$	50 5	0 0	0 0	-	-	30 5	52 42

a) mice, S-180, testing procedure given in (20)
b) mice (CBA x C57 Bl/6)F_1. The sera assayed on day 34 (passive hemagglutination)
c) for method see (22)

Table 5. Protective Effect of Synthetic Glycopeptides in Experimental Infections (Percent of Survived Animals).

Microorganism	Dose µg/mouse	Glycopeptide 2		3		6		7	
		a	b	a	b	a	b	a	b
E. coli	0.1	25	20	10	20	80	60	20	60
	1.0	25	40	50	40	100	80	20	80
	10	25	60	50	50	100	80	20	60
	50	40	50	70	50	60	100	60	100
Control	-	0	0	0	0	10	10	10	10
Ps. aeruginosa	0.1	50	80	70	70	20	40	60	100
	1.0	20	70	50	80	40	40	20	60
	10	40	50	50	80	20	80	40	100
	50	40	60	40	60	80	40	40	100
Control	-	0	0	0	0	0	0	0	10
St. aureus	0.1	0	0	0	0	0	20	0	40
	1.0	0	0	0	0	0	20	0	0
	10	0	0	10	0	20	20	0	40
	50	0	0	0	0	40	80	0	40
Control	-	0	0	0	0	0	10	0	10

a - single dose, b - double dose

Results of the preliminary biological study of the compounds obtained are presented in Tables 4, 5. In all tests the tetrasaccharide containing glycopeptides showed very high biological activity exceeding in some cases the most potent disaccharide containing glycopeptide (GMDP). For instance, the glycopeptide (6) exibited higher antitumour and espesially adjuvant activity than GMDP. The higher adjuvant activity was evident both from the 5-fold increase in specific antibody titer to T-dependent antigen BSA and from considerable reduction of glycopeptide dose required for pronounced adjuvant effect: injection of 5 mg/kg of glycopeptide (6) resulted in higher antibody titer than injection of 50 mg/kg of GMDP.
The benzyl ester of glycopeptide (6) (compound 7) also exibited high antitumour activity, but was devoid of adjuvant activity. Consequently, its antitumour effect has to be attributed to cellular immunity: T-killer cell, NK-cells and/or macrophages. It is of in-

terest that the benzyl ester (9) differing from (6) in D-configuration of the alanine residue also showed antitumour potency. Protective action of glycopeptides (6) and (7) was studied in experimental infections caused by intraperitoneal introduction of a 90-100% lethal dose of 24 hr. culture of *E. coli. Ps. aeruginosa* or *St. aureus*. Relevant data for MDP (3) and GMDP (2) are showed for comparison; all results are obtained by Nikitin A.V., Lobuseva A.N., Navashin S.M. (All Union Institute of Antibiotic Research, Moscow). The drug was introduced intraperitoneally in a single dose 24 hr. before the infection or in two doses of the same size 48 hr. and 24 hr. before the infection. As seen from Table 5 in all cases a pronounced effect was observed with gram-negative strains, the tetrasaccharide containing glycopeptides being somewhat more active than the reference samples. With *St. aureus* the glycopeptides (6) and (7) also showed some effect, especially when

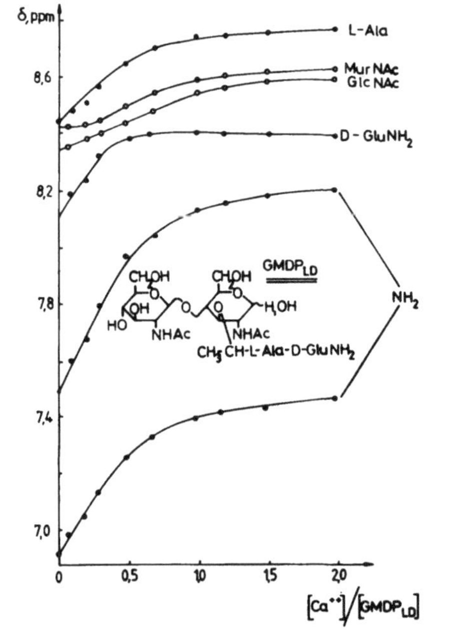

Fig. 5. Titration curves of GMDP in 96% ethanol by 0.1 M $CaCl_2$. Starting concentration of GMDP is $1.2 \cdot 10^{-2}$ M.

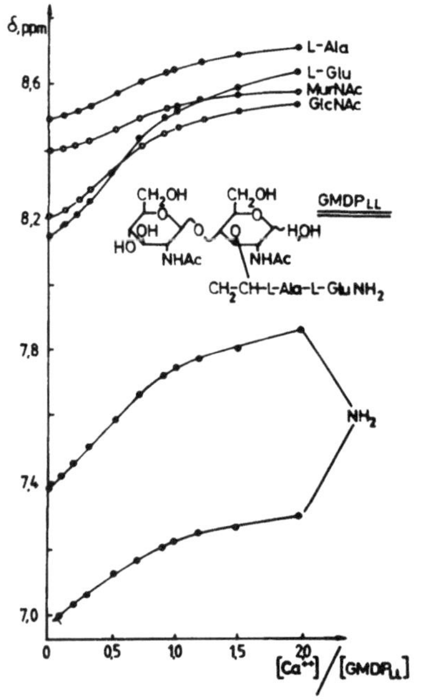

Fig. 6. Titration curves of LL-diastereoisomer of GMDP. Conditions are given in caption to Fig. 5.

applied twice while MDP and GMDP were totally inactive. Preliminary data related to the molecular principles of glycopeptide action will be considered in the concluding part of the paper. E.Lederer in his review (23) mentioned the possible participation of MDP in the transmembrane transfer of calcium. In order to probe that hypothesis we studied the calcium binding of GMDP in solution. CD spectra of GMDP were calcuim independent in water and calcium dependent in ethanol in the 0-10^{-2} M range of $CaCl_2$ concentrations. Calcium binding in the latter solvent was confirmed by NMR followed titration of GMDP by $CaCl_2$. In particular, considerable shifts occured in the region of NH signals. As seen from Fig. 5, the initial slope of the titration curves most sensitive to the presence of Ca^{2+} (amide NH_2 protons as well as the peptide NH of Ala and D-Glu-NH_2) point to the presence of a rather stable 2:1 GMDP:Ca^{2+} complex in equilibrium with usual equimolar complex. Analogous experiment with the inactive LL-analogue also showed considerable Ca dependence of the NH shifts (Fig. 6). However, there was no indication of 2:1 complex formation in this case. Further studies are required to find whether this fact is relevant to sharp differences in biological properties of these diastereoisomers.

References

1. Mathe, G. 1971. Adv. Cancer Res. 14, 1.

2. Milas, L., M.T. Scott. 1978. Adv. Cancer Res. 26, 251.

3. Silobrcic, V. 1980. Folia Biologica (Praha) 26, 145.

4. Ermolyeva, Z.V., G.E. Vaisberg. 1974. Soviet Medicine (Russian), No.2, 12.

5. Bogdanov, I. 1982. In: Observations on the therapeutic effect of the anti-cancer preparation from *Lactobacillus bulgaricus* LB-51 tested on 100 oncologic patients (Sofia), 1.

6. Bogdanov, I.G., V.T. Velichkov, A.J. Gurevich, P.G. Dalev, M.N. Kolosov, V.P. Mal'kova, I.B. Sorokina, L.N. Khristova. 1977. Bull. Exp. Biol. and Med. (Russian) No. 12, 709.

7. Bogdanov, I.G., P.G. Dalev, A.I. Gurevich, M.N. Kolosov, V.P. Mal'kova, L.A. Plemyannikova, I.B. Sorokina. 1975. FEBS Lett. 57, 259.

8. Andronova, T.M., L.I. Rostovtseva, E.P. Dobrushkina, Yu.D. Gavrilov, T.N. Deshko, V.T. Ivanov. 1980. Bioorg. Khim. (Russian) 6, 1830.

9. Lipkin, V.M., N.A. Aldanova, M.Yu. Feigina, E.B. Zhigulina, E.I. Vinogradova. 1972. Biochemistry (Russian) 37, 410.

10. Schleifer, K.H., O. Kandler. 1972. Bacteriol. Rev. 36, 407.

11. Ghuysen, J.-M. 1968. Bacteriol. Rev. 32, 425.

12. Jackson, L.K., E. Slodk, R.D. Platner, K.A. Burton, M.C. Cadmus. 1982. Carbohydr. Res. 110, 273.

13. Halls, T.D.J., M.S. Ragu, E. Wenkert, M. Zuber, P. Lefrancier, E. Lederer. 1980. Carbohydr. Res. 81, 173.

14. Klaic, B. 1982. Carbohydr. Res. 110, 320.

15. DcBoer, W.R., F.J. Kruyssen, J.T.M. Wouters, C. Kruk. 1976. Eur. J. Biochem. 62,1; De Boer e.a.Eur.J.Biochem.,1978,85,N2,p.433-436.

16. Dalev, P.G., P.N. Tsoneva, I.G. Bogdanov. 1977. Compt. rend. Acad. bulg. Sci. 30, 1305.

17. Reshetova, O.S., B.V. Rosynov, M.V. Bezrukov, I.A. Bogdanova: Bioorg. Khim. (Russian) (in press).

18. Rosynov, B.V., V.V. Onoprienko, V.B. Muratov, O.S. Reshetova, Yu.V. Volgin. 1980. V All-Union Symposium on Chemistry and Physics of Proteins and Peptides, Baku, Abstracts, p. 227.

19. Ivanov, V., L. Rostovtseva, T. Andronova, I. Sorokina, V. Mal'kova, Yu. Gavrilov, F. Noscov, E. Fridman. 1981. In: Peptides 1980 (E.Brunfeldt, ed.). Scriptor, Copenhagen, p. 494.

20. Rostovtseva, L.I., T.M. Andronova, V.P. Mal'kova, I.B. Sorokina, V.T. Ivanov. 1981. Bioorg. Khim. (Russian) 7, 1843.

21. McOmie, J.F.W. 1973. Protective groups in organic chemistry, Plenum Press, London and New York

22. Denisov, N. 1981. Lab. Delo (Russian) 6, 321.

23. Adam, A., E. Lederer. 1984. Med. Res. Rev. 4, 111.

ENZYMATIC-CHEMICAL EXCHANGE OF TYROSINE (B16) IN THE B-CHAIN OF INSULIN

H.-G. Gattner, M. Leithäuser
Deutsches Wollforschungsinstitut
Veltmanplatz 8, D-5100 Aachen, FRG

Introduction

Chemical and enzymatic semisynthesis have facilitated the demanding task of preparing insulin analogues for structure-function studies, but their application have so far been limited to terminal regions of the molecule.
We have now begun to investigate the possibilities to modify positions within the bicyclic disulfide structure. We want to report our results with the B-chain to replace tyrosine (B16). This amino acid is an invariant residue, it is involved in dimer formation and constitutes part of the receptor binding region (1). We therefore expect that replacements of tyrosine (B16) would have a marked influence on biological activity and physico-chemical properties.

Results

Preparation of B-chain derivatives (2). di-S-sulfonate-B-chain. The sulfitolysis of insulin was carried out according to Bailey and Cole (3). The di-S-sulfonate-B-chain was separated by ion exchange chromatography on SP-Sephadex equilibrated with 7M urea, 1,5M acetic acid and eluted with a linear sodium chloride gradient, 0,0 - 0,2M (4).
B-chain-disulfide (BSS).
500 mg (0,14 mMol) di-S-sulfonate-B-chain were dissolved in 30 ml glycine buffer (8M urea, 0.05 glycine adjusted to pH 10,6 with NaOH). Two drops of octanol were added and the solution was flushed with nitrogen for 10 min. The reduction was performed with 3 ml

Chemistry of Peptides and Proteins, Vol. 3

(43 mMol) 2-mercaptoethanol for 30 min. Then the solution was acidified with 20 ml formic acid. The reduced B-chain was separated by gel filtration on Sephadex G-25 equilibrated with 30% acetic acid (column 4 x 60 cm). The protein containing fractions were diluted with acetic acid to a concentration of about 2 mg/ml B-chain. Oxidation of di-thiol-B-chain was performed by titration with 0,1 N iodine until the solution was faintly yellow. The excess of iodine was destroyed by dropwise addition of sodiumthiosulfate solution. The colourless solution was dialysed against demineralized. Polymers were removed by gel filtration on Sephadex G-50 in 10% acetic acid (yield 75%).

(B1),(B29)-di-Msc-B-chain-disulfide(Msc_2BSS). 200 mg (57 µMol) BSS were dissolved in 20 ml dimethylsulfoxide, 60 mg (228 µMol) methylsulfonylethyloxycarbonyl succinimide ester (Msc-ONSu) and 100 µl triethylamine were added. The reaction mixture was stirred for 4 hrs at RT. The reaction was stopped by addition of 2 ml glacial acetic acid and diluted with 5 ml water. The peptide was separated by gel filtration on Sephadex G-25 in 10% acetic acid and lyophilized (90%).

The amino protected B-chain was purified by ion exchange chromatography on SP-Sephadex (2,5 x 25 cm) equilibrated with 7M urea and 1,5M acetic acid with a linear salt gradient from 0,0 - 0,2 M NaCl. The peptide fractions were desalted on Sephadex G-25 in 10% acetic acid and lyophilized (yield 75%).

Msc_2-B-chain-disulfide-trimethyl ester ($Msc_2BSS(OMe)_3$) (5). 200 mg Msc_2BSS were dissolved in 50 ml methanol and 2,5 ml of a solution of 20% BF_3 in methanol (5). After 24 hrs at RT the solution was evaporated. The residue was desalted on Sephadex G-25 in 10% acetic acid and lyophilized (yield 90%).

Digestion of $Msc_2BSS(OMe)_3$ with proteinase K. 150 mg $Msc_2BSS(OMe)_3$ were dissolved in 9 ml 4M urea solution. The pH was adjusted to 4,5 with acetic acid. A solution of 1,5 mg proteinase K (Merck) in 1 ml 4M urea was added and incubated for 2 hrs at RT. The reaction was stopped by lowering the pH to about 2 with formic acid. Removal of the enzyme was achieved by gel filtration on Sephadex G-50 (2,6 x 100 cm) in 10% acetic acid. The B-chain containing fractions were pooled and lyophilized. To separate unchanged B-chain derivative from split B-chain the mixture was chromatographed on SP-Sephadex (1,6 x 30 cm) in 1,5M acetic acid and 7M urea with a

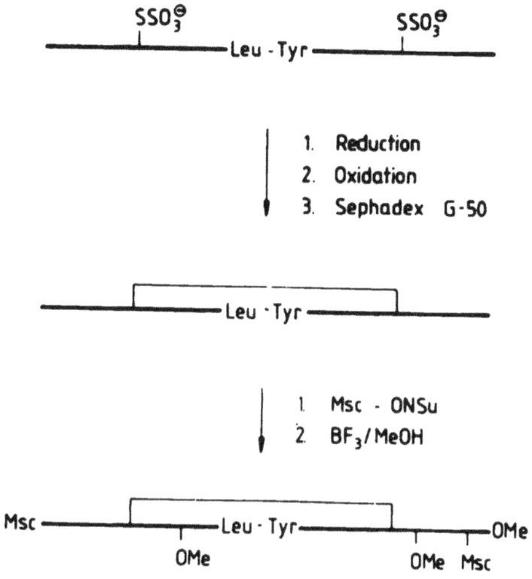

Fig. 1. Preparation of protected B-chain.

linear salt gradient (250/250 ml 0,0 - 0,25 M NaCl). Split B-chain derivative was eluted behind the unreacted B-chain (Fig. 2).

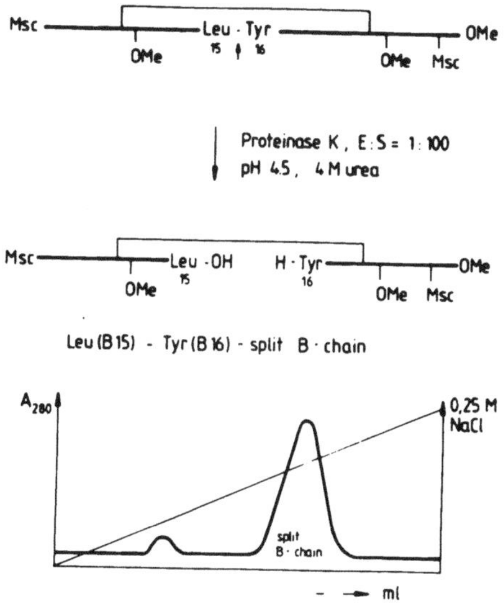

Fig. 2. Cleavage of protected B-chain with proteinase K and purification of the split B-chain derivative by ion exchange chromatography on SP-Sephadex.

The peptide containing fractions were desalted on Sephadex G-25 in 10% acetic acid and lyophilized (yield: 80 mg).
The split B-chain derivative was characterized by end group analysis, amino acid analysis and thin layer electrophoresis at pH 2,2 (2,4 M formic acid, 4 M urea, Pauly staining).
Removal of tyrosine (B16) by Edman-degradation.
175 mg of split Leu(B15)-Tyr(B16)-Msc_2-BSS$(OMe)_3$ were dissolved in 15,6 ml pyridine and 1,56 ml water. The solution was flushed with nitrogen for 15 min and 300 ul phenylisothiocyanate (PICT) were added. Reaction was performed for 5,5 hrs in the dark. The solution was concentrated nearly to dryness and triturated with ether. Excess of reagent was removed by gel filtration on Sephadex G-25 in 10% acetic acid. Purification was achieved by ion exchange chromatography on SP-Sephadex as described above (yield: 150 mg). The purity of the product was controlled by end group analysis and thin layer electrophoresis at pH 2,2.
150 mg of the (B16)-Ptc-B-chain were dissolved in 5 ml trifluoroacetic acid (TFA) and incubated for 2 hs at RT. After evaporation of the TFA the peptide was precipitated with ether and purified by ion exchange chromatography on SP-Sephadex (see above). Amino acid analysis: Tyr 1,07 (1,00).

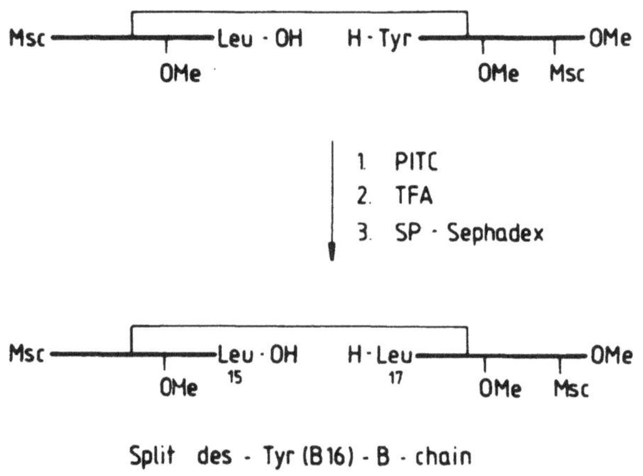

Fig. 3. Preparation of split des-Tyr(B16)-B-chain.

Insertion of phenylalanine at B16. 180 mg (51 µMol) of Msc_2-des Tyr(B16)-BSS$(OMe)_3$ were dissolved in 10 ml dimethylsulfoxide (DMSO) and 56 ml N-methylmorpholine. 92 mg Boc-Phe-ONSu (255 µMol) in 2,5 ml DMSO were added. After 5 hrs the reaction was complete according to thin layer electrophoresis. The solution was acidified with 10 ml 10% acetic acid and desalted. After removal of the Boc-group with TFA electrophoresis indicated on O-acylation of about 5%. The overreacted product was removed by SP-Sephadex chromatography (see above). Yield of Phe(B16) open chain B-chain derivative: 79 mg.

Coupling of Leu(B15) with Phe(B16). To remove acetate ions from desalting the open chain B-chain derivative was dissolved in water by addition of 1N HCl to pH 2,5 - 3,0 and lyophilized. 80 mg were then dissolved in 3,2 ml dimethylformamide with 3,2 µl N-methylmorpholine and 2,96 mg 1-hydroxybenzotriazole. The solution was cooled to -5°C and 5,2 mg dicyclohexylcarbodiimide were added. The solution was kept for 5 days at +2°C. The reaction could be followed by thin layer electrophoresis at pH 2,2. After desalting and lyophilization 77 mg were obtained. For purification ion exchange chromatography on SP-Sephadex was used (yield 50 mg).

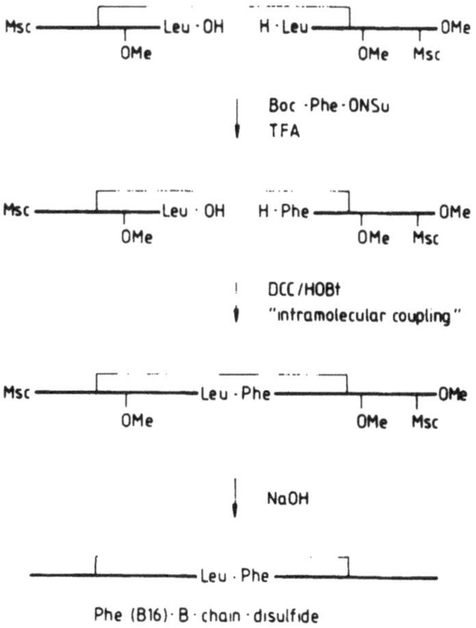

Fig. 4. Insertion of phenylalanine in position (B16).

Removal of the protecting groups.
Msc-groups and methyl ester could be removed in one step. Protected B-chain was dissolved at +2°C in 0,1 N NaOH (5 mg/ml) and kept at this temperature for 30 min (2). Then the pH of the solution was adjusted to 3-4 with acetic acid, and the crude B-chain analogue was isolated by gel filtration on Sephadex G-50 in 10% acetic acid. It was further purified by ion exchange chromatography on CM-cellulose equilibrated with 7M urea, 0.1M acetic acid, pH 4,8, linear salt gradient 0.0 - 0.15 M NaCl.

Conclusions

1. Under selected conditions proteinase K cleaves only the Leu(B15)-Tyr(B16) peptide bond of the protected B-chain disulfide.
2. Edman degradation removes Tyr(B16).
3. The two following chemical coupling steps yield an insulin B-chain with changed sequence in the important position B16.
4. During the synthesis amino- and carboxyl-component are connected by a disulfide bridge which facilitates purification and coupling steps.

References

1. Pullen, R.A., D.G. Lindsay, S.P. Wood, I.J. Tickle, T.L. Blundell, A. Wollmer, G. Krail, D. Brandenburg, H. Zahn, J. Gliemann, S. Gammeltoft. 1976. Nature (London), 259, 369.
2. Büllesbach, E.E., E.W. Schmitt, H.-G. Gattner. 1982. Int. J. Peptide Prot. Res. 20, 207.
3. Bailey, J.L., R.D. Cole. 1959. J. Biol. Chem. 234, 1733.
4. Brandenburg, D. 1972. Hoppe-Seyler's Z. Physiol. Chem. 353, 263.
5. Gattner, H.-G., E.W. Schmitt, V.K. Naithani. 1975. Hoppe-Seyler's Z. Physiol. Chem. 356, 1465.

SEMISYNTHETIC CONVERSION OF THE BOVINE TRYPSIN INHIBITOR (KUNITZ) INTO AN EFFICIENT LEUKOCYTE-ELASTASE INHIBITOR BY SPECIFIC VALINE FOR LYSINE SUBSTITUTION IN THE REACTIVE SITE

H.R. Wenzel, J. Beckmann, A. Mehlich, E. Schnabel*, H. Tschesche

Universität Bielefeld, Fakultät für Chemie, D-4800 Bielefeld 1
and *Bayer AG, Institut für Biochemie, D-5600 Wuppertal 1

Proteinases and their naturally occurring inhibitors are known to play a crucial role in a variety of biologically important processes like food digestion, blood coagulation and fibrinolysis, blood pressure regulation, fertilization, phagocytosis and complement immune reactions (1).

Moreover, several pathological situations have been related to the action of proteolytic enzymes. Pulmonary emphysema for example is a disease characterized by a decrease in physiological lung function due to the loss of elastic recoil (2). A proteinase/proteinase inhibitor imbalance hypothesis has been proposed as a model for the development of the disease: It suggests that either an increase in destructive proteolytic activity, mainly due to the release of elastase from leukocytes, or a decrease in protective serum proteinase inhibitors can lead to emphysema. This finally involves the irreversible destruction of lung connective tissues. A similar model may apply to other related connective tissue diseases such as rheumatoid arthritis (3, 4).

Thus, the development of specific inhibitors especially for leukocyte elastase is of particular pharmacological importance (5, 6). Low-molecular weight substances, for example peptide chloromethyl ketones (7) or certain heterocyclic compounds (8), which have been synthesized for this purpose, often exhibit cytotoxic effects. Therefore, naturally occurring inhibitors or derivatives thereof, which are proteins, could turn out to be the antagonists of choice. In order to understand the function of such inhibitors or to remodel them for a special concern, specific modification of their structure is sometimes required.

During the last decade we have developed methods for tailoring protein inhibitors against distinct target proteinases with particular emphasis on elastase.

Chemistry of Peptides and Proteins, Vol. 3

Our efforts have mainly concentrated upon derivatives of aprotinin, the trypsin inhibitor (Kunitz) from bovine mast cells (9). The single polypeptide chain of aprotinin consists of 58 amino acid residues which are cross-linked by three disulfide bridges. It is folded in such a way that a pear-shaped molecular structure results (10). The hatched area in Fig. 1, including the reactive-site peptide bond Lys^{15}-Ala^{16}, displays the region of the inhibitor that is in close contact with the enzyme in the trypsin/aprotinin complex (11). The five negatively charged residues (Asp^{3}, Glu^{7}, Glu^{49}, Asp^{50}, Ala^{58}) are concentrated at the opposite end of the molecule.

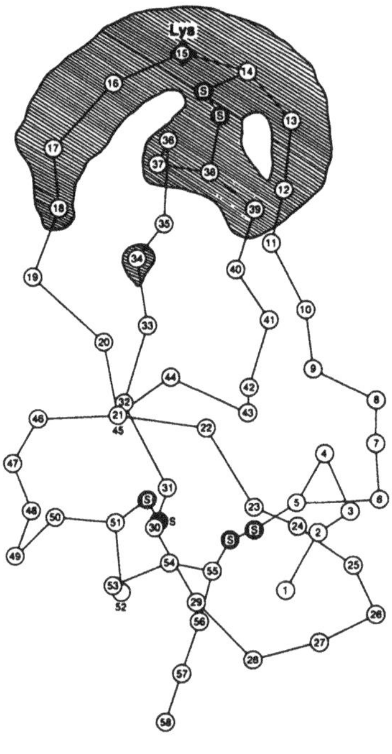

Fig. 1: Projection of the three-dimensional structural model of aprotinin (10). The open circles indicate the positions of the α-carbon atoms of the corresponding amino acid residues.

Aprotinin is an excellent inhibitor of trypsin, mainly due to the positively charged side chain of its Lys^{15} residue, which is accomodated in the specificity pocket of the proteinase. The complex-dissociation constant is extremely low: $K = 6 \times 10^{-14}$ M (12). Inhibition of chymotrypsin with $K = 9 \times 10^{-9}$ M (13) is relatively weak and that of leukocyte elastase with $K = 4 \times 10^{-6}$ M (14) is hardly detectable. Because chymotrypsin preferentially splits peptide bonds following aromatic amino acid residues and leukocyte elastase those behind aliphatic residues, especially valine, it was surmised that replacement of Lys^{15} in aprotinin by suitable other amino acids would change the inhibitory specificity in a predictable way. The first semisynthetic pathway which allows for such replacements is the so-called 'enzymatic mutation', originally developed for the soybean trypsin inhibitor (15). Its application to aprotinin led to Phe^{15}- and Trp^{15}-homologues (16) with anti-chymotryptic activity and the Arg^{15}-homologue (17) with conserved antitryptic activity as expected. Briefly, the 'enzymatic mutation' consists of the following steps: a) Enzymatic hydrolysis of the reactive-site peptide bond Lys^{15}-Ala^{16} to yield 'modified' aprotinin. b) Removal of Lys^{15} by carboxypeptidase B. c) Coupling of the new amino acid to Cys^{14} by an enzymatically catalyzed reaction (carboxypeptidase B for Arg, carboxypeptidase A for Phe and Trp) giving a 'modified' aprotinin homologue in minute amounts. d) Trapping of this product by complex formation with a proteinase (kallikrein for Arg, chymotrypsin for Phe und Trp) thus shifting the carboxypeptidase-catalyzed reaction towards synthesis. e) Kinetic control complex dissociation to yield single-chain aprotinin homologues.

There have been several attempts to apply this series of enzymatic reactions to the introduction of aliphatic amino acids in order to render aprotinin an efficient elastase inhibitor. However, so far all efforts have remained unsuccessful, mainly because a proper enzyme system is lacking. We therefore decided to develop methods which overcome these difficulties and allow for introduction of almost any amino acid.

In a preliminary series of experiments the new amino acid, as methyl ester, was coupled to Cys^{14} of 'modified' aprotinin lacking Lys^{15} with water-soluble carbodiimides (18, 19). Upon complex formation with a proper proteinase the reactive-site peptide bond was closed again whereby methanol was eliminated. There are two inherent drawbacks, however: 1) The need for a proteinase in the final reactive-site peptide bond synthesis questions the general applicability of the method. 2) During the introduction step, coupling of the amino acid methyl ester also takes place to some extent at the five carboxyl groups already present in native aprotinin. Consequently, a generally applicable chemical condensation step is required and reversible carboxyl protection has to be employed.

Our strategy, leading from naturally occurring aprotinin with Lys^{15} to its homologue with Val^{15} in nine or ten steps respectively, is schematically depicted in the following four figures.

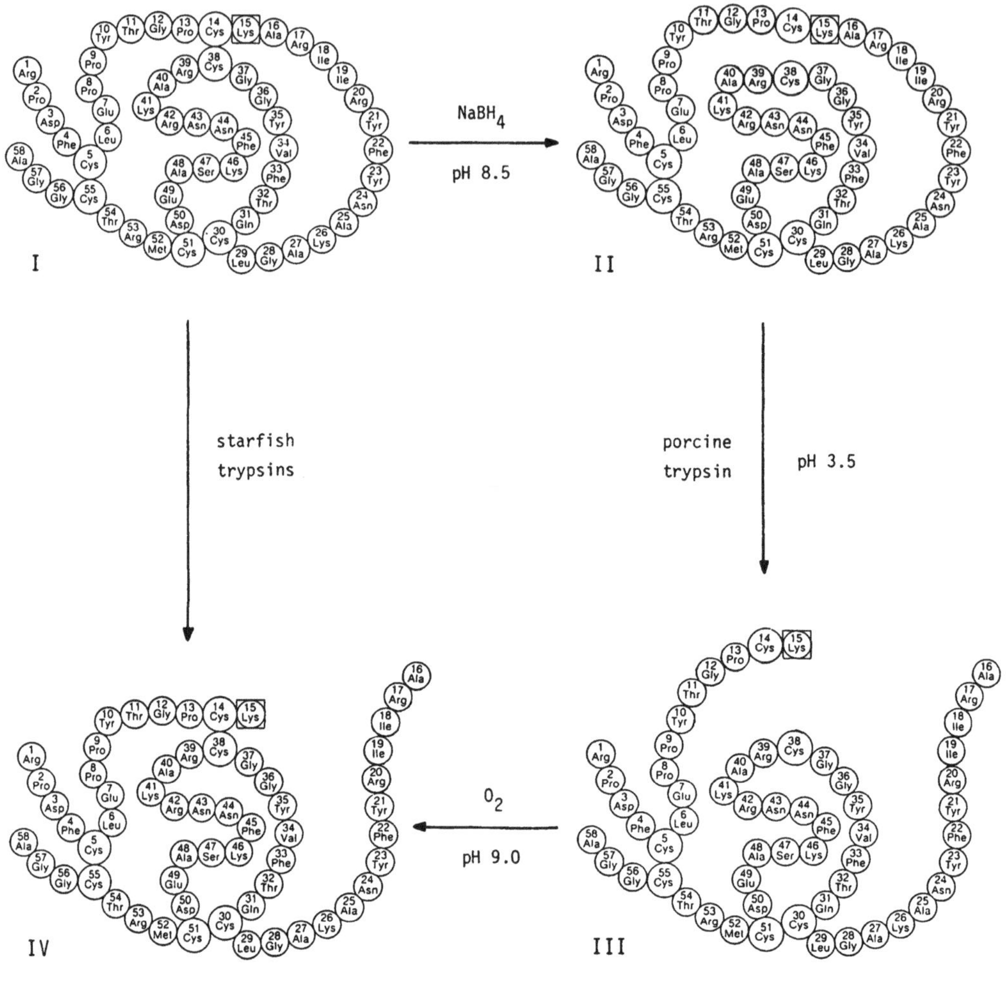

Fig. 2: Preparation of 'modified' aprotinin with the reactive-site peptide bond, Lys^{15}-Ala^{16}, cleaved either by direct proteolysis (I → IV) or by a reaction sequence also comprising reduction and reoxidation of the adjacent disulfide bond, Cys^{14}-Cys^{38} (I → II → III → IV).

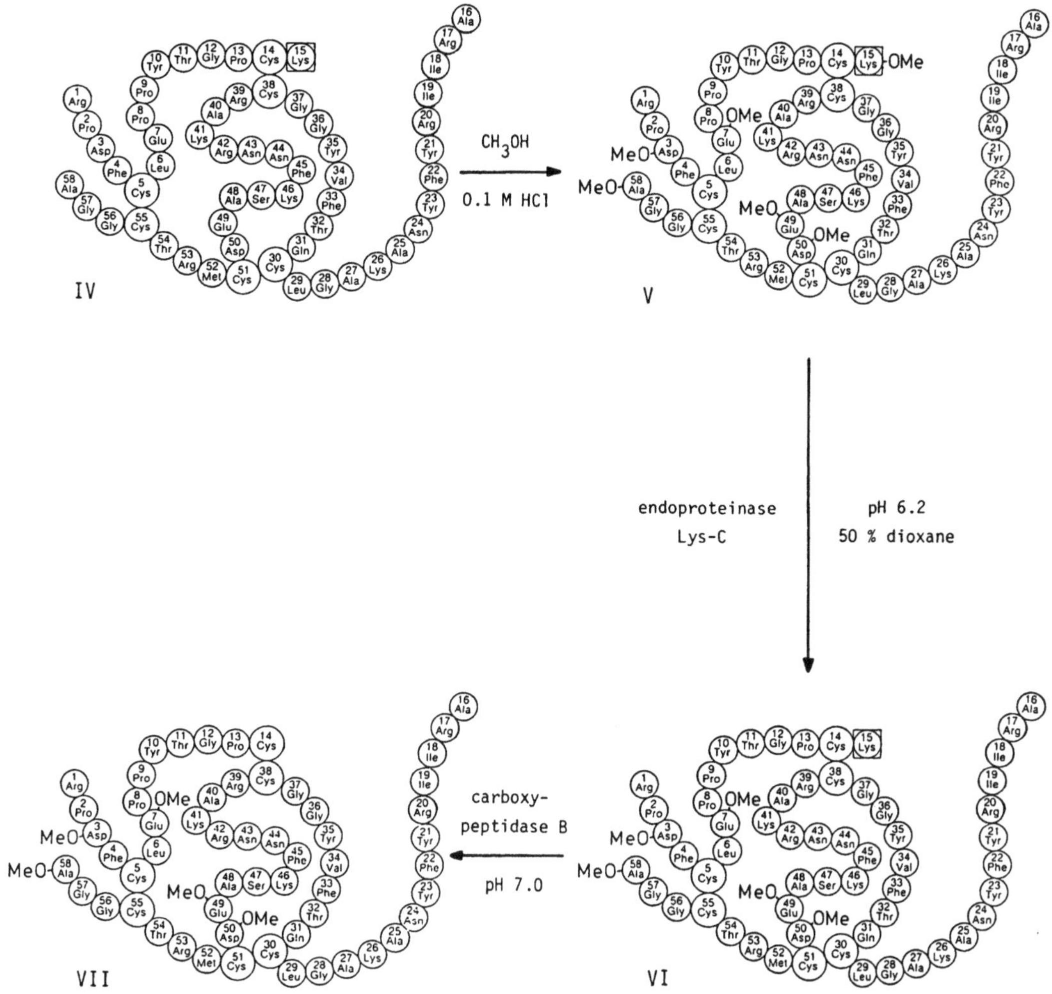

Fig. 3: Preparation of the key intermediate for the chemical introduction of new reactive-site amino acids:
a partially protected des-Lys^{15} 'modified' aprotinin with only the carboxyl group of Cys^{14} deprotected.

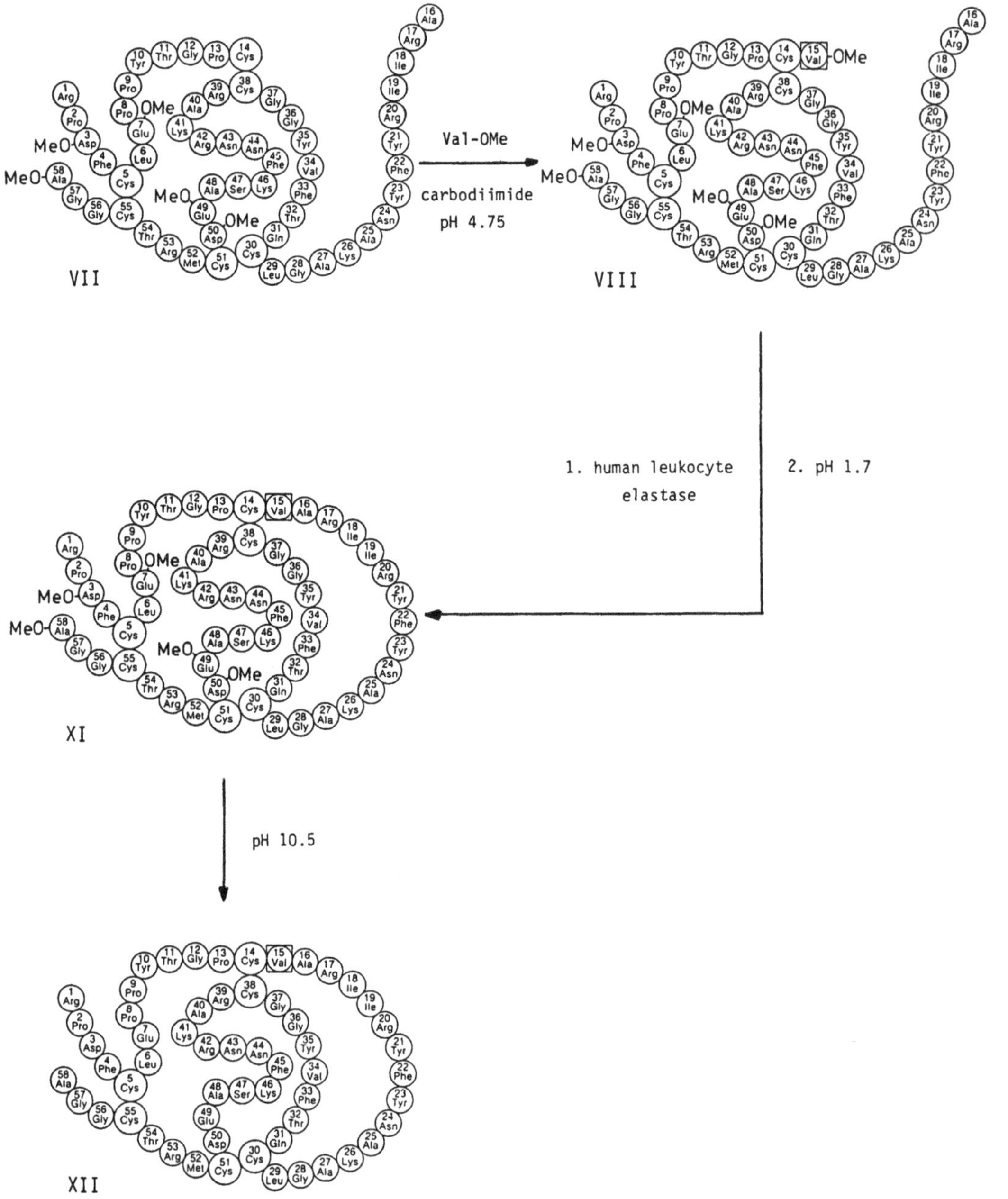

Fig. 4: Synthesis of Val^{15}-aprotinin involving carbodiimide-coupling of valine methyl ester and enzymatic reactive-site peptide bond closure.

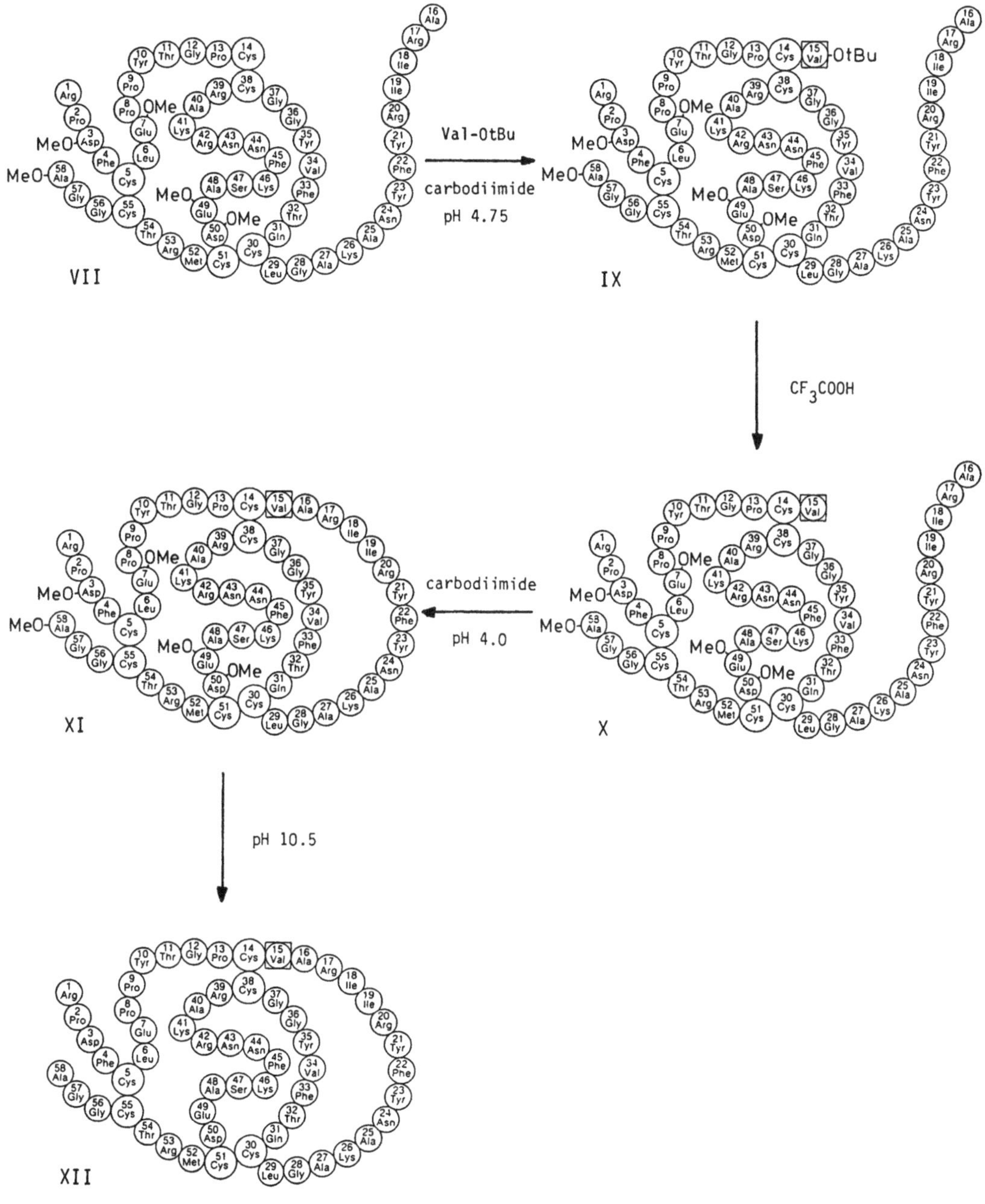

Fig. 5: Synthesis of Val^{15}-aprotinin using carbodiimides for coupling of valine t-butyl ester and reactive-site peptide bond closure.

The first derivative of aprotinin I (Fig. 2), which is isolated in a homogeneous form is the 'modified' inhibitor IV with the reactive-site peptide bond Lys^{15}-Ala^{16} hydrolyzed; it is also a key intermediate during the 'enzymatic mutation' reaction sequence. Bovine trypsin, bovine chymotrypsin and porcine plasmin catalyse this specific cleavage only very slowly; the equilibrium at neutral pH, in which about half of the inhibitor molecules are 'modified' is reached after several months (20). This difficulty can be overcome by selective borohydride reduction of the Cys^{14}-Cys^{38} disulfide bond (→ II), followed by fast tryptic cleavage of the reactive-site peptide bond (→ III) and air reoxidation of the half-cystine residues (→ IV) (21). After cation-exchange chromatography, pure 'modified' aprotinin IV is obtained in about 40% yield even on a 10 gram scale. In the meantime trypsin-like proteinases from the starfish species Dermasterias imbricata (22) and Asterias rubens (Tschesche, H., Schmuck, R., Wenzel, H.R., unpublished) have been isolated, which establish the equilibrium in some days without disulfide reduction.

The five carboxyl groups remote from the reactive site have now to be protected specifically and reversibly to avoid their irreversible derivatisation during the following peptide-chemical coupling steps. A combination of chemical and enzymatic reactions proved to be the method of choice (Fig. 3): 'Modified' aprotinin IV is almost quantitatively esterified during 24 h in methanol containing 0.1 M HCl (23) to yield the hexamethyl ester V. The ester group of Lys^{15} can be specifically hydrolyzed in four days using the endoproteinase Lys-C from Lysobacter enzymogenes (Boehringer) in a mixture of 50% 0.1 M acetic acid and 50% dioxane with the pH adjusted to 6.2. Under these conditions the proteinase, while keeping its side-chain specificity, mainly exhibits esterolytic activity. The same behaviour was described for bovine trypsin (24, 25), which can be used as an alternative enzyme, however, with inferior results as to reaction time and purity of the product. From the 'modified' aprotinin pentamethyl ester VI Lys^{15} is completely removed by carboxypeptidase B within some hours at pH 7.0. Cation-exchange chromatography provides pure des-Lys^{15} 'modified' aprotinin pentamethyl ester VII in 25% overall yield from 'modified' aprotinin IV.

We exploited two alternative pathways to introduce valine, the first of which is shown in Fig. 4. Valine methyl ester is coupled to the only free carboxyl group of VII at pH 4.75 by using water-soluble carbodiimides, N-(3-dimethylaminopropyl)-N'-ethylcarbodiimide-hydrochloride or N-cyclohexyl-N'-(2-morpholinoethyl)carbodiimide-methyl-p-toluenesulfonate, to yield 'modified' Val^{15}-aprotinin hexamethyl ester VIII. Stoichiometric amounts of human leukocyte elastase are then needed to form, at about neutral pH, the proteinase/inhibitor complex, which releases Val^{15}-aprotinin pentamethyl ester XI upon kinetic control dissociation after lowering the pH to 1.7. Raising the pH to 10.5 for some hours removes all methyl ester groups thus leading to Val^{15}-aprotinin XII. Because of limited supply of leukocyte elastase only small amounts of the Val^{15}-homologue can normally be synthesized in this way. Moreover, the extension of the method to the introduction of other amino acids critically depends on the availability of proteinases capable of closing the various reactive-site peptide bonds.

The second pathway (Fig. 5), which only includes peptide-chemical reactions, is more general. The carbodiimide-condensation is carried out with valine t-butyl ester to yield derivative IX. The t-butyl ester protecting group is then specifically split off in trifluoroacetic acid for one hour at room temperature leading to derivative X with one free carboxyl group. Synthesis of the peptide bond between Val^{15} and Ala^{16} is effected with water-soluble carbodiimides at pH 4.0. Saponification of the resulting Val^{15}-aprotinin pentamethyl ester XI at alkaline pH leads to the desired Val^{15}-homologue XII, which is purified by cation-exchange chromatography. The overall yield from the key intermediate VII is about 12%.

Val^{15}-aprotinin was characterized as follows:

During high-performance cation-exchange chromatography (Fig. 6), which we recently presented as a powerful tool to characterize a variety of structurally related proteins (26), the homologue, devoid of the positively charged Lys^{15} side-chain, elutes in front of aprotinin. The sharp, symmetrical peak obtained strongly suggests homogeneity of the preparation. Moreover, polyacrylamide gel electrophoresis at pH 9.3 (21) shows only a single band migrating about 25% slower than aprotinin.

Amino acid analysis is in good agreement with the expected integral values and clearly indicates the Lys→Val substitution. The specific introduction of valine in position 15 and the single-chain nature of the semisynthetic inhibitor is unambigously demonstrated by automatic sequencing, which was carried out up to residue 31.

The Val^{15}-homologue has inhibitory properties quite different from those of aprotinin. At enzyme/Val^{15}-aprotinin ratios of 1:2 there is no appreciable inhibition of the following proteinases: bovine trypsin, porcine kallikrein, bovine chymotrypsin and human leukocyte cathepsin G, a chymotrypsin-like proteinase. The

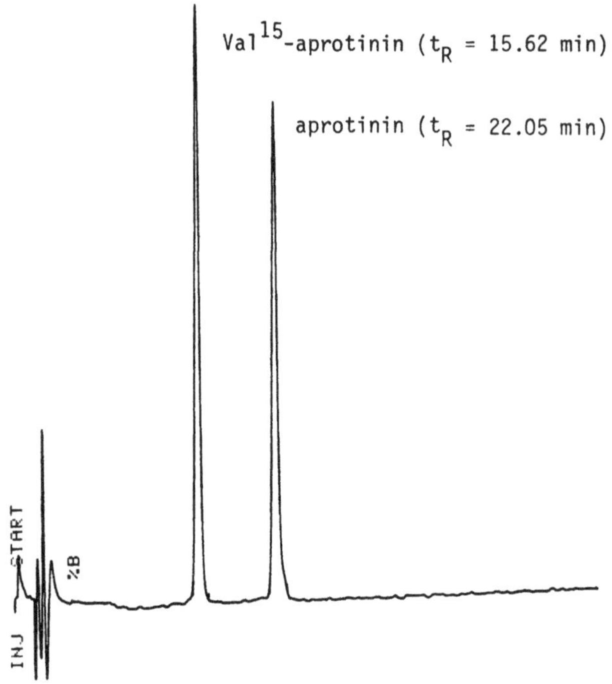

Fig. 6: Separation of 10 µg aprotinin and 10 µg semisynthetic Val^{15}-aprotinin by high-performance cation-exchange chromatography (26) (HP 1084 A chromatograph with detection at 220 nm, LKB Ultropac TSK SP-5PW column, 30°C, 1 ml/min; elution: 10% B for 5 min, then, insert "%B", linear gradient from 10% B to 100% B over 45 min, solvent A: 0.01 M phosphate buffer, pH 7.0, solvent B: 1 M NaCl in 0.01 M phosphate buffer, pH 7.0).

activity of porcine pancreas elastase is reduced to about 70% by a twofold molar amount of the inhibitor. Fig. 7 shows the titration curves of human leukocyte elastase with Val^{15}-aprotinin and aprotinin respectively. It is evident that the Lys→Val exchange in the reactive site is accompanied by a dramatic increase in affinity of leukocyte elastase for the Kunitz inhibitor. From the Val^{15}-aprotinin curve, which is linear down to 10% residual activity, the existence of a strong 1:1 complex can be deduced with a dissociation constant of about 10^{-10} M. This value is almost five orders of magnitude lower than the constant for the elastase/aprotinin complex and one of the smallest values reported for leukocyte-elastase inhibitors so far.

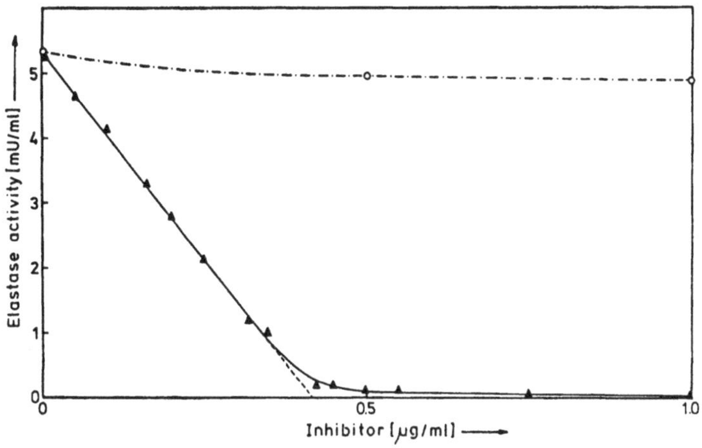

Fig. 7: Inhibition of human leukocyte elastase by aprotinin (o) and semisynthetic Val^{15}-aprotinin (▲). Enzymatic activity was determined using the synthetic chromogenic peptide substrate 3-carboxypropionyl-alanyl-alanyl-valine-4-nitroanilide (27) as described previously (28).

In conclusion, the semisynthetic reaction sequences outlined in this paper, virtually allow any single amino acid replacement within the reactive site of aprotinin. This corresponds to a post-translational 'chemical mutation' of the completed protein. The homologues of aprotinin with different amino acid substitutions at their reactive site allow to assay the individual contribution of a single side chain to the inhibition of a particular proteinase and to design specific inhibitors for it. This approach could provide an excellent rationale for the application of laborious site directed mutagenesis techniques for tailor-made inhibitor production. It can be anticipated that modern genetic engineering could play a decisive role in finally providing us with new protein inhibitors in sufficient amounts for therapeutic control of pathological destructive proteinase action.

Acknowledgement

Financial support by the 'Deutsche Forschungsgemeinschaft' and the 'Fonds der Chemischen Industrie' is gratefully acknowledged.

References

1. Holzer, H., Tschesche, H., eds. 1979. Biological Functions of Proteinases. Springer-Verlag, Berlin-Heidelberg-New York.

2. Mittman, C., ed. 1972. Pulmonary Emphysema and Proteolysis. Academic Press, New York.

3. Barrett, A.J. 1978. Agents Actions 8, 11-18.

4. Menninger, H., Mohr, W. 1981. Therapiewoche 31, 2134-2148.

5. Powers, J.C. 1983. Am. Rev. Respir. Dis. 127, S54-S58.

6. Burkhardt, H., Menninger, H., Wenzel, H., Tschesche, H., Hoss, R., Berner, W., Mierau, H.D. 1983. In: Theoretische und klinische Befunde der Knorpelforschung (Dettmer, N., Lindner, J., Kleesiek, K., Mohr, W., Puhl, W., eds.). Eular Verlag, Basel, pp. 153-159.

7. Powers, J.C., Gupton, B.F., Harley, A.D., Nishino, N., Whitley, R.J. 1977. Biochim. Biophys. Acta 458, 156-166.

8. Teshima, T., Griffin, J.C., Powers, J.C. 1982. J. Biol. Chem. 257, 5085-5091.

9. Fritz, H., Wunderer, G. 1983. Arzneim.-Forsch./Drug Res. 33, 479-494.

10. Huber, R., Kukla, D., Rühlmann, A., Epp, O., Formanek, H. 1970. Naturwissenschaften 57, 389-392.

11. Huber, R., Kukla, D., Bode, W., Schwager, P., Bartels, K., Deisenhofer, J., Steigemann, W. 1974. J. Mol. Biol. 89, 73-101.

12. Vincent, J.-P., Lazdunski, M. 1972. Biochemistry 11, 2967-2977.

13. Vincent, J.-P., Lazdunski, M. 1973. Eur. J. Biochem. 38, 365-372.

14. Lestienne, P., Bieth, J.G. 1978. Arch. Biochem. Biophys. 190, 358-360.

15. Sealock, R.W., Laskowski, Jr., M. 1969. Biochemistry 8, 3703-3710.

16. Jering, H., Tschesche, H. 1976. Eur. J. Biochem. 61, 453-463.

17. Bode, W., Walter, J., Huber, R., Wenzel, H.R., Tschesche, H. 1984. Eur. J. Biochem. 144, 185-190.

18. Wenzel, H.R., Tschesche, H. 1980. Hoppe-Seyler's Z. Physiol. Chem. 361, 345.

19. Wenzel, H.R., Tschesche, H. 1981. Angew. Chem. 93, 292-293.

20. Tschesche, H., Kupfer, S. 1976. Hoppe-Seyler's Z. Physiol. Chem. 357, 769-776.

21. Jering, H., Tschesche, H. 1976. Eur. J. Biochem. 61, 443-452.

22. Estell, D.A., Wilson, K.A., Laskowski, Jr., M. 1980. Biochemistry 19, 131-137.

23. Quast, U., Engel, J., Steffen, E., Tschesche, H., Kupfer, S. 1978. Biochemistry 17, 1675-1682.

24. Coletti-Previero, M.-A., Previero, A., Zuckerkandl, E. 1969. J. Mol. Biol. 39, 493-501.

25. Büllesbach, E.E., Schmitt, E.W., Gattner, H.-G. 1982. Int. J. Peptide Protein Res. 20, 207-217.

26. Wenzel, H.R., Beckmann, J., Mehlich, A., Siekmann, J., Tschesche, H., Schutt, H., 1985. In: Modern Methods in Protein Chemistry, Vol. 2 (Tschesche, H., ed.). Walter de Gruyter, Berlin-New York, pp. 173-184.

27. Wenzel, H.R., Engelbrecht. S., Reich, H., Mondry, W., Tschesche, H. 1980. Hoppe-Seyler's Z. Physiol. Chem. 361, 1413-1416.

28. Tschesche, H., Engelbrecht, S., Wenzel, H.R. 1984. In: Methods of Enzymatic Analysis, Vol. V (Bergmeyer, H.U., ed.). Verlag Chemie, Weinheim, pp. 176-184.

UNIFIED SYSTEM FOR HIGH SPEED ANALYSIS OF Dns- AND PTH AMINO ACIDS

I.V. Nazimov, N.B. Levina
Shemyakin Institute of Bioorganic Chemistry, USSR Academy of Sciences, Moscow, USSR

Introduction

The system common for high speed separation of both Dns and PTH amino acid derivatives is proposed. Separation time for a mixture of 22 Dns amino acids is 26 min, sensitivity 10-20 picomoles (fluorimetry); for PTHs 9.2 min, 50-100 picomoles (UV, 254 nm). Conditions of peptide dansylation and hydrolysis of Dns-peptides with the following HPLC of Dns-amino acids were elaborated. N-terminal analysis takes 2-3 hrs.

Results

Determination of PTH and Dns amino acids is a very widely used procedure in modern protein chemistry. A great number of publications deals with methodological improvements of separation process and identification of PTH and Dns amino acids. However, there are only few examples of complete separation of mixtures of Dns derivatives in reasonable time.
The purpose of present investigation is development of a unified system of high speed N-terminal amino acid and PTH amino acid analysis. The same separation system (chromatograph, column, eluents) is employed in both cases which adds to the practicability of the method.
Analysis of N terminal amino acids as Dns derivatives by RP HPLC provides useful criteria (qualitative and quantitative) of purity of peptides (natural and synthetic). Still the advantages of HPLC for routine analysis could not be utilized because separation of standard mixtures rather than the more complicated mixtures of N-terminal amino acids obtained by dansylation of peptides followed

Chemistry of Peptides and Proteins, Vol. 3

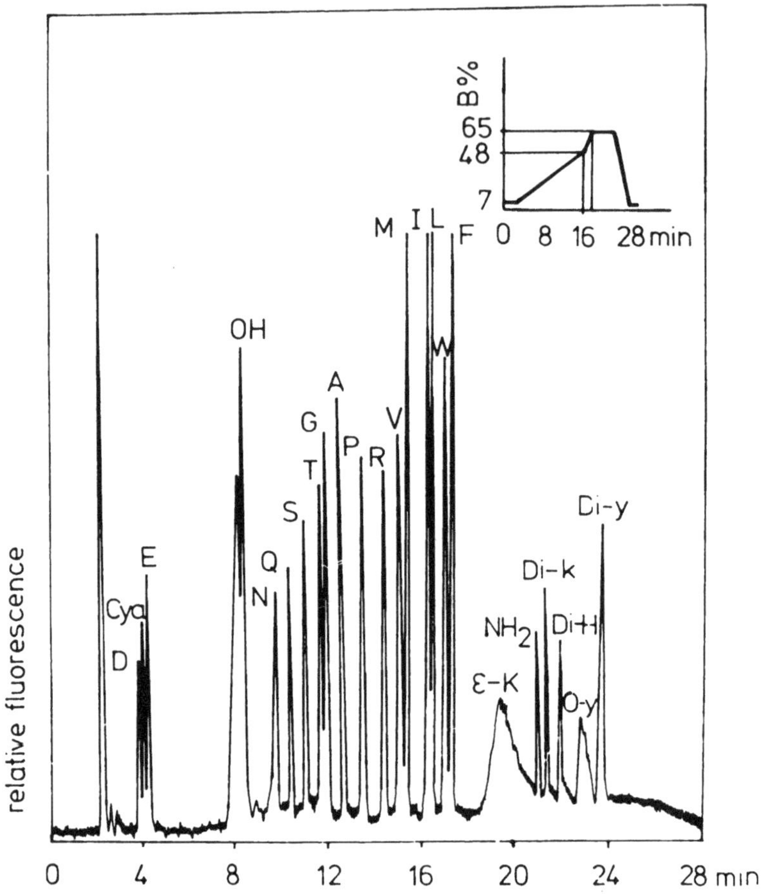

Fig. 1. RP HPLC of Dns amino acids. 322 MP Altex chromatograph, 4.6x250 mm Ultrasphere ODS column, particle size 5 µ. Solvent A: CH_3CN-Na-TFA buffer (25 mM, pH 7.6), 10:90. Solvent B: CH_3CN-Na-TFA buffer (25 mM, pH 7.6), 70:30. Solvent program: 0 min after injection, 7% B; 1 min, 7%; 16 min, 48%; 23 min, 65%; 25 min, 7%. Flow rate: 1.2 ml/min, temperature 39°C. 20 picomoles of each derivative except for di-derivatives which were applied in 10 picomoles amounts. Detector:fluorimeter FS 970 (Kratos). λ_{ex} 330 nm, λ_{em} >470 nm. One letter code for amino acids is used for most mono-dansyl derivatives; Cya=Dns-cysteic acid; Di-K= Di-Dns-Lys; O-Y=O-Dns-Tyr; OH=dansylic acid; NH_2=Dns-amide.

by Dns-peptide hydrolysis is usually discussed in the literature. We aimed at developing a rapid quantitative N-terminal analysis at subnanomol level of peptides composed of up to 30 amino acid residues, in particular, hydrophobic fragments of membrane proteins. The procedure could be of value for the purity control of peptides to be further subjected to manual or automatic sequencing. Three procedures were optimized: (1) separation of the mixture of standard Dns-amino acids; (2) dansylation of the peptide; (3) hydrolysis of the dansyl peptide. Besides, we developed a rapid and sensitive method for identification of PTHs cleaved on a sequenator.

Separation of Dns-amino acids. In order to obtain reproducible results with the mixture of 22 Dns derivatives (19 mono-derivatives of usual amino acids plus di-Dns derivatives of His, Tyr, Lys), separation conditions with respect to buffer composition, its molarity and pH were optimized.

Of different buffer systems examined, the sodium trifluoroacetate (Na-TFA) buffer appeared to be the most selective. With this buffer the components of the mixture were eluted as sharp symmetric peaks, an important precondition for rapid and highly sensitive analysis. The capacity factor (k') of Dns-amino acids proved highly dependent upon the buffer pH and Na-TFA molarity in case of Asp, Glu, Gln, Ser, Thr, Gly, His, Arg, the 25-30 mM buffer, with 7.0-7.6 pH being optimal for separation (1,2).

Fig. 1 illustrates the separation pattern of the standard mixture of the 22 mono- and di-Dns derivatives (10-20 picomoles each) as well as dansylic acid and dansyl amide. Complete separation of this mixture takes 26 min; equilibration of the column requires additional 5 min. It should be mentioned that O-Dns-Tyr and ε-Dns-Lys give diffuse peaks and intensity of the Dns-Asp and Dns-Glu peaks is twice lower than for other derivatives.

Dansylation of peptides. When dansylation of amino acids and peptides is performed according to conventional methods the chromatographic pattern differs considerably from that in Fig. 1. The usual 100-1000 fold excess of Dns chloride gives rise to considerable amounts of side products which complicate identification of many amino acids. We found that 5-10-fold excess of Dns chloride in the lithium carbonate buffer of Tapuhi and coworkers (3) applied in a small volume (3 - 5 μl) resulted in quantitative dansylation and was accompanied by minimal side product formation.

Hydrolysis of Dns peptides. Dns peptides are usually hydrolyzed with 5.7 M HCl for 4-12 hr at 105° which results in substantial destruction of Dns derivatives of Met, Gly and Pro. To accelerate the hydrolysis of Dns peptides we increased the temperature and employed a mixture of 5.7 M HCl and TFA. A similar mixture (11 M HCl and TFA) and temperature were used for hydrolysis of unprotected peptides by Tsugita et al. (4). Fig. 2 provides a comparison of the yields obtained by hydrolysis of a Dns peptide under classical conditions and the conditions proposed in this paper; 50 min duration

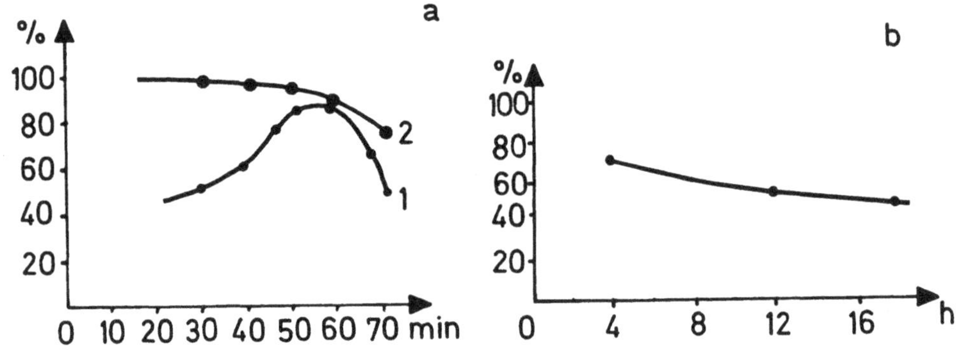

Fig. 2. Dependence of the yield of Dns Phe on hydrolysis time of Dns insulin B-chain (a_1 - 5.7 M HCl/TFA, 2:1, 166°C; b - 5.7 M HCl, 105°C). Hydrolysis of Dns Phe is shown for comparison (a_2 - 5.7 M HCl/TFA, 2:1, 166°C).

and 166°C proved optimal for hydrolysis of Dns peptides of about 30 residues in length. At 190° the hydrolysis time drops to 15 min but the yield of the N-terminal amino acids is lower. As usual, correction should be made for partial destruction of Dns amino acids formed in the course of hydrolysis (Table 1). Hydrolysis with 5.7 M HCl/TFA allowed us not only to shorten the hydrolysis time as compared to hydrolysis with 5.7 M HCl but also to obtain higher yields of Dns amino acids. With slowly hydrolyzable sequences such as Val-Val, Ile-Val, Ile-Leu, frequently found in hydrophobic fragments of the membrane proteins hydrolysis takes 100 min (Fig. 3). Under usual conditions the yield of Dns Val or Dns Ile is 1.4-2.2 times lower (Table 2).

Table 1. Destruction of Dns Amino Acids (5.7 M HCl/TFA, 2:1, 160°, 50 min)

Derivative	Survival after hydrolysis, %
Dns Asp	105
Dns Arg	95
Dns Ala	91
Dns Val	91
Dns Ile	90
Dns Leu	85
Dns Phe	85
Dns Gly	83
α,O-di-Dns Tyr	80
Dns Thr	80
Dns Ser	75
α,ε-di-Dns Lys	75
Dns Met	65
α-di-Dns His	60
Dns Pro	58

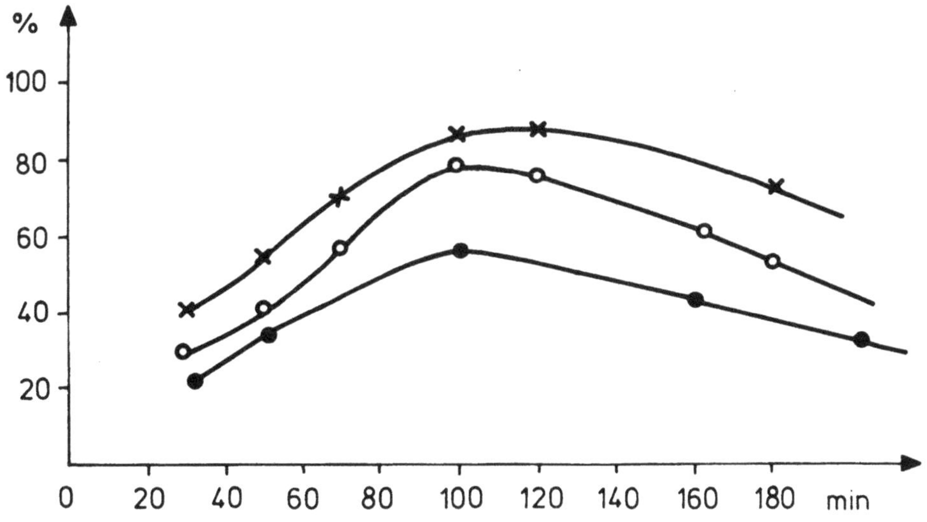

Fig. 3. Dependence of the yield of N-terminal amino acids of Dns dipeptides on hydrolysis time: x - Val-Val, o - Ile-Val, ● - Ile-Leu.

Table 2. Dependence of the Yield of N-Terminal Dns Amino Acids of Slowly Hydrolyzable Dns-Dipeptides on Hydrolysis Conditions

Dipeptide	Yield of N-terminal amino acid, %*			Relative increase of yield, (III/I)
	I 5.7 M HCl, 105°, 18 hr	II 5.7 M HCl/TFA, 166°, 50 min	III 5.7 M HCl/TFA 166°, 100 min	
Val-Val	60	70	87	1.45
Ile-Val	57	58	80	1.40
Ile-Leu	25	45	57	2.20

* mean result of three determinations

Separation of PTH amino acids. A successful use of TFA-containing buffers for separation of Dns amino acids prompted us to try that system for analysis of PTH amino acids. With an Ultrasphere ODS column and Na-TFA the highly selective separation of 20 PTH amino acids was achieved in 9.2 min (Fig. 4).

Conclusion. The procedures proposed here for the N-terminal analysis of peptides made possible rapid and quantitative evaluation of peptide purity. Of special importance is application of TFA-containing buffer resulting in considerable reduction of analysis time of Dns and PTH amino acids. The conditions suggested are superior to those described in the literature with respect of speed and accuracy.

Protein hydrolysates containing short and middle size peptides are currently separated with TFA and acetonitrile (CH_3CN) on columns with C_{18} coated phase. Therefore this standard separation system is now applicable to great majority of chromatographic separations involved in analysis of peptides up to 30 amino acid residues in length.

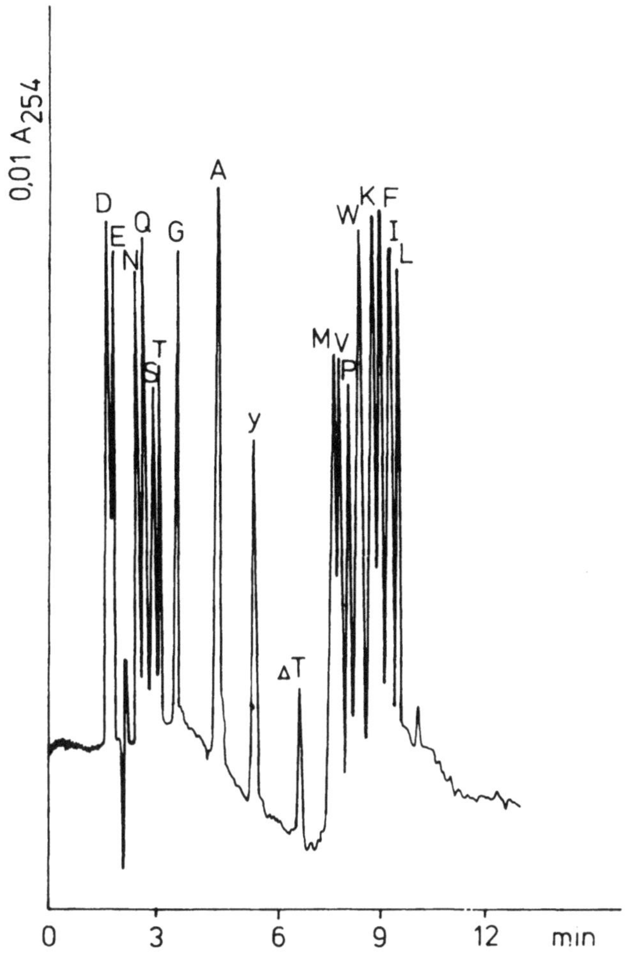

Fig. 4. Separation of standard mixture of PTH amino acids. Instrument, column - as in Fig. 1. Solvent A - 30 mM Na-TFA (pH 5.4) containing 10% CH_3CN; solvent B - CH_3CN. Gradient B: time 0 min, 35% B; 1 min, 35%; 8 min, 75%; 10 min, 35 %. Flow rate - 1.3 ml/min; temperature 39°. Detector - UV-photometer, 254 nm, Altex). PTH derivatives: 50-100 picomoles each.

References

1. Levina, N.B., I.V. Nazimov. 1984. J. Chrom. 286, 207-216.

2. Levina, N.B., Ch.G. Muradov, I.V. Nazimov. 1986. Bioorg. Khim. (Russian) 11, N 2, in press.

3. Tapuhi, Y., D.E. Schmidt, W. Lindner, B.L. Karger. 1981. Anal. Biochem. 115, 123-129.

4. Tsugita, A., J.J. Scheffler. 1982. Eur. J. Biochem. 124, 585-588.

CONFORMATION OF GRAMICIDIN A IN SOLUTION AND MICELLES: TWO DIMENSIONAL ^{1}H NMR STUDY

Alexander S. Arseniev, Igor L. Barsukov, Vladimir F. Bystrov

Shemyakin Institute of Bioorganic Chemistry,
USSR Academy of Sciences, 117437 Moscow, USSR

Introduction

Gramicidin A (GA) is a linear pentadecapeptide antibiotic consisting of alternating L- and D-amino acid residues. Its most abundant isomer [Val^1] GA has the structure: HCO-L-Val^1-Gly^2--L-Ala^3-D-Leu^4-L-Ala^5-D-Val^6-L-Val^7-D-Val^8-L-Trp^9-D-Leu^{10}--L-Trp^{11}-D-Leu^{12}-L-Trp^{13}-D-Leu^{14}-L-Trp^{15}-NHCH$_2$CH$_2$OH. Ability of GA to form well characterized ion permeable pores in the natural and artificial lipid bilayer membranes made it popular as a model of transmembrane molecular channel (1). Although the dimeric nature of the GA channel was established with confidence (2), its molecular conformation evoked several controversial assumptions based on indirect evidence (3,4). Indeed the problem is dramatically complicated by diversity of energetically allowed conformations of GA and its metal complexes (5,6), including single- and double-stranded helices with different relative orientation of N(head)- and C(tail)-terminals, handedness and number of residues per turn.

In some organic solvents GA forms conformationally defined structures. In dioxane, for instance, four basic species were detected, which interconvert into each other so slowly that they were isolated separately (6). From CD and IR spectra (6,7) it followed that those species were double helices of different types. The X-ray study (8) of crystals grown from methanol and ethanol solutions of GA complexes with Cs^+, K^+ and Tl^+ showed a molecule of cylindrical shape with one metal cation per GA monomer, but was consistent with both single- and double-helical dimer conformations. From IR and laser Raman spectra of GA and

Chemistry of Peptides and Proteins, Vol. 3

Cs^+-GA crystals it was inferred that they had double-helical structure (9). Conformational energy calculations (10,11) yielded almost identical results for single- and double-stranded helices. For GA incorporated into lysolecithin micelles the ^{13}C NMR chemical shifts induced by thallium cations were interpreted in favour of a head-to-head dimer formed by left-handed, single-stranded $\overrightarrow{\pi}^{6.3}_{LD}$ $\overleftarrow{\pi}^{6.3}_{LD}$ helices (3). Similar conclusion on a head-to-head dimer was reached in the ^{13}C and ^{19}F NMR study of modified GA in phosphatidylcholine liposomes with use of paramagnetic probes (12). In contrast, the CD and IR spectra of GA in liposomes were considered as evidence for formation of double-helical channels at high lipid/GA ratio and extended β-structures at the low ratio (4,13).

In this situation the high-field two dimensional proton NMR spectroscopy is the choice for direct evaluation of conformation of GA and its metal complexes in different milieus.

Materials and Methods

$[Val^1]$ GA was obtained from gramicidin D (Serva) by countercurrent distribution (14). The sample used for NMR studies also contained 4-5% of $[Ile^1]$ GA. The solvents dioxane-D_8 (99% deuterium, Stohler), $CDCl_3$ ("100%" deuterium, Stohler), D_2O ("100%" deuterium, Stohler), CF_3CD_2OD (98% deuterium, Stohler), CD_3OD and CD_3OH (99% deuterium, Isotop USSR) were used without further purification. Sodium dodecyl-D_{25} sulphate (CEA, France) was a generous gift of Drs. L. Zetta and A. De Marco (Milano).

The NMR spectra were recorded on a Bruker WM-500 spectrometer interfaced to an ASPECT-2000 computer. Proton chemical shifts for organic solvents and water media were reported relative to tetramethylsilane and sodium 3-trimethylsilyl-(2,2,3,3-D_4)-propionate as internal standards, respectively. Samples were placed in the 5 mm NMR tubes.

The 500 MHz proton COSY and NOESY 2D NMR spectra were obtained in a combined COSY-NOESY experiment (15) with the nonselective 90°-pulse sequence:

$$(90^o-t_1-90^o-t_{2,C}-\Delta-90^o-t_{2,N}-T_p)_n, \quad [1]$$

where t_1 - evolution time period 0.1-51.2 ms (512 increments); $t_{2,C}$ and $t_{2,N}$ - periods of 102.4 ms for aquisition of COSY and NOESY spectra, respectively; $t_{2,C}+\Delta$ - mixing time τ_m = 120 ms for NOESY spectrum; T_p - preparation period of 1.2 s during which the system was allowed to reach equilibrium. The two free induction decay (FID) interferrograms (2048 data points each) were recorded during $t_{2,C}$ and $t_{2,N}$ and stored in different parts of the computer memory.

The 2D RELSY spectra were obtained as in (16) with the pulse sequence:

$$(90^o-t_1-90^o-\tau-180^o-\tau-90^o-t_2-T_p)_n \qquad [2]$$

where t_1 = 0.1-51.2 ms, t_2 = 102.4 ms, and the time period τ= 32 ms was set to optimize relayed coherence transfer involving vicinal coupling constants of ~7-8 Hz in the $C^{\alpha}H-C^{\beta}H-C^{\gamma}H_3$ fragment of the valine amino acid residue.

The NOERELSY spectra were recorded as in (17) with the pulse sequence:

$$(90^o-t_1-90^o-\tau_m-90^o-\tau-180^o-\tau-90^o-t_2-T_p)_n \qquad [3]$$

where t_1 = 0.1-51.2 ms, t_2 = 102.4 ms, and τ_m = 150 ms and τ = 25 ms were set to optimize longitudinal magnetization transfer due to NOE between the C^{α}_iH and $N_{i+1}H$ protons, and coherence transfer involving vicinal coupling constant of ∿ 10 Hz for the $HN_{i+1}-C^{\alpha}_{i+1}H$ fragment, respectively. The double-quantum transition (2QT) 2D-spectra were obtained as in (18) with the pulse sequence:

$$(90^o-\tau-180^o-\tau-90^o-t_1-90^o-t_2-T_p)_n \qquad [4]$$

in which the time period τ= 25 ms was set to facilitate optimum creation of double-quantum coherence between vicinal protons in the $NH-C^{\alpha}H-C^{\beta}H$ fragment involving coupling constants of ~10 Hz. The t_1 was varied from 9 μs to 46 ms in 50 μs intervals (1024 data points), t_2 = 102.4 ms.

The COSY, NOESY, RELSY, NOERELSY and 2QT spectra were acquired n times with different phases of the receiver and carrier radiofrequency (r.f.) during 90°-pulses both to achieve the desired signal-to-noise ratio and eliminate experimental artifacts. The time domain matrices were expanded to 2048 points in t_1 and 4096 points in t_2 by "zero-filling" to end up with a 1024x1024 points frequency domain (ω_1 and ω_2) data matrices. To enchance the spectral resolution, prior to Fourier transformation the time domain data matrix was multiplied (19) in the t_1 direction with a phase-shifted sine bell, $\sin(\pi(t_1+t_{01})/t_{s1})$, and in the t_2 direction with a phase-shifted sine-squared bell, $\sin^2(\pi(t_2+t_{02})/t_{s2})$. The length of the window functions t_{s1} and t_{s2} was adjusted for the functions to reach zero at the last experimental data point in the t_1 or t_2 direction, respectively. The phase shifts t_{01}/t_{s1} and t_{02}/t_{s2} were 1/64 and 1/16 for t_1 and t_2 directions, respectively. The two-dimensional Fourier transformation of the matrix was performed by a Bruker FTNMR-2D software for the absolute value presentation of the spectra.

The phase-sensitive absorption mode COSY spectra were aquired as in (20) with the pulse sequence:

$$(90^o-t_1-90^o-t_2-T_p)_n \qquad [5]$$

where t_1 was varied from 9 μs to 102.4 ms in 100 μs intervals (1024 data points), t_2 = 409.6 ms. The time domain matrices were expanded by "zero-filling" to reach the digital resolution 2.44 and 0.76 Hz/point for ω_1 and ω_2 directions, respectively.

The phase-sensitive absorption mode NOESY spectra were obtained as in (21) with the pulse sequence:

$$(90^o-t_1-90^o-\tau_m-90^o-t_2-T_p)_n \qquad [6]$$

where t_1 was varied from 9 μs to 51.2 ms (512 increments), t_2 = 204.8 ms, τ_m = 100 ms. The time domain matrices were zero-filled to reach the digital resolution 5 Hz/point both for ω_1 and ω_2 directions. The Fourier transformation of phase-sensitive absorption mode spectra was performed by a modified version of a Bruker FTNMR-2D software. Quadrature detection was used with the carrier r.f. at the low-field end of a spectrum.

The residual solvent resonances were suppressed by selective irradiation at the corresponding frequency with relatively weak r.f. field (γH_2 ∿1 Hz) at all times except during data aquisition t_2.

CD spectra were recorded on Jobin Ivon Dichrograph III at the ambient temperature.

Gramicidin A Double Helices in Organic Solvents

Species 3 in dioxane. After crystallization from ethanol at 20°C as in (6) of the purified by countercurrent distribution GA and subsequent dissolving in dioxane-D_8, the proton NMR spectrum of the 10 mM solution (Fig. 1) was found to correspond to a single conformer (22,23), named species 3 after Veatch et al. (6). The ^{1}H-NMR spectra of prepared GA species 3 display one signal for each chemically equivalent proton without any indication of conformational inhomogeneity for at least 4 days required for recording a set of 2D NMR spectra at 30°C. 10 days later at 30°C or 4 h at 90°C minor signals appeared (5-7% of the spectrum intensity), presumably due to other conformational species of GA. Absorption of moisture from air caused changes in the chemical shift of some NH and $C^{\alpha}H$ signals. Special care was taken to avoid these processes.

Proton signal assignments to specific positions in the amino acid sequence were carried out in two steps. The spin systems of most amino acid residues were identified by analysis of COSY spectra (23). However, in some cases this procedure did not allow unambiquous assignment. For instance, the chemical shifts of $C^{\beta}H$ protons of three valine residues (D-Val^6, Val^7 and D-Val^8) practically coincided and therefore the spin systems of the NH-$C^{\alpha}H$-$C^{\beta}H$ protons of these residues could not be correlated with the $C^{\gamma}H_3$ protons. In this and other similar situations RELSY pulse sequence [2] was used. In case of the valine residue the consecutive coherence transfer from $C^{\alpha}H$ proton to $C^{\beta}H$ proton and then from $C^{\beta}H$ to $C^{\gamma}H_3$ protons gives rise to a cross-peak in a RELSY spectrum with coordinates ω_1 and ω_2 corresponding to chemical shifts of $C^{\alpha}H$ and $C^{\gamma}H_3$ protons (Fig. 2). This technique allows assignment of NH-$C^{\alpha}H$-$C^{\beta}H(C^{\gamma}H_3)_2$ spin systems of all GA valine residues. The RELSY spectrum was similarly employed to overcome the COSY spectrum problem in assigning the NH-$C^{\alpha}H$ J-connectivity for interferring $C^{\alpha}H$ signals of Ala^5 and Val^7 as well as of the D-Val^8 and Trp^{15} residues.

Spin systems of NH-$C^{\alpha}H$-$C^{\beta}H_2$ protons of tryptophan residues were linked to spin systems of indole protons through nuclear Overhauser effect (NOE) connectivities of the $C^{\beta}H$ protons with the $C_{(4)}H$ and $C_{(2)}H$ ring protons. The corresponding cross-peaks are normally well observed in the NOESY spectra of polypeptides.

The next step was the assignment of a spin system to specific position of the amino acid residue in the primary structure of GA. This could be done by the NOESY spectra sequential resonance assignment through ${}^1d_{\alpha N}$, ${}^1d_{NN}$ and ${}^1d_{\beta N}$ NOE-connectivities (Table 1) for the amide $N_{i+1}H$ proton of residue $i+1$ with, respectively, $C_i^{\alpha}H$, N_iH and $C_i^{\beta}H$ protons of the preceding residue i. The assignment of the [Val[1]] GA backbone resonances were based on ${}^1d_{\alpha N}$ connectivities shown in the upper left part of Fig. 3. The starting point

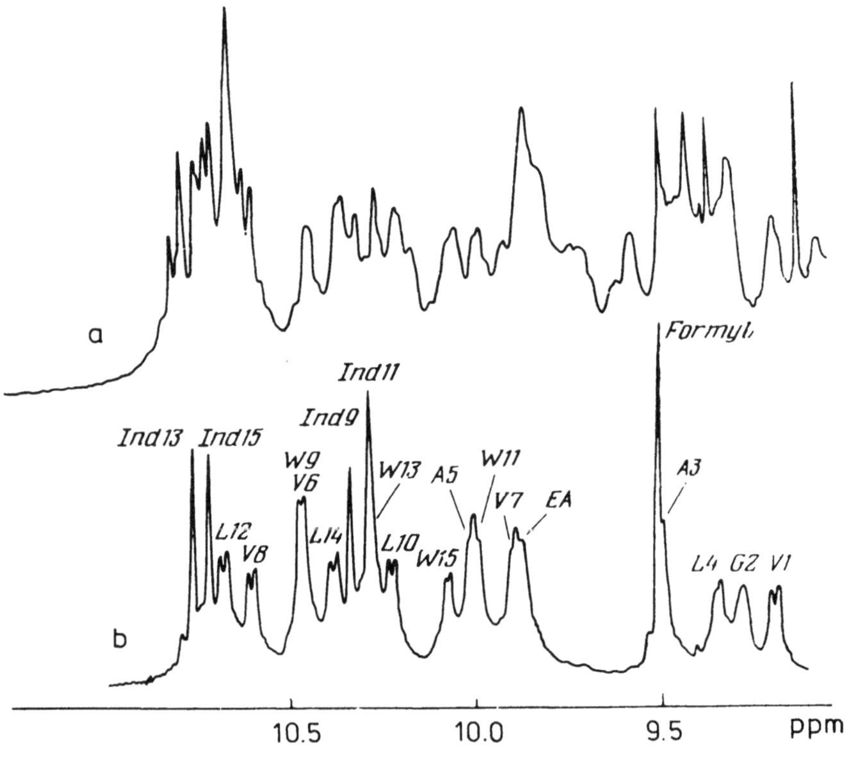

Fig. 1. Low-field region of the 500 MHz proton 1D NMR spectra of a freshly prepared 10 mM solution of [Val[1]] gramicidin A in dioxane-D_8 at 30°C: (a) as obtained by countercurrent distribution; (b) after crystallization from ethanol (species 3). The backbone NH resonance assignments are indicated by one-letter notation for amino acid residues and the serial number in the primary structure. EA - amide proton of ethanolamine moiety; Ind - indole NH protons of corresponding tryptophan residues.

was the COSY cross-peak of the C-terminal ethanolamine NH proton, which was connected via $^1d_{\alpha N}$ NOE with the $C^{\alpha}H$ proton of Trp^{15}. The last proton revealed COSY cross-peak with the neighbour NH due to the vicinal coupling $^3J(H\text{-}NC^{\alpha}\text{-}H)$. Next, the NH proton of Trp^{15} demonstrated $^1d_{\alpha N}$ NOE with C H of D-Leu^{14} and so on towards

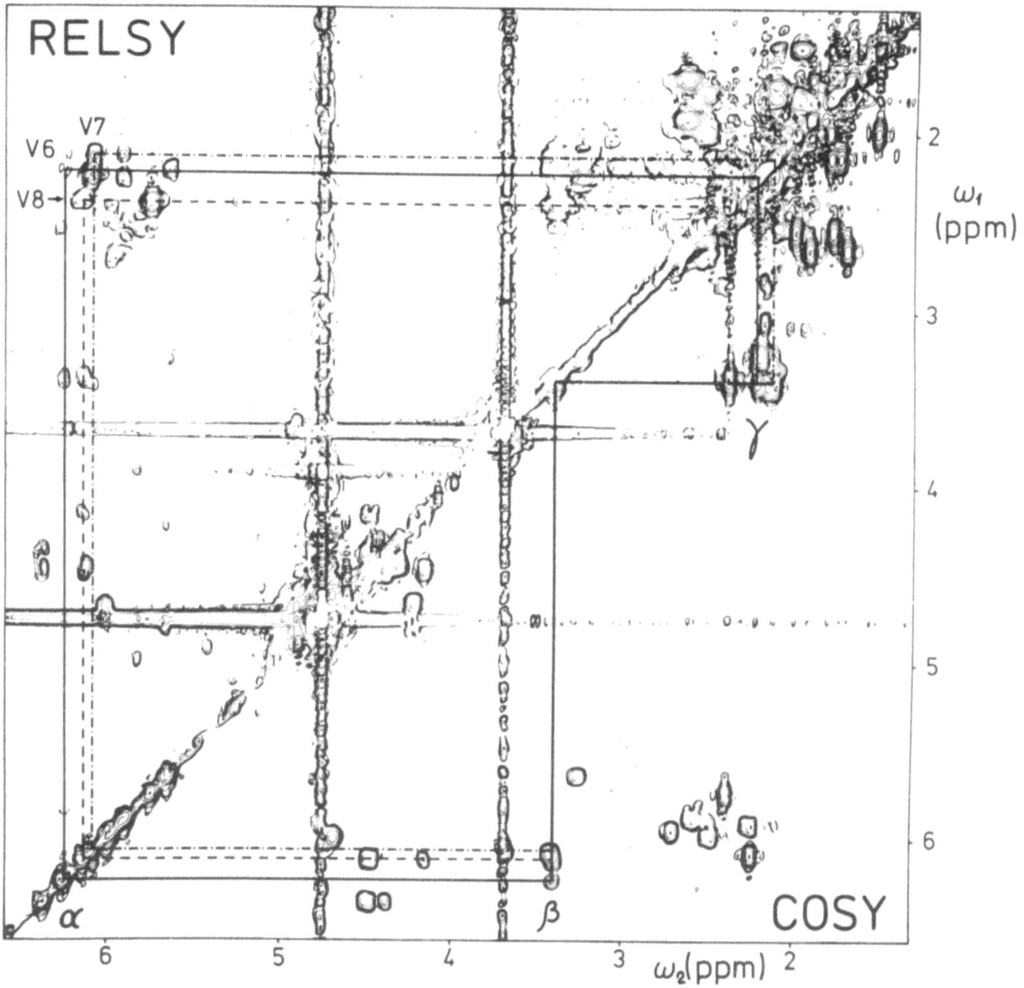

Fig. 2. Combined RELSY-COSY contour plot of the aliphatic spectral region of a 10 mM solution of the [Val^1] gramicidin A species 3 in dioxane-D_8 at 30°C. The pulse sequences [1] and [2] were used for COSY and RELSY (τ = 32 ms) 2D-spectra, respectively, with total aquisition times 24 h and 48 h. The proton-proton J-connectivities in the $C^{\alpha}H\text{-}C^{\beta}H(C^{\gamma}H_3)_2$ fragments are shown for the D-Val^6 (———), Val^7 (-·-·-) and D-Val^8 (----) residues.

the N-terminal [Val[1]] residue, as shown in Fig. 3 by the clockwise straight line with arrows.
The pattern of d-connectivities (all ${}^{1}d_{\alpha N}$ were observed, whereas no ${}^{1}d_{NN}$ was revealed) and relatively high, L-D alternating values of the vicinal spin-spin coupling constants H-NC^{α}-H (Table 2) indicate that the GA polypeptide backbone of species 3 in dioxane has an extended conformation.

Table 1. Notation for interproton NOE-connectivities in polypeptides:

i residue ...... *i+n* residue

Residue *i* \ Residue *i+n*	NH	$C^{\alpha}H$	$C^{\beta}H$	. . .
NH	${}^{n}d_{NN}$	${}^{n}d_{N\alpha}$	${}^{n}d_{N\beta}$	
$C^{\alpha}H$	${}^{n}d_{\alpha N}$	${}^{n}d_{\alpha\alpha}$	${}^{n}d_{\alpha\beta}$	
$C^{\beta}H$	${}^{n}d_{\beta N}$	${}^{n}d_{\beta\alpha}$	${}^{n}d_{\beta\beta}$	
⋮				

In addition, analysis of the NOESY spectrum, obtained with mixing time τ_m = 120 ms preventing spin diffussion, reveals over 50 through space proton connectivities between residues which are non-adjacent to each other in the amino acid sequence (23). The appearance of a NOE cross-peak in the spectrum evidence that the distance between corresponding protons is less than 0.45 nm. The concise map of NOE contacts, shown in Fig. 4b, clearly demonstrates the dimeric nature of the GA species 3 in dioxane solution. This conclusion is in agreement with the previous osmometric determination of molecular mass of this species in the same solvent (24).

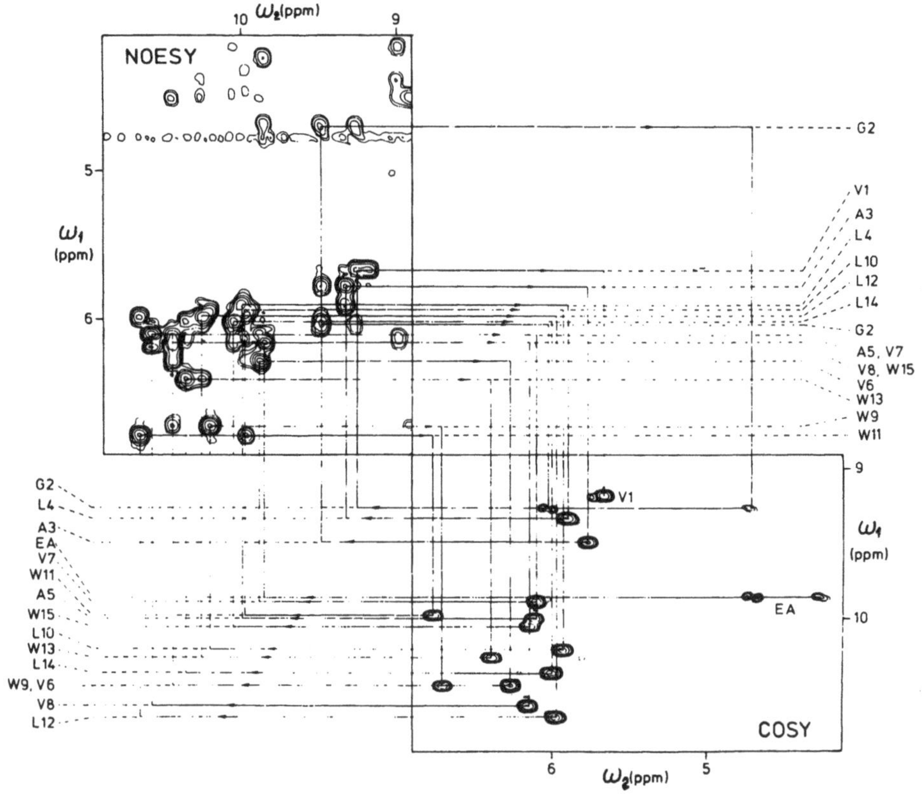

Fig. 3. [Val[1]] Gramicidin A species 3 combined COSY-NOESY connectivity diagram for sequential resonance assignments via NOE's between amide protons and $C^{\alpha}H$ protons of the preceding residue ($^{1}d_{\alpha N}$ connectivity). In the upper left, the region (ω_1 = 4.1-6.9 ppm, ω_2 = 8.9-10.9 ppm) of a ^{1}H-NOESY 2D-spectrum of a 10 mM solution in dioxane-D_8 at 30°C is presented. To the lower right, the region (ω_1 = 8.9-10.9 ppm, ω_2 = 4.1-6.9 ppm) of a ^{1}H-COSY 2D-spectrum obtained under identical sample conditions is shown. Both 2D-spectra were recorded simultaneously with the pulse sequence [1], τ_m = 120 ms. Total aquisition time was 24 h. The straight lines and arrows indicate the connectivities between neighbouring residues from the C-terminal ethanolamine towards the Val[1] residue. The amide proton chemical shifts are indicated by the assignments in the lower lefthand corner, and those of the $C^{\alpha}H$ protons - in the upper righthand corner.

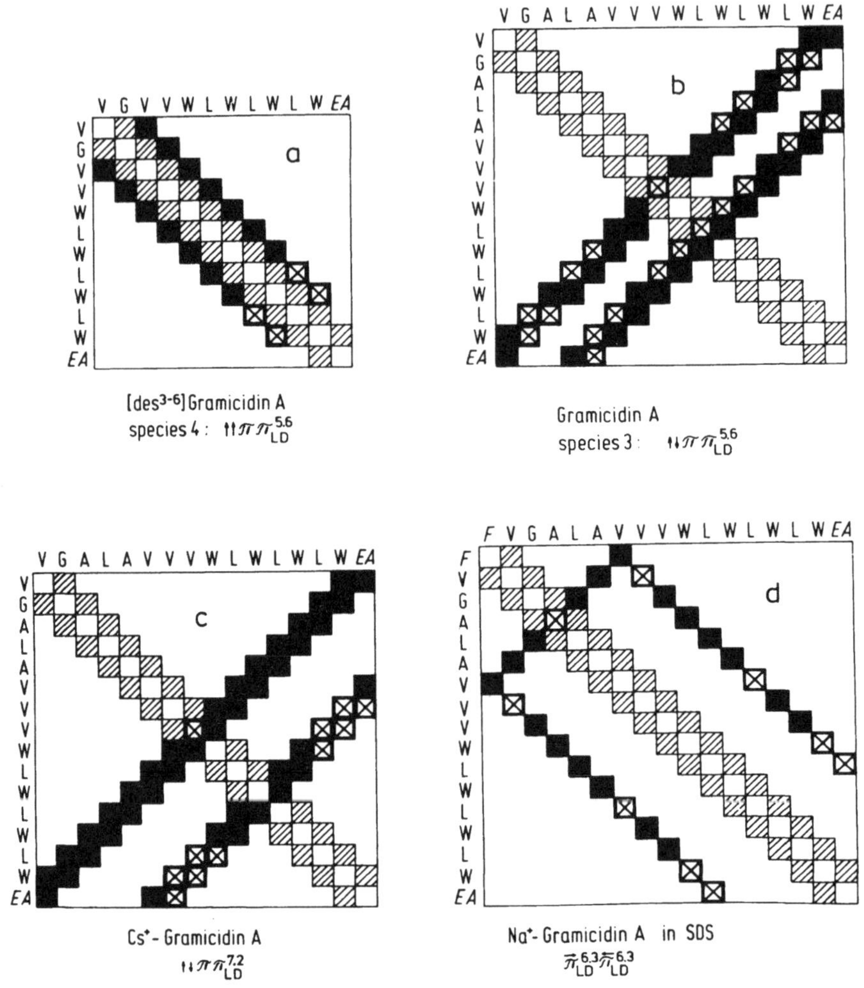

Fig. 4. Schematic maps of NOE-connectivities between backbone protons of: (a) des[Ala^3, D-Leu^4, Ala^5, D-Val^6]GA in dioxane (species 4); (b) [Val^1]GA species 3 in dioxane; (c) 1:1 Cs^+-[Val^1]GA complex in the $CD_3OH/CDCl_3$ solution; (d) [Val^1]GA in SDS micelles. The filled and crossed squares correspond to, respectively, the identified and impended NOE cross-peaks for dimer interchain connectivities; the hatched squares-to the identified intrachain connectivities ($^1d_{\alpha N}$)

Table 2. Vicinal proton H-NC^{α}-H coupling constants (^3J, Hz) and the deuterium exchange half-times ($t_{1/2}$, hours) of the backbone NH groups of the [Val^1]gramicidin A species 3 (GA-3) in dioxane-D_8 (22,23), des[Ala^3,D-Leu^4,Ala^5,D-Val^6]gramicidin A (des^{3-6}GA) in dioxane-D_8 (species 4) (25), the Cs^+-[Val^1]gramicidin A complex 1:1 (Cs^+-GA) in the $CD_3OH/CDCl_3$ solution at 30°C (26), and [Val^1] gramicidin A incorporated into sodium dodecyl-D_{25} sulphate micelles in D_2O/CF_3CD_2OD (Na^+-GA-SDS) at 55°C (27). EA - ethanolamine moiety.

Residue	GA-3		des^{3-6}GA		Cs^+-GA		Na^+-GA-SDS	
	^3J	$t_{1/2}$	^3J	$t_{1/2}$	^3J	$t_{1/2}$	^3J	NH Exchange*
L-Val 1	9.2	1.5	10.3	2.5	9.0	<0.2	9.4	Slow
Gly 2		1.5		15		0.2		Slow
L-Ala 3	7.8	8	-	-	8.4	0.6	9.4	Slow
D-Leu 4	8.0	8	-	-	8.8	23	8.8	Slow
L-Ala 5	6.6	6	-	-	6.0	0.2	9.2	Slow
D-Val 6	9.2	27	-	-	7.2	64	8.8	Slow
L-Val 7	7.5	46	9.6	>100	9.4	>100	9.9	Slow
D-Val 8	9.2	33	8.8	>100	8.4	>100	8.7	Slow
L-Trp 9	7.0	35	8.8	>100	8.4	>100	9.1	Slow
D-Leu 10	9.3	83	5.9	>100	7.6	>100	9.7	Slow
L-Trp 11	7.4	49	9.9	>100	9.2	>100	9.0	Slow
D-Leu 12	9.0	59	6.6	>100	6.4	>100	9.8	Fast
L-Trp 13	7.0	11	8.8	>100	9.6	>100	9.8	Slow
D-Leu 14	9.2	28	7.4	>100	8.8	>100	9.8	Fast
L-Trp 15	7.2	26	8.1	>100	9.6	>100	9.4	Slow
EA		46		10		7.7		Fast

* Estimated from pH dependence of the backbone NH peak intensity when the solvent H_2O sighal was saturated by an r.f. field.

Important aspect of the polypeptide spatial structure is the intra- and intermolecular hydrogen bonds of the backbone NH groups. They can be revealed by the rate of deuterium exchange with solvent. For species 3 of GA the majority of amide NH groups slowly exchange with deuterium when 4% of CD_3OD is added to dioxane-D_8 solution (Table 2) indicating their participation in hydrogen bonding. Only the NH of Val^1 and Gly^2 exchange somewhat faster.

To fit the spatial structure of GA species 3 with the experimental NMR data, two equivalent, extended polypeptide chains have to form an antiparallel β-structure. However, in this case only part of the detected intermolecular NOE contacts and hydrogen bonds are produced. Thus the antiparallel structure has to be rolled into the left-handed double helix $\uparrow\downarrow\pi\pi_{LD}^{5.6}$ with 5.6 residues per turn as shown in Fig. 5a. It has to be noted that this structure comprises a few intermolecular NOE-connectivities marked in Fig. 4b by crossed squares, which can not be revealed because the positions of corresponding cross-peaks are too close to the strong diagonal ridge of a NOESY spectrum.

The double helix is reinforced by 26 interstrand NH...O=C hydrogen bonds. The conformation appears to be reasonably rigid as far as the peptide backbone is concerned, but some side chains are subjected to a considerable degree of flexibility as manifested by the $H\text{-}C^{\alpha}C^{\beta}\text{-}H$ proton coupling constants (22,23). The dimer is 3.6 nm long (as estimated from the distance between the formyl oxygens), possesses a C_2 symmetry axis perpendicular to the helix axis, and has an axial cavity with average diameter of ∿ 0.3 nm. Thus 2D NMR spectroscopy corroborates the formerly proposed $\uparrow\downarrow\pi\pi_{LD}^{5.6}$ structure of GA species 3 in dioxane (6,7) and refines its conformation and hydrogen bonding pattern. On the other hand the NMR spectra do not reveal any indication of the minor, single-stranded $\vec{\pi}_{LD}\overleftarrow{\pi}_{LD}$ dimeric forms postulated in (7).

Species 4 in dioxane. Other conformational forms of GA in organic solvents are not easy to isolate in enough for NMR study quantity. Thus to imitate the GA species 4 conformation use was made of the synthetic shortened derivative - des[L-Ala^3,D-Leu^4,L-Ala^5,D-Leu^6]-gramicidin A ($des^{3-6}GA$), CD and IR spectra of which are similar to those of species 4.

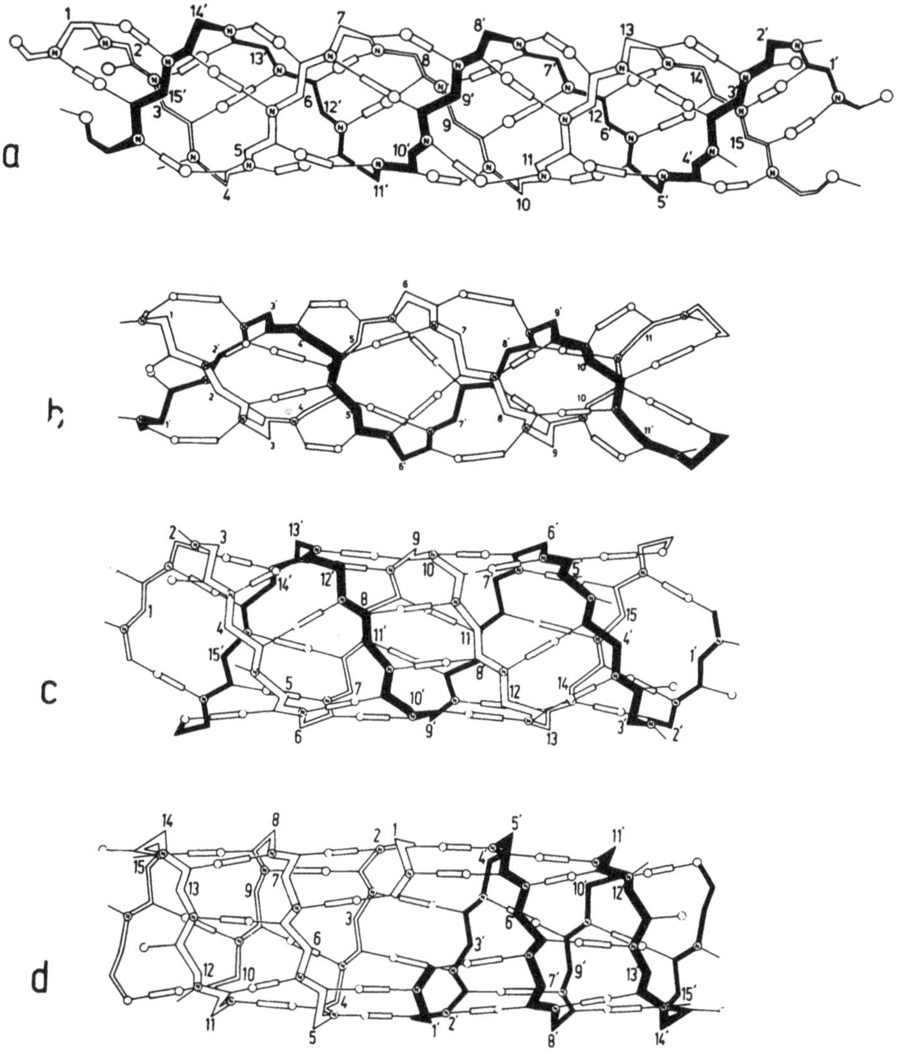

Fig. 5. Conformations of [Val^1] gramicidin A: (a) left-handed antiparallel double helix $\uparrow\downarrow\pi\pi\,^{5.6}_{LD}$ of species 3 in dioxane; (b) right-handed parallel double helix $\uparrow\uparrow\pi\pi\,^{5.6}_{LD}$ of des[Ala^3,D-Leu^4, Ala^5,D-Val^6]GA in dioxane (species 4); (c) right-handed antiparallel double helix $\uparrow\downarrow\pi\pi\,^{7.2}_{LD}$ of the Cs^+-complex of GA in CD_3OH-$CDCl_3$, and (d) head-to-head right-handed, single-stranded helical dimer $\vec{\pi}\,^{6.3}_{LD}\ \overleftarrow{\pi}\,^{6.3}_{LD}$ of GA incorporated into SDS micelles.

Freshly dissolved in dioxane-D_8, this compound contains a mixture of slowly interconverting conformational forms (Fig. 6a), but after 12 hours at 30°C the sample reaches equilibrium with one highly predominant (>90%) conformer (Fig. 6b). This form, which according to the IR and CD spectra (7) has a stable conformation similar to that of GA species 4, is subjected to the 2D NMR study. Most of the spin systems of amino acid residues are assigned in the absolute mode and phase-sensitive COSY spectra (25). However, when proton signals of different residues overlap or interfere with strong diagonal peaks, the COSY technique of assignment is failed. For instance, the $C^{\alpha}H$ signals of D-Leu^8 and D-Leu^{10} are superimposed in the des^{3-6}GA spectrum. This prevents assignment of the corresponding NH-$C^{\alpha}H$-$C^{\beta}H$ spin systems. The 2QT 2D NMR pulse sequence [4] is used to resolve this problem (Fig. 7). The double-quantum coherence of two spins NH and $C^{\alpha}H$ is transferred to the observable single-quantum transition of the third $C^{\beta}H$ spin if the coupling constants of $C^{\alpha}H$ with NH and $C^{\beta}H$ protons are high enough to be observable in a spectrum. In case of leucine residues the 2QT 2D-spectrum reveals cross-peaks with coordinates $\omega_1=\omega_{NH}+\omega_{C^{\beta}H}$ and $\omega_2=\omega_{C^{\alpha}H}$. By combining the available chemical shifts of the NH and $C^{\alpha}H$ protons of D-Leu^8 and D-Leu^{10} it becomes evident that ω_1 = 13.15 ppm (Fig. 7) corresponds to D-Leu^8 and ω_1 = 12.53 ppm - to D-Leu^{10}. The 2QT 2D-spectrum was also used for assignment of the leucine $C^{\gamma}H$-$C^{\delta}H_3$ J-connectivities (25), which were not possible to delineate in the COSY spectrum due to interference of corresponding cross-peaks with diagonal resonances.

Analysis of the NOESY spectrum of des^{3-6}GA put forward another problem. While all $^1d_{\alpha N}$ connectivities along the backbone were observed, their assignment in the D-Leu^8-Trp^9-D-Leu^{10}-Trp^{11} sequence was uncertain due to overlapping of NH signals of Trp^9, D-Leu^{10} and Trp^9 as well as of $C_i^{\alpha}H$ signals of the D-Leu^8 and D-Leu^{10} residues. In such cases the NOERELSY pulse sequence [3] for the $C_i^{\alpha}H$-CO-$N_{i+1}H$-$C_{i+1}^{\alpha}H$ fragment was used, by which the magnetization transfer was first produced during τ_m due to NOE from the initial spin $C_i^{\alpha}H$ to the spatially close proton $N_{i+1}H$ and then coherence transfers from the $N_{i+1}H$ proton to the $C_{i+1}^{\alpha}H$ proton due to J-coupling. As a result the cross-peaks for $C_i^{\alpha}H$ and $C_{i+1}^{\alpha}H$ were observed in the 2D-spectra, indicating the spatial proximity of the $N_{i+1}H$ and $C_i^{\alpha}H$ protons. Fig. 8 illustrates assignment of all

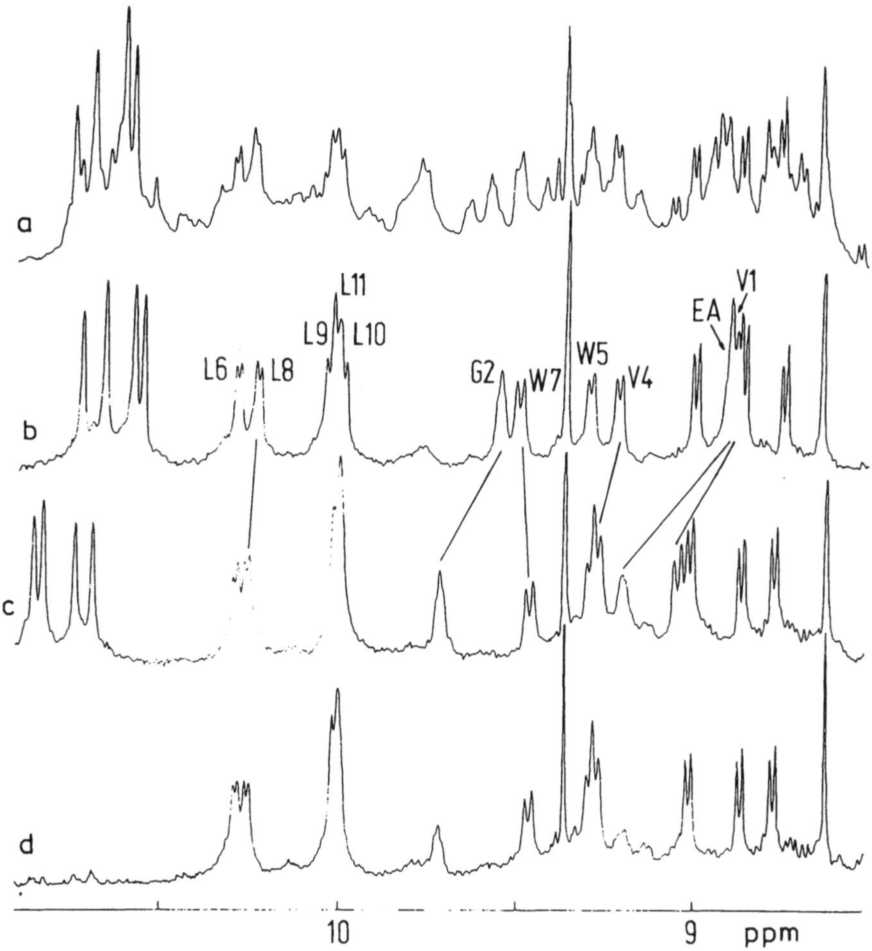

Fig. 6. Low-field region of the 500 MHz proton 1D NMR spectra of a 5 mM solution of des$^{3-6}$GA in dioxane-D_8 at 30°C: (a) immediately after preparation; (b) after 12 hours; (c) when 3% of D_2O was added to the sample (b); (d) - after 10 hours of 3% D_2O addition to the sample (b). See caption in Fig. 1 for notation.

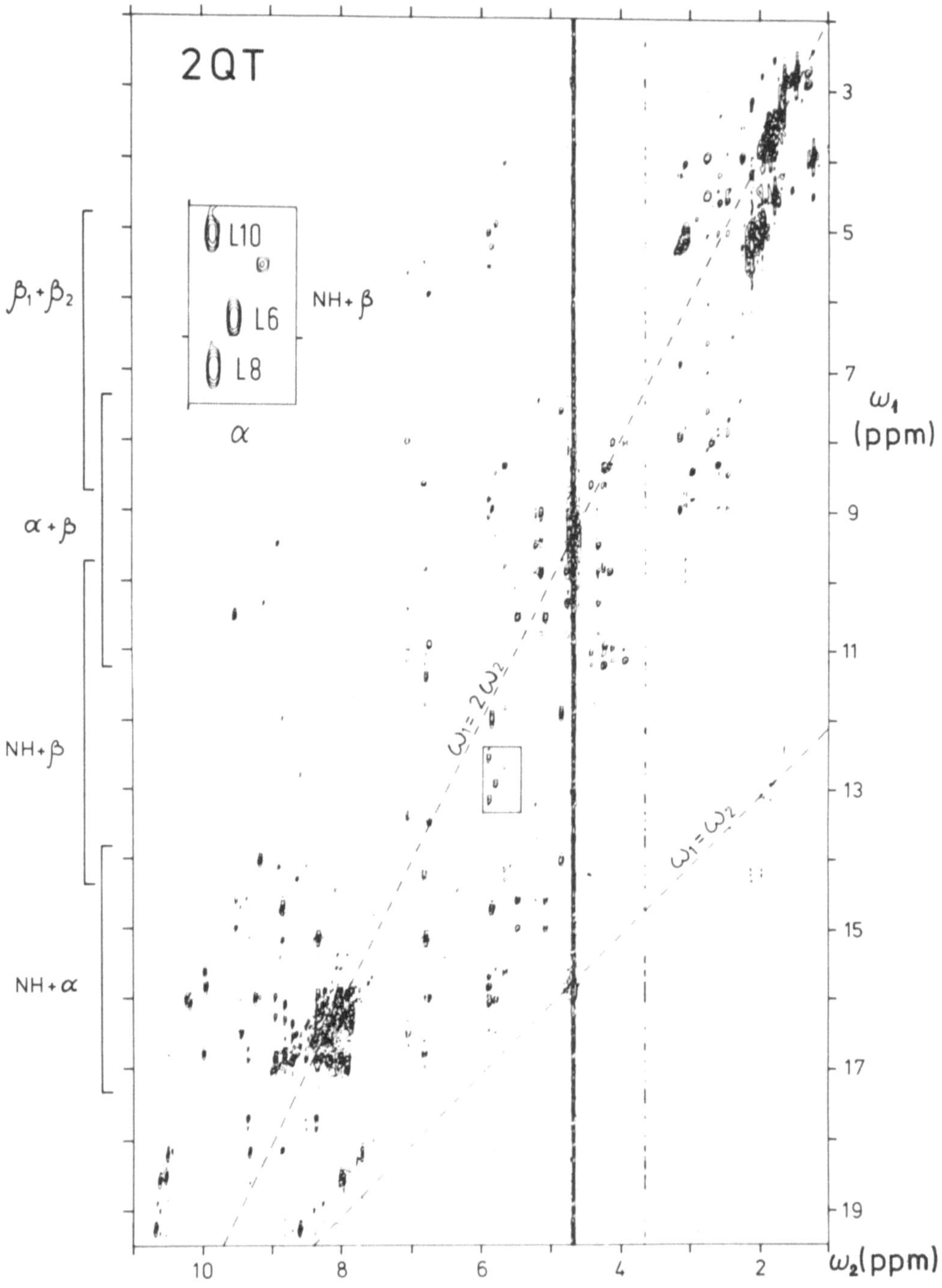

Fig. 7. Contour plot of the 2QT 2D NMR spectrum of a 10 mM solution of des$^{3-6}$GA in dioxane-D_8 at 30°C, obtained with pulse sequence [4], τ = 25 ms. Regions of double-quantum transitions $\omega_{NH} + \omega_{C^{\alpha}H}$, $\omega_{NH} + \omega_{C^{\beta}H}$, $\omega_{C^{\alpha}H} + \omega_{C^{\beta}H}$ and $\omega_{C^{\alpha\beta 1}H} + \omega_{C^{\beta 2}H}$ are indicated on the left-hand side of ω_1 axis. The insert depicts enlarged spectrum part with cross-peaks, corresponded to transfer of double-quantum coherence $\omega_{NH} + \omega_{C^{\beta}H}$ to single-quantum transition $\omega_{C^{\alpha}H}$ in the D-Leu^{6,8,10} spin systems. Dashed lines show diagonals for $\omega_1 = 2\omega_2$ and $\omega_1 = \omega_2$

$C^{\alpha}H$ protons of des$^{3-6}$GA performed by NOERELSY technique. This experiment confirms $^{1}d_{\alpha N}$ connectivities between all $C_i^{\alpha}H$ and $N_{i+1}H$ protons and completes signal attribution to the residues in the des$^{3-6}$GA sequence.

The majority of spin-spin H-NC^{α}-H and H-$C^{\alpha}C^{\beta}$-H coupling constants were directly measured from the 1D NMR spectrum (Fig. 6b). However, overlapping of NH resonances of the Trp^{9}, D-Leu^{10} and Trp^{11} residues in the whole temperature range of 25-80°C forced to use the phase-

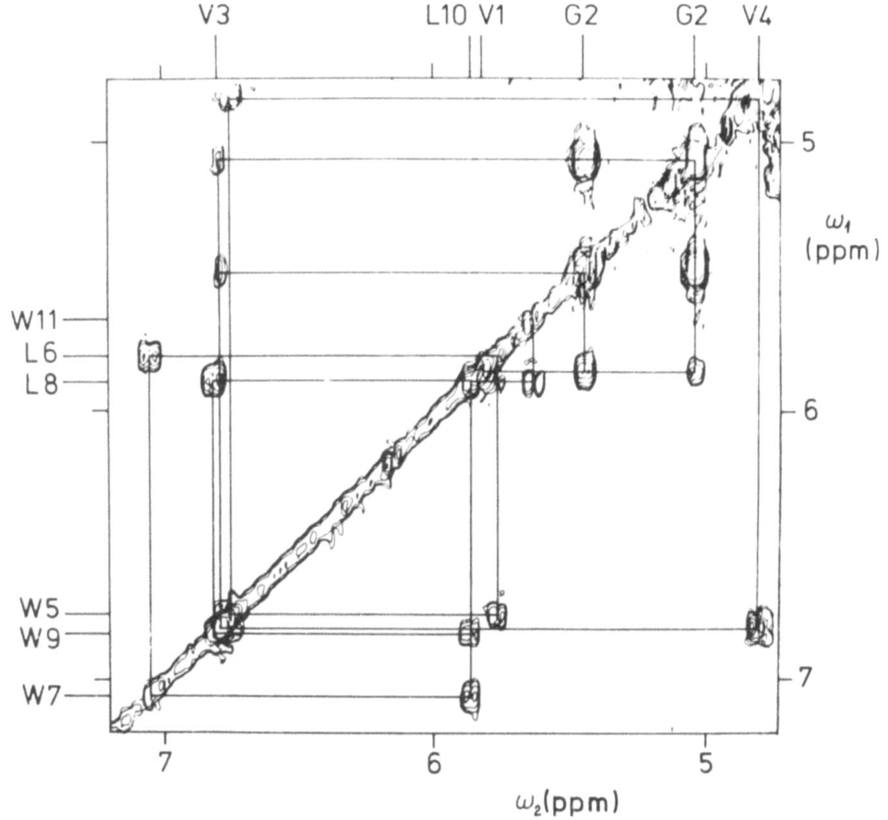

Fig. 8. Contour plot of the NOERELSY 2D-spectrum (ω_1, ω_2 = 4.7-7.2 ppm) of a 10 mM solution of des$^{3-6}$GA in dioxane-D_8 at 30°C, obtained with pulse sequence [3], τ = 25 ms and τ_m = 150 ms. Total aquisition time was 48 h. The straight lines connect cross-peaks originated by the NOE-*J* connectivities between $C^{\alpha}H$ protons of neighbouring residues from Val^{1} towards Trp^{11}. The assignments are shown in the plot margins.

sensitive COSY pulse sequence [5] and absorption mode of spectrum presentation. The obtained relatively high $^3J(H\text{-}NC^{\alpha}\text{-}H)$ values (Table 2) indicate extended conformation of the des$^{3-6}$GA backbone in conformity with observation of the full set of the $^1d_{\alpha N}$ connectivities and the absence of any $^1d_{NN}$ contact (25).
To evaluate the degree of accessibility of backbone NH groups by solvent medium, the 3% D_2O was added to the dioxane-D_8 solution of des$^{3-6}$GA and deuterium exchange was observed during 48 h by 1D NMR spectra (Fig. 6 b-d). For the NH signals of Val^1, Gly^2 and ethanolamine residues low-field shift (Fig. 6 b,c) and higher exchange rates (Table 2) were detected, indicating that the NH groups of the Val^1 residue and ethanolamine moiety as well as carbonyl group of Gly^2 are exposed to the solvent, whereas NH and C=O groups of other residues are involved in formation of the hydrogen bond network of des$^{3-6}$GA.
Important results on the des$^{3-6}$GA spatial structure followed from the NOESY 2D-spectrum, which revealed more than 40 NOE-connectivities between protons of non-adjacent residues in the amino acid sequence (25). Whereas the vicinal $H\text{-}NC^{\alpha}\text{-}H$ couplings and the NH deuterium exchange half-times for des$^{3-6}$GA are similar to those for the GA species 3 (Table 2), the regularity of NOE-connectivities of the backbone protons (Fig. 4a,b) is completely different. For des$^{3-6}$GA it is remarkable that amide N_iH proton is proximate to the $C^{\alpha}_{i+2}H$ proton if i is even ($^2d_{N\alpha}$ contact) and to $C^{\alpha}_{i-2}H$ if i is odd ($^2d_{\alpha N}$ contact).
The observed NOE-connectivities and NH hydrogen bondings could be combined with the extended conformation of the backbone only if two des$^{3-6}$GA molecules form a parallel β-structure rolled into a right-handed double helix $\uparrow\uparrow\pi\pi^{5.6}_{LD}$ with 5.6 residues per turn, as shown in Fig. 5b. The dimensions of this helix, when adjusted to the full length of the [Val^1]GA amino acid sequence, are the same as for the above considered GA species 3 structure. Hydrophobic external surface and axial cavity lined by polar carbonyl oxygens also make both structures compatible, in principle, with transmembrane channel requirements.
Cs^+-Gramicidin A complex. We failed in preparing metal complexes directly with the above evaluated conformational forms of GA in dioxane. Thus the mixed solvent $CD_3OH/CDCl_3$ (1:1) was chosen. Methanol facilitated solubility of an alkali metal salt while

chloroform was added to prevent precipitation of the GA complex due to aggregation.

The proton NMR spectrum of salt-free solution of [Val[1]]GA in this medium demonstrates a mixture of various forms of GA (Fig. 9a) presumably differing in conformation. When progressively increasing amounts of 0.4 M solution of CsCNS in CD_3OH are added to the sample (simultaneously corresponding volume of $CDCl_3$ is introduced to keep the 1:1 ratio of methanol/chloroform) a new set of signals appears and gradually grows in intensity (Fig. 9 b-c). Chemical

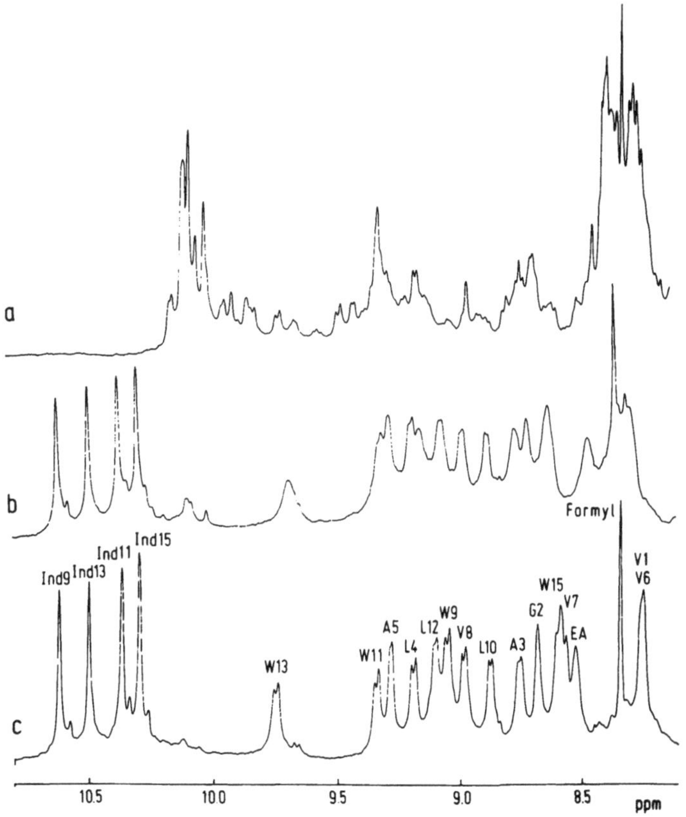

Fig. 9. Low-field region of the 500 MHz ^{1}H-NMR 1D-spectra of 30 mM solution of [Val1] gramicidin A in $CD_3OH/CDCl_3$ (1:1), 30°C at different $[Cs^+]/[GA]$ molar ratio: (a) 0.0, (b) 0.8 and (c) 1.1. See Fig. 1 caption for notation.

shifts of the new signals also change and their line widths narrow until the molar ratio $[Cs^+]/[GA]$ becomes ~1. The final spectrum corresponds to a single major conformational form of the Cs^+-GA 1:1 complex with less than 15% admixture of minor components (Fig. 9d).

The behaviour of NMR spectra in the presence of caesium salt can be rationalized by the two-step equilibria:

$$Cs^+ + GA_2 \overset{K_1}{\rightleftarrows} (Cs \cdot GA_2)^+$$

$$Cs^+ + (Cs \cdot GA_2)^+ \overset{K_2}{\rightleftarrows} (Cs_2 \cdot GA_2)^{2+}$$

in which the stable dimeric form of $(Cs_2 \cdot GA_2)^{2+}$ complex is demonstrated by NMR conformational analysis (see below). Both sites of cation binding are tight and identical for single occupancy. Presumably the binding of the first cation induces the GA dimer formation and its proper molecular structure. The $(Cs \cdot GA_2)^+$ complex is stable; its lifetime exceeds 30 ms, as indicated by the line width of NMR signals, and equilibrium binding constant $K_1 \approx 10^4\ M^{-1}$. The second equilibrium ($K_2 \approx 3 \cdot 10^3\ M^{-1}$) is faster - lifetime of $(Cs_2 \cdot GA_2)^{2+}$ is < 1 ms and therefore only the averaged NMR signals of $(Cs \cdot GA_2)^+$, and $(Cs_2 \cdot GA_2)^{2+}$ are observed. It can be assumed that when the second Cs^+ is accommodated, then both the identical sites become less tight due to electrostatic repulsion of the cations.

The conformation of the $(Cs_2 \cdot GA_2)^{2+}$ complex was studied by 2D ^{1}H-NMR spectroscopy at $[Cs^+]/[GA] = 1.1$. First, the chemical shifts of most protons were assigned to specific positions in amino acid sequence of GA (26) by the conventional procedure of sequential resonance assignments: proton spin systems of amino acid residues were identified in COSY spectra via *J*-connectivities and then they were aligned, via NOE-connectivities between neighbouring residues into the amino acid sequence of [Val^1] GA. All the assignments from the C-terminal ethanolamine moiety towards the N-terminal valine residue were based on NOE $^1d_{\alpha N}$ connectivities. The $^1d_{NN}$ connectivities were not observed and only a few $^1d_{\beta N}$ connectivities were identified in NOESY spectra (26). Analysis of the NOESY spectra of the Cs^+-GA complex in $CD_3OH/CDCl_3$ and in $CD_3OD/CDCl_3$ (with amide protons exchanged for deuterium) also revealed over 50 NOE-connectivities between protons of non-adjacent residues in the amino acid sequence (26). For example, Fig. 10 shows a fragment of the phase-sensitive absorption mode NOESY 2D-spectrum with NOE cross-peaks between NH and C H protons, i.e. $^{n}d_{N\alpha}$ and $^{n}d_{\alpha N}$ connectivities.

Additional information on the spatial structure of the Cs^{+}-GA complex was obtained from spin-spin coupling constants of vicinal H-NC^{α}-H protons and from deuterium exchange of amide NH groups (Table 2). The coupling constants were measured directly from the resolution enchanced 1D NMR spectrum. For deuterium exchange measurments the Cs^{+}-GA complex was lyophilized from $CD_3OH/CDCl_3$ solution and then redissolved in $CD_3OD/CDCl_3$. The exchange half-times $t_{1/2}$ were estimated from the time dependence of peak intensity of corresponding NH signals. The majority of backbone NH

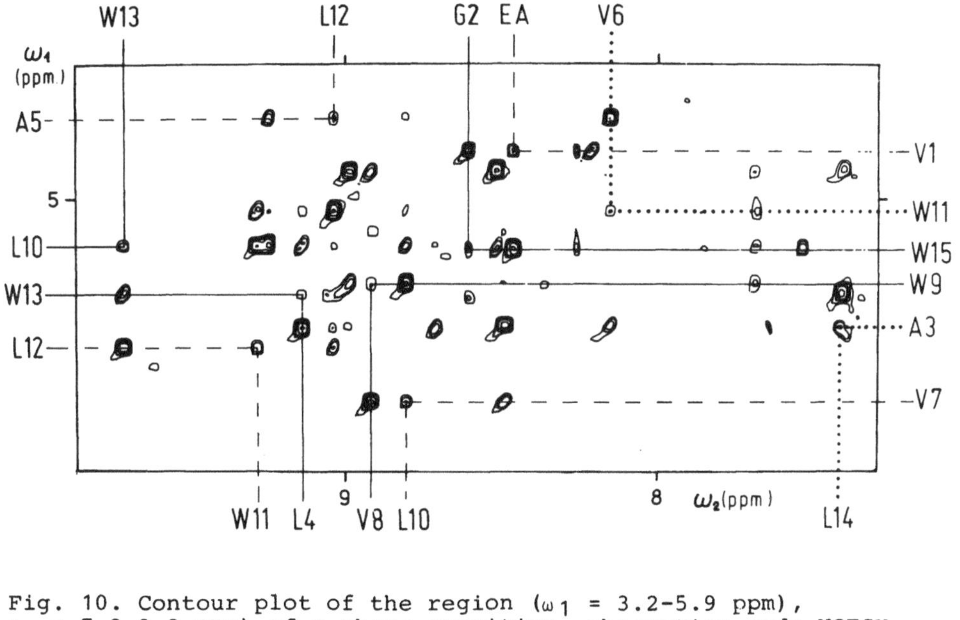

Fig. 10. Contour plot of the region (ω_1 = 3.2-5.9 ppm), ω_2 = 7.3-9.9 ppm) of a phase sensitive, absorption mode NOESY proton 2D-spectrum (obtained with pulse sequence [6], τ_m = 200 ms) of a 30 mM [Val^1] gramicidin A complex with caesium (Cs^+ / GA = 1.1) in $CD_3OH/CDCl_3$ (1:1) at 30°C. The digital resolution is 5 Hz/point for both ω_1 and ω_2 directions. The spectrum was recorded in 22 h. The amide NH proton chemical shifts are indicated by the assignments at the top and at the bottom, those of C^{α}H protons on the left- and right-hand sides of the plot.

groups exchanged slowly enough to indicate their participation in hydrogen bonding. At 30-50°C, the NH signal of D-Leu^{14} was hidden in the tryptophan aromatic region and thus its $t_{1/2}$ values were not measured.

Observation of $^{1}d_{\alpha N}$ connectivities over the entire amino acid sequence as well as of relatively large values of spin-spin coupling constants of vicinal H-NC$^{\alpha}$-H protons confirms an extended conformation of the polypeptide backbone of the Cs^{+}-GA complex. The data on tertiary NOEs (between protons of non-adjancent residues in the amino acid sequence) unequivocally prove (Fig. 4c), that the Cs^{+}-GA complex is formed by two extended, equivalent chains adjusted in antiparallel manner and rolled into right-handed $\uparrow\downarrow\pi\pi^{7.2}_{LD}$ double helix with 7.2 residues per turn (Fig. 5c). In accordance with deuterium exchange of amide NH groups (Table 2), the molecular structure is stabilized by 26 interstrand hydrogen bonds. The amide NH groups of Val^{1}, Ala^{3} and Ala^{5}, as well as the carbonyl groups of Val^{1} and Ala^{3} do not participate in hydrogen bonds. Although the exchange of the Gly^{2} NH is also rapid, nevertheless, according to NOE-connectivities and the overall molecular structure it is likely that this NH amide group could participate in interstrand hydrogen bonding.

The $\uparrow\downarrow\pi\pi^{7.2}_{LD}$ dimer is 2.7 nm long (as estimated by the distance between the Val^{1} amide nitrogens) and has a C_2 symmetry axis perpendicular to the helix axis. The central axial cavity is lined by the polar peptide group, while the nonpolar side chains project from the exterior surface of the dimer. The luminal diameter ~0.4 nm (as estimated from the distance between the carbonyl oxygens and the helix axis minus the van der Waals radius of the carbonyl oxygen) suits well to accommodate the caesium cation (0.33 nm ion diameter).

Thus, the determined conformations of [Val^{1}] GA in organic solvents are double-helical dimers with a different backbone alignment, handedness and a number of residues per turn. However, the diversity of the GA conformation in organic solvents, certainly, is not exhausted, at least because GA species 1 and 2 in dioxane possess alternative structures (6,7). Nevertheless, their relation to the structure of the GA ion channel in membranes is still under discussion (4,6-11,13,24) and thus the next section is dedicated to direct elucidation of conformation of GA complex with alkali cation in membrane-like environment.

Gramicidin A Single-stranded Dimer in Micelles

The membrane-like environment of the GA ion channel is imitated by micelles of sodium dodecyl sulphate (SDS). In water those roughly spherical in shape micelles consist of 60-80 SDS molecules each forming an internal hydrocarbon core with polar groups at the surface (28).

[Val^1] GA insoluble in water was incorporated into micelles by dropwise addition of its 0.05 M solution in 2,2,2-trifluoroethyl alcohol-D_3 into a 0.5 M sodium dodecyl-D_{25} sulphate (SDS-D_{25}) dispersion in H_2O or D_2O. The sample was diluted with water to final concentration of 0.005 M of gramicidin A and 0.25 M of SDS-D_{25} in water/2,2,2-trifluoroethyl alcohol-D_3 (molar ratio 16/1). These experimental conditions correspond to incorporation of not more than two GA molecules into one SDS micelle.

The CD spectrum of GA in SDS micelles is almost identical to that of GA in dipalmitoyl phosphatidylcholine liposomes (Fig. 11) and lysolecithin micelles (3). The red shift of the liposome spectra is ascribed to light scattering of lipid suspensions [see (29)]. The unique CD pattern observed evidences that the system contains a channel state of GA (3). Furthermore, by means ^{7}Li-, ^{23}Na-, ^{13}C- and ^{1}H-NMR spectra we have proved that GA embeded into SDS micelles does indeed interact with alkali metal ions and the interactions are competitively blocked by Tl^+ ions. Therefore, one can conclude that GA in SDS micelles forms a channel with a similar molecular structure to that in lysolecithin micelles and in liposomes.

The proton NMR spectrum of GA incorporated into SDS-D_{25} micelles (Fig. 12) displays features, which make it acceptable for high resolution 2D NMR study of the detailed molecular structure.

It should be specifically noted, that the system in SDS contains an abundance of sodium cations relative to the GA concentration. Further increase in Na^+ concentration by addition of NaCl does not change the NMR spectrum, testifying that cation binding sites of GA are saturated under the conditions used. Thus, the results obtained corresponded to the Na^+ complex of GA incorporated into SDS micelles.

Chemical shifts of all protons were assigned to specific positions in the amino acid sequence of GA by a conventional procedure of

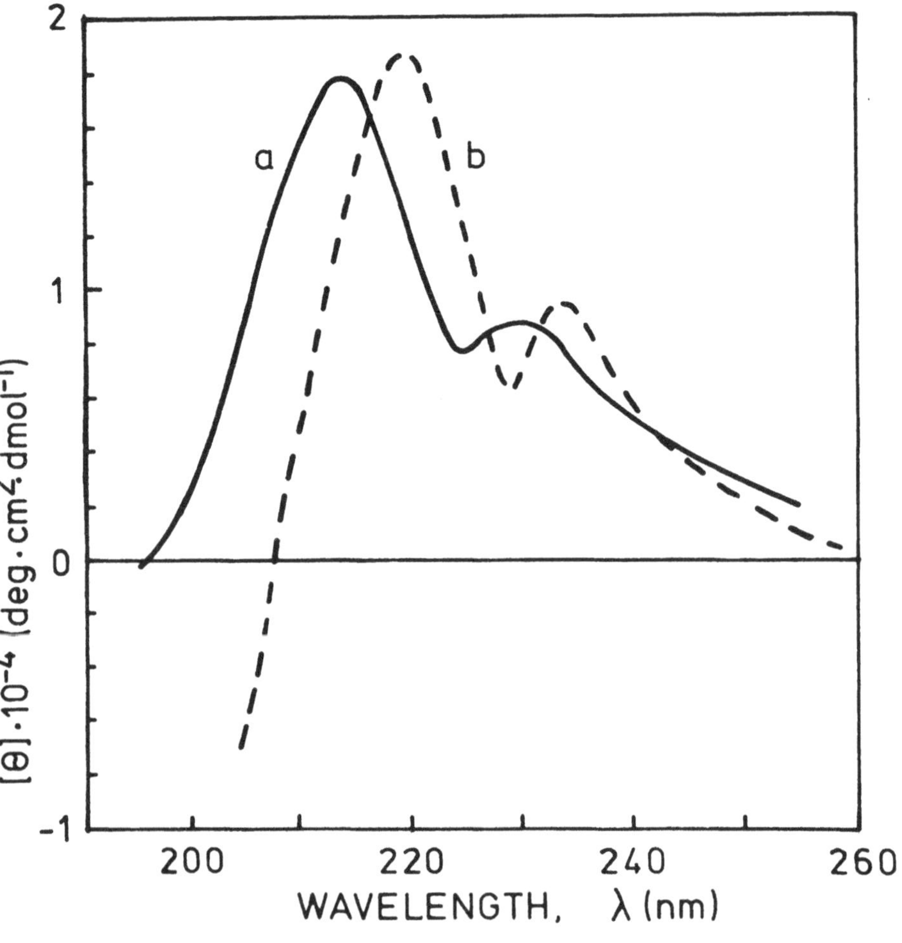

Fig. 11. CD spectra of (a) 5 mM [Val[1]] gramicidin A incorporated into SDS micelles (peptide/detergent molar ratio 1/50) in D_2O/CF_3CD_2OD (molar ratio 16/1) and (b) 0.5 mM gramicidin A incorporated into dipalmitoyl phosphatidylcholine liposomes (peptide/lipid molar ratio 1/330) after heat incubation, 15 hr at 70°C.

sequential resonance assignments: proton spin systems of amino acid residues were identified in COSY 2D-spectra via *J*-connectivities and then by NOESY 2D-spectra they were aligned via NOE-connectivities between protons of neighbouring residues into amino acid sequence (27). In particular, all NOE $^1d_{\alpha N}$ connectivities were revealed (Fig. 13), while only few $^1d_{\beta N}$ connectivities were detected and no $^1d_{NN}$ connectivities were observed at all. NOESY 2D-spectra also reveal many NOE-connectivities between protons of sequentially distant residues (27). The most significant of them for molecular structure analysis are presented in Fig. 4d. It is noteworthy that the regularity of the NH...C$^{\alpha}$H contacts for GA incorporated into SDS micelles is completely different from the observed one for the double helical conformations of GA in organic solvents (Fig. 4 a-c). In particular the NOE-connectivities between the N_iH and $C^{\alpha}_{i+6}H$ protons ($^6d_{N\alpha}$) for

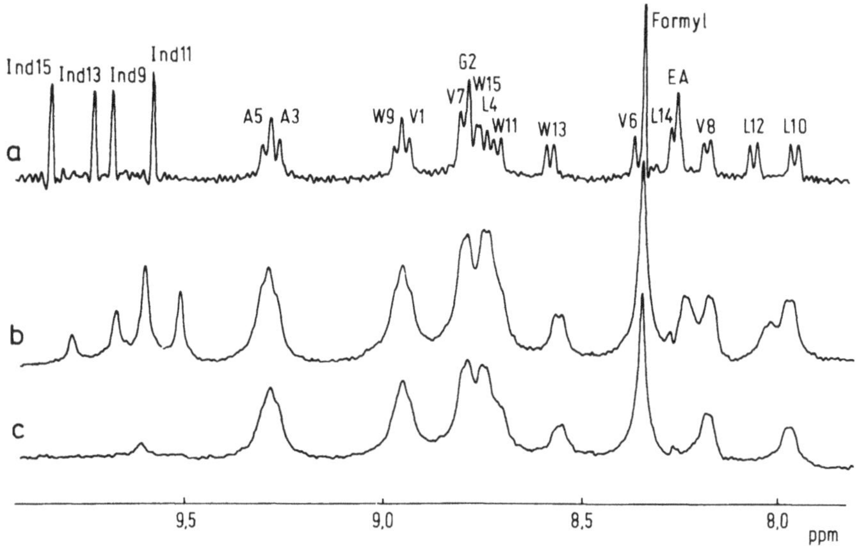

Fig. 12. Low-field region of the 500 MHz ^{1}H-NMR 1D-spectra of 5 mM [Val^1] gramicidin A in SDS-D_{25} micelles (peptide/detergent molar ratio 1/50) in H_2O/CF_3CD_2OD (molar ratio 16/1), 55°C at different pH values: (a) pH 6.5, (b) pH 8.5 and (c) pH 9.7. Spectrum (a) is resolution enhanced by Lorentzian to Gaussian transformation (30). See Fig. 1 caption for notation.

even i and between the $C_i^{\alpha}H$ and $N_{i+6}H$ protons ($^6d_{\alpha N}$) for odd i, recorded by a phase-sensitive absorption mode NOESY 2D-spectrum (Fig. 14), are inherent to a right-handed single stranded $\beta^{6.3}$-helices with 6.3 residues per turn. The alternative handedness of the helix is excluded because specific for the left-handed helix NOE-connectivities $^6d_{N\alpha}$ for odd i and $^6d_{\alpha N}$ for even i are not observed in the NOESY 2D-spectrum of GA in SDS micelles. Some of the observed NOE-connectivities are attributed to intermolecular contacts: NH(Val[1])... NH(Ala[5]), NH(Val[1])... $C^{\beta}H_3$(Ala[5]) and C^{α}H(Gly[2])... C^{α}H(D-Val[4]).

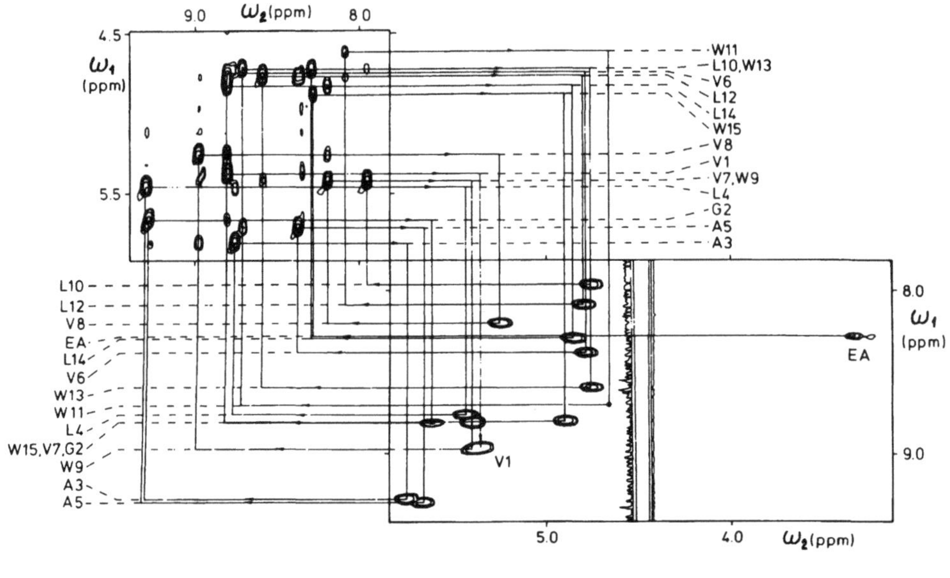

Fig. 13. Combined COSY-NOESY connectivity diagram for sequential resonance assignments via NOEs between backbone amide protons and $C^{\alpha}H$ protons of the preceding residue ($^1d_{\alpha N}$ connectivity) for 5 mM [Val[1]] gramicidin A in SDS-D_{25} micelles (peptide/detergent molar ratio 1/50) in H_2O/CF_3CD_2OD (molar ratio 16/1). The phase-sensitive absorption mode NOESY, τ_m = 100 ms (upper left) and absolute mode COSY (lower right) 2D-spectra were recorded for the same sample at identical conditions, pH 6.5, 55°C. The straight lines and arrows indicate the connectivities between neighbouring residues from the C-terminal ethanolamine moiety towards the Val[1] residue. The amide proton chemical shifts are indicated by the assignments in the lower lefthand corner, and those of the $C^{\alpha}H$ protons - in the upper righthand corner.

The helical structure of GA in SDS micelles was also supported by spin-spin coupling constants of vicinal H-NC^{α}-H protons (Table 2) and by limited solvent accessibility of backbone NH groups. The rates of NH proton exchange for GA in SDS micelles

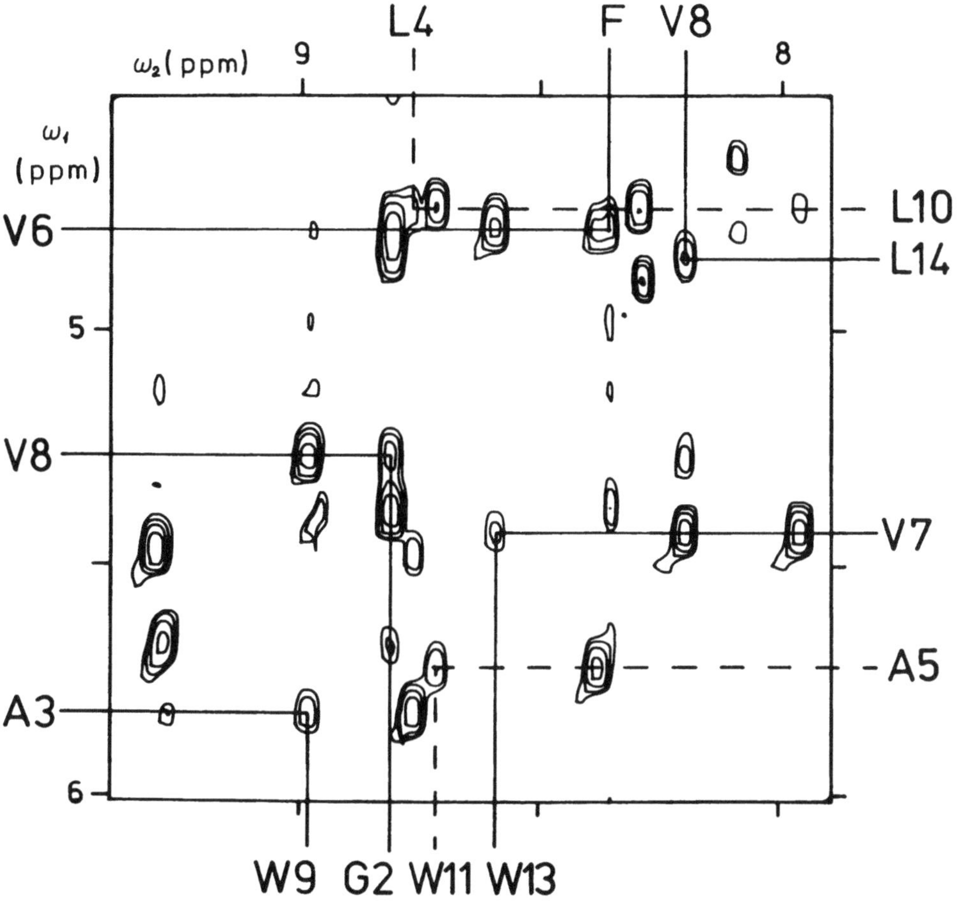

Fig. 14. Spectral region (ω_1 = 4.5-6.0 ppm, ω_2 = 7.9-9.4 ppm) of a phase-sensitive absorption mode NOESY proton 2D-spectrum (obtained with pulse sequence [6], τ_m = 100 ms) of a 5 mM [Val^1] gramicidin A in SDS-D_{25} micelles (peptide/detergent molar ratio 1/50) in H_2O/CF_3CD_2OD (molar ratio 16/1) at 55°C. The digital resolution is 5 Hz for both ω_1 and ω_2 directions. The spectrum is recorded in 22 h. The amide NH proton chemical shifts are indicated by the assignments at the top and at the bottom, those of the C^{α}H protons - on the left- and right-hand sides of the plot. F - formyl HCO proton.

were qualitatively estimated from the pH dependence of the peak intensity of corresponding NH signals when the H_2O signal was saturated by an r.f. field (Fig. 12). At high pH values the rate of the base-catalysed exchange with water molecules increases and signals of solvent accessible NH groups diminish and finally disappear from the spectrum due to saturation transfer. The majority of backbone NH groups exchange slowly enough to indicate their participation in the hydrogen bonding. Only the NH groups of the D-Leu^{12} and D-Leu^{14} residues and of the ethanolamine moiety are solvent accessible.

All the above NMR data fit only to one molecular structure - a head-to-head dimer $\vec{\pi}_{LD}^{6.3}\overleftarrow{\pi}_{LD}^{6.3}$ formed by two right-handed, single-stranded $\beta^{6.3}$-helices with 6.3 residues per turn. The dimer has a C_2 symmetry axis perpendicular to the channel axis in accordance with NMR observation of only one signal for each chemically equivalent proton. The corresponding wire model was constructed, adjusted in accordance with NMR data, and refined by means of conformational energy minimization, which was done using fixed bond angles and bond lengths and planar rigid trans-peptide units (31,32). Energy parameters were identical to those used in (33), except that an electrostatic term was discarded. Initial values of the torsion angles were deduced from a wire model constructed in accordance with experimental NMR parameters (spin-spin coupling constants of vicinal H-NC^α-H and H-$C^\alpha C^\beta$-H protons and NOE-connectivities between distinct protons). Conformational energy was minimized in three steps. First for a monomer - by varying the torsion angles ϕ , ψ , χ^1 and χ^2. Then the torsion angles were fixed and mutual arrangement of monomers into the dimer was adjusted, C_2 symmetry axis taken into account. Finally, complete energy minimization was performed by varying both torsion angles and mutual arrangement of monomers. Thus obtained molecular structure of the $\vec{\pi}_{LD}^{6.3}\overleftarrow{\pi}_{LD}^{6.3}$ dimer is shown in Fig. 5c and its conformational parameters are in Table 3. Twenty intramolecular and six intermolecular N-H...O=C hydrogen bonds stabilize the structure. The dimer is 2.6 nm length; its central axial cavity has an average luminal diameter of $\sim$ 0.4 nm (obtained by subtracting the van der Waals radius of the oxygen

Table 3. Torsion angles ϕ, ψ, χ^1 and χ^2 (in degree) for the energy refined conformation of [Val^1]gramicidin A incorporated into SDS micelles.

Residue	ϕ	ψ	χ^1	χ^2
L-Val 1	-141	90	178	
Gly 2	153	-142		
L-Ala 3	-99	124		
D-Leu 4	141	-107	-171	-178
L-Ala 5	-125	116		
D-Val 6	141	-142	53	
L-Val 7	-100	127	-166*	
D-Val 8	143	-90	61	
L-Trp 9	-143	154	169	98
D-Leu 10	103	-127	-179	159
L-Trp 11	-104	155	-46	-68
D-Leu 12	108	-110	47	67
L-Trp 13	-129	161	-60	-99
D-Leu 14	104	-117	172	88
L-Trp 15	-106	140	-73	-66
EA	-69		165	-66

* This torsion angle has considerable degree of freedom as estimated from spin-spin coupling constant of vicinal protons and NOE-connectivities between distinct protons.

from the distance between the carbonyl oxygens and the helix axis) forming the water-permeable cation-selective transmembrane channel.

Although the demonstrated right-handed dimeric structure of the Na^+ complex of GA in micelles differs from the widely accepted Urry's structure (3) in handedness of the helices, it is still in full agreement with single-channel investigations of numerous appropriately modified gramicidines (1,34). Presumably significance of the handedness might be established by further biophysical and theoretical study of ion-conductance dynamic processes in the GA channel.

Conclusion

Direct conformational study, performed by high-field 2D NMR spectrocsopy, indicates that GA can adopt several double-helical structures in organic solvents and the single-stranded helical dimer structure in membrane-like environment. The vast conformational diversity is especially due to the presence of D- and L-residues in alternation. All the structures reported in this paper fit principal requirenments of the transmembrane channel in dimensions (length and diameter of the axial cavity), hydrophobic external surface formed by the non-polar side chains and polar central cavity lined by the carbonyl groups. However this study reveals that water-lipid interface and relatively organized nonpolar environment promote the formation of a single-stranded helical dimer as a molecular structure of the GA transmembrane ion channel.
Of special interest is the operational structure of GA responsible for its biological action as endogenous inhibitor of RNA polymerase of the producing microorganism, which is still to be discovered.

References

1. Andersen, O.S.. 1984. Ann. Rev. Physiol. 46, 531-548.

2. Veatch, W.R., L. Stryer. 1977, J. Mol. Biol. 113, 89-102.

3. Urry, D.W., T.L. Trapane, K.U. Prasad. 1983. Science 221, 1064-1967.

4. Ivanov, V.T., S.V. Sychev. 1983. In: Peptides: Structure and Function (V. Hruby and D.H. Rich, eds.). Pierce Chem. Co., Rockford, pp. 451-462.

5. Chandrasekaran, R., B.V.V. Prasad. 1978, CRC Crit. Rev. Biochem. 5, 125-161.

6. Veatch, W.R., E.T. Fossel, E.R. Blout, 1974. Biochemistry 13, 5249-5256.

7. Sychev, S.V., A.N. Nevskaya, S. Jordanov, E.N. Shepel, A.I. Miroshnikov, V.T. Ivanov. 1980. Bioorg. Chem. 9, 121-151.

8. Koeppe II, R.E., J.M. Berg, K.O. Hodgson, L. Stryer. 1979. Nature 279, 723-725.

9. Naik, V.M., S. Krimm. 1984. Biophys. J. 45, 109-112.

10. Venkataram Prasad, B.V., R. Chandrasekaran. 1977. Int. J. Pept. Protein Res. 10, 129-138.

11. Colona-Cesari, F., S. Premilat, F. Heitz, G. Spach, B. Lotz. 1977, Macromolecules 10, 1284-1288.

12. Weinstein, S., B.A. Wallace, J.S. Morrow, W.R. Veatch. 1980. J. Mol. Biol. 143, 1-19.

13. Sychev, S.V., V.T. Ivanov. 1984. Biol. Membr. (USSR) 1, 1109-1124.

14. Ramachandran, L.R.. 1963. Biochemistry 2, 1138-1142.

15. Gurevich, A.Z., I.L. Barsukov, A.S. Arseniev, V.F. Bystrov. 1984. J. Magn. Res. 56, 471-478.

16. Eich, G., G. Bodenhausen, R.R. Ernst. 1982. J. Amer. Chem. Soc. 104, 3731-3732.

17. Wagner, G.. 1984. J. Magn. Res. 57, 497-505.

18. Braunschweiler, L., G. Bodenhausen, R.R. Ernst. 1983. Mol. Phys. 48, 535-560.

19. Wagner, G., K. Wüthrich, H. Tschesche. 1978. Eur. J. Biochem. 86, 67-76.

20. Marion, D., K. Wüthrich. 1983. Biochem. Biophys. Res. Commun. 113, 967-974.

21. States, D.J., Haberkorn R.A., D.J. Ruben. 1982. J. Magn. Res. 48, 286-292.

22. Arseniev, A.S., V.F. Bystrov, V.T. Ivanov, Yu.A. Ovchinnikov. 1984, FEBS Lett. 165, 51-56.

23. Arseniev, A.S., I.L. Barsukov, S.V. Sychev, V.F. Bystrov, V.T. Ivanov, Yu.A. Ovchinnikov. 1984. Biol. Membr. (USSR) 1, 5-17.

24. Veatch, W.R., E.R. Blout. 1974. Biochemistry. 13, 5257-5264.

25. Arseniev, A.S., I.L. Barsukov, E.N. Shepel, V.F. Bystrov, V.T. Ivanov. 1985. Bioorg. Khim. (USSR) 11, 5-20.

26. Arseniev, A.S., I.L. Barsukov, V.F. Bystrov. 1985. FEBS Lett. 180, 33-39.

27. Arseniev, A.S., I.L. Barsukov, V.F. Bystrov, A.L. Lomize, Yu.A. Ovchinnikov, 1985. FEBS Lett. (in press).

28. Dill, K.A., D.E. Koppel, R.S. Cantor, J.D. Dill, D. Bendedouch, S.-H. Chen. 1984. Nature 309, 42-45.

29. Long, M.M., D.W. Urry. 1981. In Membrane Spectroscopy (E. Grell, ed). Springer Verlag, West Berlin, pp. 143-171.

30. Ferrige, A.G., Lindon, J.C. 1978. J. Magn. Reson. 31, 337-340.

31. Momany, F.A., R.F. McGuire, A.W. Burgess, H.A. Scheraga. 1975. J. Phys. Chem. 79, 2361-2381.

32. Nemethy, G., M.S. Pottle, H.A. Scheraga. 1983. J. Phys. Chem. 87, 1883-1887.

33. Lomize, A.L., E.M. Popov. 1982. Bioorg. Khim. (USSR) 8, 1474-1485.

34. Urry, D.W., S. Alonso-Romanowski, C.M. Venkatachalam, R.D. Harris, K.U. Prasad. 1984. Biochem. Biophys. Res. Commun. 118, 885-893.

AN IMPROVED ^{19}F NMR RACEMIZATION TEST

M.Baumeister, K.Seeholzer, I.Ugi (Speaker), B.Westinger

Organisch-Chemisches Institut der Technischen Universität München, D-8046 Garching

W.E.Hüll

Bruker Analytische Meßtechnik GmbH 7512 Rheinstetten 4-Fo. Present adress: Institut für Biochemie DKFZ, 6900 Heidelberg

Since the customary racemization tests are based on the chemical degration of peptides (1), the derivatization of α-amino acids, and the separation of enantiomers (2) and diastereomers (3), such tests are relatively time-consuming. A fast and convenient racemization test could reduce considerably the work involved.

Racemization tests by spectroscopic methods are generally less cumbersome. Due to signal overlap, the analysis of racemized peptide derivatives through ^{1}H NMR (4) and ^{13}C NMR (5) is feasible only in a few cases. Bayer et al. (6) have demonstrated that the ^{19}F NMR spectra of diastereomeric N-TFA peptides differ in ^{19}F chemical chift. The observed differences were too small ($\Delta\delta$ = 0.02 - 0.03 ppm) for a convenient quantitative ^{19}F NMR racemization test, however, they indicate the possibility of analyzing mixtures of ^{19}F-labelled peptide derivatives by ^{19}F NMR.

Mosher et al. (7) introduced chiral α-methoxy-α-trifluoromethylphenylacetic acid 1 (MTPA, Mosher's acid) as a reagent for ^{19}F NMR analysis of enantiomer purity of chiral amines.

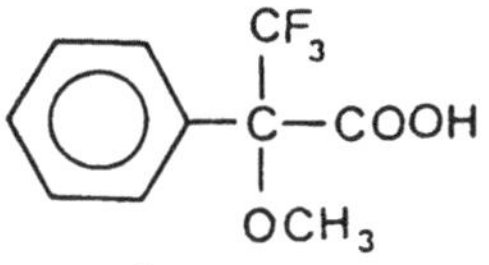

1

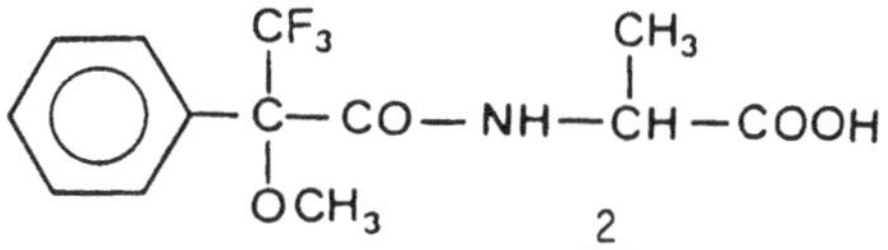

2

Chemistry of Peptides and Proteins, Vol. 3

The N-MTPA derivatives of diverse amino acids were studied by Breuer and Ugi (8) as ^{19}F probes; it was found that the MTPA derivatives of aliphatic amino acids, e.g.2, differ sufficiently in their ^{19}F chemical shift ($\Delta\delta$ = 0.32 - 0.40 ppm) in order to serve in a ^{19}F NMR racemization test. The use of various methods in the synthesis of peptide derivatives from 2-(R,S) and ^{19}F analysis of the products demonstrated that 1 is indeed a very suitable reagent for ^{19}F NMR racemization assays. Such tests are convenient to carry out; but until recently they usefulness was severely limited, because sensitivity was no better than 2 - 3 %, when the customary ^{19}F NMR techniques and equipment were used (8). The present paper deals with an improvement in sensitivity of epimer detectability level. Accordingly this method now becomes a racemization test with specific advantages. Since the epimers of 2, and related compounds have very similar ^{19}F NMR properties, it suffices to confine these sensitivity studies to the model compound 2.

It has been demonstrated that ^{19}F NMR provides a convenient means of assaying diastereomer purity in peptide synthesis by virtue of the distinguishable ^{19}F chemical shifts for (R,R) and (R,S) N-MTPA derivatives of amino acids. The minor diastereomer can readily be detected at levels as low as 0.05 % with an accuracy of $\pm$ 0.005 % in 0.4 ml samples of 0.082 solution of MTPA derivative.

A high-field NMR spectrometer (9.4 Tesla) was used in the present experiments. Any superconducting magnet system of the 200 - 360 MHz class will however suffice for such ^{19}F NMR measurements. Even electromagnets of the 80 - 100 MHz class will do if a sensitivity of ca. 0.5 % and prolonged recording times are acceptable. With this type of equipment ^{1}H decoupling may be necessary. The ^{19}F NMR spectra were measured with a Bruker AM-400 FT spectrometer at 376,5 MHz and 297° K. The samples had a concentration of 25 mg/ml in $CDCl_3$ and a volume of 0.4 ml in the 5 mm tubes used.

The ^{19}F MNR spectra were measured without ^{1}H decoupling. A standard 5 mm ^{1}H probe was simply tuned to the ^{19}F frequency, using the external tuning control.

The ^{1}H-^{19}F long range coupling between CF_3 and C_6H_5 groups of MTPA resulted in a linewidth of 5.4 Hz. Customary Fourier transform NMR techniques were used: spectral width 5.4 Hz for 16 K data points with quadrature detection, 16 bit A/D converter for better dynamic range behaviour, pulse angle ca 30° (3 μsec), aquisition time 1.64 sec, relaxation time 1.0 sec, number of transients per FID 64 - 256. The homogeneity of the magnetic field was carefully adjusted, in order to ensure optimum line shape and minimal rotational side bands.

For improved accuracy on ^{19}F signal integration, each FID was multiplied by a Lorentz-Gauss line shape transformation function (9). This improves resolution without influencing the ratios of the integrals. After Fourier transformation one finds a main signal of the MTPA derivatives **2-(R,S)** surrounded by two $^{13}CF_3$-^{13}C-satellites ($^{J}CF_3$ = 290.1 Hz, isotope shift $\Delta\delta$ = 0.127 ppm). The signal of **2-(R,R)** is 0.319 ppm downfield.

Mixtures of **2-(R,S)** as the main product, and **2-(R,R)** as the contaminant were used in model experiments. The reference compound **2-(R,S)** contained 0.04 % **2-(R,R)** as an impurity. The samples contained 0.00 %, 0.125 %, 0.500 % and 1.00 % of **2-(R,R)** by weight. Their ^{19}F NMR analyses indicated 0.0039 ± 0.004, 0.124 ± 0.001, 0.052 ± 0.02 and 0.98 ± 0.02 % of **2-(R,R)**. The respective acquisition times were 11.3, 7.0, 2.8 and 2.8 min. These experiments demonstrated that the present method meets the current requirements of a routine racemization test (1).

This ^{19}F NMR racemization test is particularly suitable for the determination of reaction conditions at which minimum racemization occurs with a given coupling method (8). In the development of peptide syntheses by four component condensations (10) the present method has the advantage that it allows to follow the four component condensation step (4CC), as well as the step in which the auxiliary group is removed. The productive stereoselectivity of 4CC and the destructive stereoselectivity (11) of the cleavage step are presently under investigation. The 4CC of **3-6** to form **7**, and the cleavage of the latter to form **8** and **9** are used as a model system.

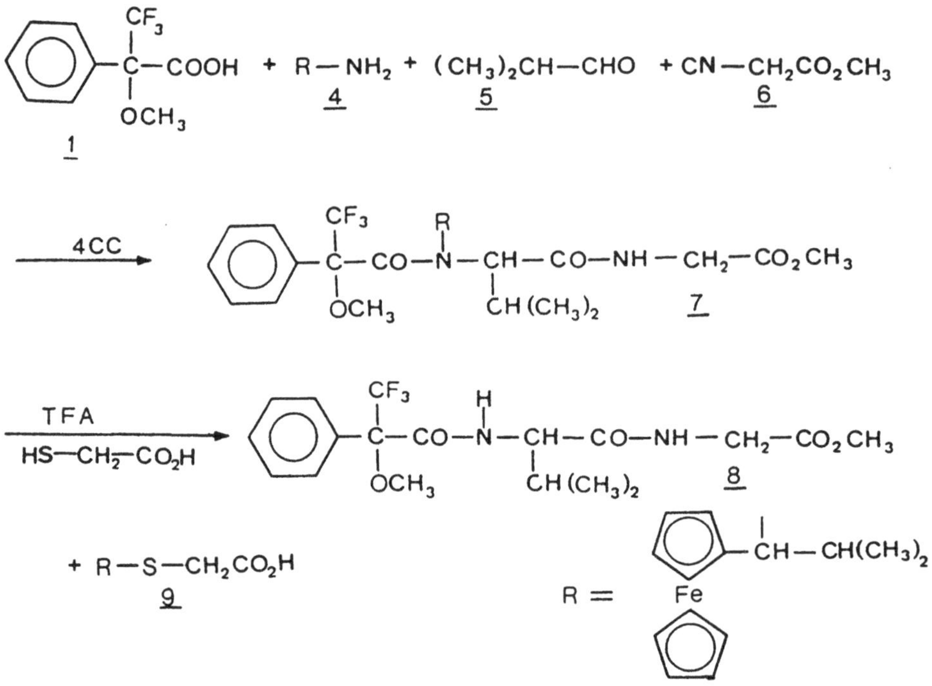

1. E. Wünsch, Angew. Chem., **83**, 773 (1971); Angew. Chem. Int. Ed. Engl., **10**, 786 (1971).

2. E. Gil-Av, J. Mol.-Evol. **1975**, 131; H. Frank, G.J. Nicholson and E. Bayer, J. Chromatog. Sci., **15**, 174 (1977); see also: V. Schurig, Angew. Chem., **96**, 733 (1984); Angew. Chem. Int. Ed. Engl., **23**, 742 (1984), and ref. therein.

3. D.H. Spackman, W.H. Steingel and S. Moore, Anal. Chem., **30**, 1190 (1958); J.M. Morning and S. Moore, J. Biol. Chem., **243**, 5591 (1968); see also: M. Bodanszky and L.E. Conklin, Chem. Comm., **1967**, 773; N. Izumiya and M. Muraoka, J. Amer. Chem. So., **91**, 2391 (1969); M.R. Bosshard, I. Schechter and A. Berger, Helv. Chim. Acta, **56**, 717 (173); E. Gil Av and A. Tishbee and P.E. Hare, J. Amer.

Chem. Soc., **102**, 5115 (1980); Houben-Weyl: Methoden der organischen Chemie, Vols. 15/I + II, ed.: E. Wünsch, Thieme Verlag, Stuttgart 1974.

4. B. Halpern, L.F. Chew, and B. Weinstein, J. Amer. Chem. Soc., **89**, 5051 (1967); B. Weinstein and A.E. Pritchard, J. Chem. Soc., Perkin Trans. I **1972**, 1015: N.L. Benoiton, K. Kuroda and F.M. Chen, Int. J. Peptide Protein Res., **13**, 403 (1979): N.L. Benoiton, Kuroda, ibid., **17**, 197 (1981); R. Nagaraj and P. Balarm, Tetrahedron, **37**, 2001 (1981).

5. J.S. Davies, R.J. Thomas and M.K. Williams, Chem. Comm., **1975**, 76; R. Deslaurierts, J.C.P. Smith, R.L. Somorjai, E. Raboton, R.C. Orlowski and R. Water, Int. J. Peptide Protein Res., **13**, 473 (1979).

6. R.E. Sievers, E.Bayer and P. Hunziker, Nature, **223**, 179 (1969); see also: E. Bayer, P. Hunziker, M. Mutter, R.E. Sievers and R. Uhmann, J. Amer. Chem. Cox., **94**, 265 (1972).

7. G.R. Sullivan, J.A. Dale and H.S. Mosher, J. Org. Chem., **38**, 2143 (1973); J.A. Dale, D.L. Dull and H.S. Mosher, J. Org. Chem., **34**, 2543 (1969).

8. W. Breuer and I. Ugi, J. Chem. Research (s) **1982**, 271; (M), **1982**, 2901.

9. R. R. Ernst, in Advances in Magnetic Resonance, ed.: J.S. Wangh, Vol 2, p.1. Academic Press, New York 1966.

10. I. Ugi, D. Marquarding and R. Urban, in Chemistry and Biochemistry of Amino Acids, Peptides and Proteins, Vol. 6, ed. Weinstein, Marcel-Dekker, New York 1982, p. 245.

11. J. Brandt, C. Jochum, I.Ugi and P. Jochum, Tetrahedron **33**, 1353 (1977).

Acknowledgement:

We acknowledge gratefully the financial support of our work by Deutsche Forschungsgemeinschaft, SFB 145, and the Fonds der Chemischen Industrie.

Structural Features of Peptides and Proteins

SEQUENCE HOMOLOGY BETWEEN PHOSPHOLIPASE AND ITS INHIBITOR IN SNAKE VENOM. THE PRIMARY STRUCTURE OF THE INHIBITOR OF VIPOXIN FROM THE VENOM OF THE BULGARIAN VIPER (VIPERA AMMODYTES AMMODYTES, SERPENTES)

I. Mancheva, B. Aleksiev

Higher Institute of Chemical Technology, Department of Organic Chemistry, Sofia 1156, Bulgaria

T. Kleinschmidt, G. Braunitzer

Max-Planck-Institut für Biochemie, Abteilung Proteinchemie, D-8033 Martinsried bei München, RFG

Introduction

The Bulgarian Viper (Vipera ammodytes ammodytes) is the most toxic snake in Europe. The neurotoxin vipoxin is the main toxic factor in its venom. A simple procedure for the isolation of homogeneous vipoxin by ion-exchange chromatography of the dialysed crude venom has been reported (1). The content of vipoxin was found to be about 15% of the crude venom. According to unpublished data geographical variations of this content are possible. Vipoxin is a neurotoxin with high toxicity (LD_{100} = 8 µg per 20 g mouse) and with weak phospholipase activity (10-12 U/mg protein) (1).

It was established that vipoxin represents a complex of an acidic nontoxic protein and a strong toxic basic protein with phospholipase A_2 activity on phosphatidylcholine (2,3). This complex is stable under neutral, basic and weak acidic conditions and could be decomposed in 7 M urea solution at pH 4 and below. The viperid phospholipase A_2 is the most potent toxic phospholipase so far isolated (LD_{100} = 1-3 µg per 20 g mouse) (3). The nontoxic acidic component

---

Abbreviations:

Quadrol = N,N,N',N'-tetrakis(2-hydroxypropyl)ethylenediamine;
Reagent IV = trisodium 7-(isothiocyanato)naphthalene-1,3,5-sulphonate;
V.a.a.Inh. = inhibitor from Vipera ammodytes ammodytes venom;
B.n.Ph.A_2 = phospholipase A_2 from Bitis nasicornis venom.

Chemistry of Peptides and Proteins, Vol. 3

has an inhibitory activity with regard to viperid phospholipase (up to 60%) and to other phospholipases from snake venoms as well (3). In the absence of the inhibitor the basic component is unstable and loses its enzymatic activity within 12 - 14 days (4).

The complete sequence of an acidic phospholipase inhibitor from snake venom has not yet been reported; only some investigations have been carried out on the N- and C-terminal part of both components of vipoxin from Vipera ammodytes ammodytes (2,3). In this paper the complete primary structure of the acidic one chain component of vipoxin is presented and its structure homology with known viperid phospholipases A_2 is discussed.

Material and Methods

1) Isolation of vipoxin and component separation

Lyophilized crude venom from the Bulgarian Viper (Vipera ammodytes ammodytes) was obtained from the Research Institute of Parasitic and Infectious Diseases, Sofia, Bulgaria. Vipoxin was isolated from the crude venom by dialysis and chromatography on SE-Sephadex C-25 according to methods described earlier (1). The vipoxin components were isolated by chromatography of 300 mg of the complex on carboxymethyl-cellulose (Whatman CM 52, column: 2.6 x 25 cm) and eluted with 0.1 M acetic acid in 7 M urea (pH 4) using a linear gradient of 0-0.4 M NaCl in 400 ml of the initial solution (2). The components were desalted by dialysis. The purity was checked by electrophoresis in the presence of urea and Triton X-100 (5,6).

2) Modification of the chain

For sequence determination the chain was modified by oxidation with performic acid (15 min at freezing point) (7), reduction followed by S-carboxymethylation (8,9,10) or reaction with citraconic anhydride (11).

3) Tryptic peptides

a) Cleavage

30 mg of modified inhibitor, oxidized with performic acid or citraconylated after oxidation, were hydrolysed with $TosPheCH_2Cl$-treated

trypsin at room temperature starting at pH 11 with an enzyme/substrate ratio 2 : 100.

b) High performance liquid chromatography

The separation and isolation of all tryptic peptides was performed by high performance liquid chromatography, using a 342 Gradient Liquid Chromatograph, Controller 421 (Beckman Instruments, Palo Alto, Calif. U.S.A.) and a column (4.6 x 250 mm) of LiChrosorb RP2 (E. Merck, Darmstadt) in 50 mM ammonium acetate, pH 6.0, with a flow rate of 1 ml/min (12), using different acetonitrile gradients.

4) Sequence determination

For sequence determination the automatic Edman degradation in a liquid phase sequenator (models 890 B and 890 C, Beckman Instruments, Palo Alto, Calif., U.S.A.) was employed. A modified Quadrol programme (13) (0.25 M Quadrol) was used for degradation of the S-carboxymethylated chain and for the lysine peptides Tp1, Tp2, Tp8 and Tp10. The lysine peptides were modified with reagent IV (14). For sequencing arginine peptides Tp3, Tp5 + 6, Tp7, Tp9, C-terminal peptide Tp11, and chymotryptical peptides Tp3a, Tp7a, and Tp11b the 3-(diethylamino)propyne programme (15) was applied. Conversion was performed for 14 min at 80 °C in 3 M trifluoroacetic acid. The derivatives were identified by thin-layer chromatography (15,16) or by high performance liquid chromatography (17) (Chromatograph 1084 B purchased by Hewlett Packard GmbH, Böblingen).

Results

The neurotoxin complex vipoxin from Vipera ammodytes ammodytes consists of an acidic component with the function of an inhibitor and a basic phospholipase A_2. Under dissociating conditions these components could be separated by acrylamide gel electrophoresis (Fig.1) or column chromatography. Whereas the isolated inhibitor proved to be homogeneous, the phospholipase A_2 seemed to consist of two isoenzymes.

We present the complete primary structure of the acidic component, which was achieved by automatic sequencing and results in a chain length of 122 amino-acid residues. The N-terminal 54 residues could be identified by degradation of the intact S-carboxymethylated chain with the exception of CM-cysteine in position 43, 44, 49 and 50. The remaining part of the sequence was mainly done on tryptic peptides.

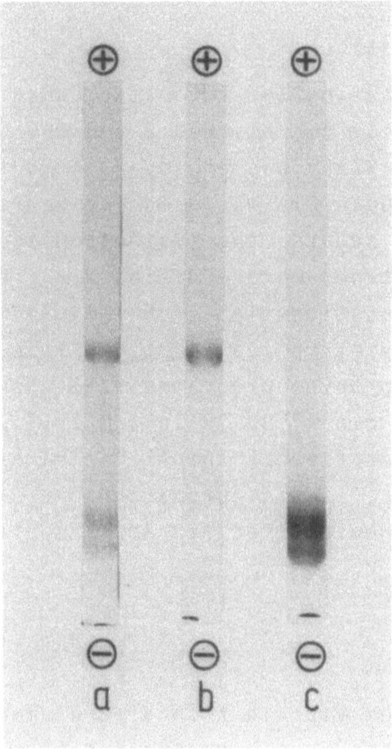

Fig. 1 Acrylamide gel elelectophoresis under dissociating conditions according to ref. (5,6). a) Vipoxin complex, b) inhibitor, c) phospholipase A_2.

Discussion

Primary structure of inhibitor

With the primary structure of the acidic component of Vipera ammodytes ammodytes venom we are presenting for the first time a complete amino acid sequence of a phospholipase inhibitor. The chain consists of 122 amino acid residues including 14 cysteine residues. The N-terminal amino acid is asparagine and the C-terminal one cysteine. The single methionine, tryptophan, and proline residues are located at positions 8, 30, and 82, respectively. The whole chain has 18 acidic amino acid residues (10 Asp and 8 Glu) and 12 basic residues (4 Lys, 6 Arg, and 2 His). The difference between acidic and basic amino acid residues (6 acidic residues) is in agreement with the earlier established acidic character of this protein with an isoelectric point at pH 4.5 - 4.7 (2).

Crotapotin - another inhibitor

Whereas an increasing number of up to 32 phospholipase sequences has been published during the last years, only little is known about the snake venom inhibitors. Some investigations on the neurotoxin complex crotoxin are reported (18). This complex from a crotalus viperid snake venom (Crotalus durissus terrificus) could be separated into two basic and a small quantitiy of an acidic phospholipase A, besides the acidic crotapotin which has inhibitor activity with regard to the basic phospholipase A isoenzymes. Crotapotin consists of three peptide chains containing 40, 34 and 14 amino acid residues, held together by disulfide bridges. The largest peptide starts with N-terminal serine, while the others are blocked (19). There seems to be no similarity between crotapotin and the inhibitor of the Bulgarian Viper venom.

Sequence homology between phospholipase A_2 and inhibitor

M.J. Dufton et al. (20) summarized 29 total sequences of phospholipases A_2 from snake venom (elapid venom, viperid venom, ß-bungarotoxin), analysed their homology and compared them with 3 phospholipases of pancreatic type. The classification according to primary structure and phospholipase evolution is well established (21). When the N-terminal sequence (18 amino acid residues) of the acidic component of vipoxin from Bulgarian Viper was determined (2), it was compared with the sequences of several phospholipases and neurotoxins and a possible homology between them was discussed (3). The primary structure of the phospholipase A_2 from Vipera ammodytes ammodytes venom is not yet known, but the sequences of phospholipases A_2 from some other Viperidae are reported. The comparison of the acidic component (phospholipase inhibitor) of Vipera ammodytes ammodytes with the basic component (phospholipase A_2) of Bitis nasicornis (CM-II) (22) and Bitis gabonica (23) showed a significant homology. We found 122 amino acid residues in the inhibitor, whereas the phospholipases have 119 and 118 residues, respectively. A homologous (24) comparison of the sequence of Vipera ammodytes ammodytes inhibitor (V.a.a. Inh.) with that of Bitis nasicornis phospholipase A_2 (B.n.Ph.A_2) is given in Fig. 2. To get the best alignment with respect to the positions of the cysteine residues four deletions had to be introduced in B.n.Ph.A_2 and one in V.a.a.Inh.. In this way 12 of the 14 cysteine residues of V.a.a.Inh. are in the same positions as the 12 cysteine residues of B.n.Ph.A_2. The cysteine resiudes in phospholipase as well as in the inhibitor are involved in disulfide bridges (25), which are supposed to be identical in both proteins, according to the homologous positions in the chains. Together with the fact that most of the cysteine residues are surrounded by completely homologous regions this points to a similar tertiary structure of phospholipase A_2 and inhibitor in vipoxin. Structural similarities between these two proteins have been reported earlier (26).

The comparison of the sequences in general shows 62 identical amino acid residues in V.a.a.Inh. and B.n.Ph.A_2, comprising 52% homology (Fig. 2). These residues are not distributed statistically, but are clustered in several regions mainly near the cysteine residues, as mentioned. The longest section of an identical sequence was found between positions 24-33 and 38-45. On the other hand, the sequence is completely different between positions 58-69 and the exchanges are not even isopolar. Both,phospholipase and inhibitor are rich in tyrosine. As in the case of cysteine residues the tyrosine homology is above the average with 8 out of 10 residues.

All phospholipases have a histidine residue located in position 47 in the active center. C.C. Viljoen et al. observed that the enzyme from Bitis gabonica venom can be completely inactivated by modifi-

```
                          10                          20
V.a.a. Inh.   Asn-Leu-Phe-Gln-Phe-Gly-Asp-Met-Ile-Leu-Gln-Lys-Thr-Gly-Lys-Glu-Ala-Val-His-Ser-Tyr-Ala-Ile-Tyr-Gly-
B.n. Ph.A2    Asp-Leu-Thr-Gln-Phe-Gly-Asn-Met-Ile-    -Asn-Lys-Met-Gly-    -Gln-Ser-Val-Phe-Asp-Tyr-Ile-Tyr-Tyr-Gly-

                       30                          40                          50
V.a.a. Inh.   -Cys-Tyr-Cys-Gly-Trp-Gly-Gly-Gln-Arg-Arg-Ala-Gln-Asp-Ala-Thr-Asp-Arg-Cys-Cys-Phe-Ala-Gln-Asp-Cys-Cys-
B.n. Ph.A2    -Cys-Tyr-Cys-Gly-Trp-Gly-Gly-Gln-Gly-Lys-Pro-Arg-Asp-Ala-Thr-Asp-Arg-Cys-Cys-Phe-Val-His-Asp-Cys-Cys-

                                      60                          70
V.a.a. Inh.   -Tyr-Gly-Arg-Thr-Ala-Thr-Tyr-Thr-Tyr-Ser-Phe-Glu-Asn-Gly-Asp-Ile-Val-Cys-Gly-Asp-Asn-Asp-Leu-    -Cys-
B.n. Ph.A2    -Tyr-Gly-Lys-Met-Gly-Thr-Tyr-Asp-Thr-Lys-Trp-Thr-Ser-Tyr-Lys-Tyr-Glu-Phe-Gln-Asp-Gly-Asp-Ile-Ile-Cys-

                          80                          90
V.a.a. Inh.   -Leu-Arg-Val-Asn-Asp-Cys-Asn-Pro-Lys-Ala-Val-Cys-Glu-Cys-Asp-Arg-Ala-Ala-Ala-Ile-Cys-Leu-Gyl-Glu-Asn-
B.n. Ph.A2    -Gly-Asp-    -Lys-Asp-Pro-Gln-Lys-Lys-Glu-Leu-Cys-Glu-Cys-Asp-Arg-Val-Ala-Ala-Ile-Cys-Phe-Ala-Asn-Ser-

               100                         110                         120
V.a.a. Inh.   -Val-Asn-Thr-Tyr-Asp-Lys-Asn-Tyr-Glu-Tyr-Tyr-Ser-Ile-Ser-His-Cys-Thr-Glu-Glu-Ser-Glu-Gln-Cys
B.n. Ph.A2    -Arg-Asn-Thr-Tyr-Asn-Ser-Lys-Tyr-Phe-Gly-Tyr-Ser-Ser-Ser-Lys-Cys-Thr-Glu-Thr-    -Glu-Gln-Cys
```

Fig. 2. Homologous (24) comparison of the vipoxin inhibitor from Vipera ammodytes ammodytes (V.a.a.Inh.) and the phospholipase A_2 (CM-II) from Bitis nasicornis venom (B.n.Ph.A_2). Identical amino acid residues are underlined.

cation of this histidine with p-bromophenacyl bromide (27). The inhibitor has glutamine in the homologous position. The measurement of the atomic parameters showed that glutamine can function in the same way as histidine (28). With the exception of one exchange in position 46 (Val → Ala) the sequence at the surroundings of this "active center" from positions 38-52 is identical in the phospholipase and inhibitor.

Tryptophan in position 30 is found in V.a.a.Inh. as well as in B.^{n}.Ph.A_2 and in the phospholipases of other viperid snakes. According to C.C. Viljoen et al. (27) it is essential for the enzymatic activity. The less important second tryptophan in some phospholipases is substituted by phenylalanine (pos. 61) in the inhibitor.

Common origin in gene duplication

F.J. Joubert et al. (22) discussed a relationship between the toxicity and the basic nature of the phospholipases A_2. The toxicity is decreasing parallel with the excess of basic residues, resulting in many acidic phospholipases A_2 which are much less toxic. The acidic component of the vipoxin complex from Vipera ammodytes ammodytes proved to have a very homologous primary structure with the phospholipases A_2, but the excess of 6 acidic residues might be responsible for its function as inhibitor.

According to the hypothesis of D. Eaker (29) the presynaptic neurotoxins have evolved from a phospholipase A structure, preserved in digestive secretes. Our findings point to the probability that not only the toxic component of the snake venom evolved from such an ancient phospholipase A, but also the acidic component with the function of an inhibitor. Investigations on other enzyme-inhibitor systems have not shown any similarities in sequence or structure up to now; for the first time we are postulating a homology between enzyme and inhibitor. Homology of proteins results from the existence of gene clusters originating in one or more gene duplications. A gene duplication has to be assumed to account for the sequence homology of phospholipase A_2 and inhibitor in the venom of Vipera ammodytes ammodytes too. The relatively late development of the reptiles and the snake venom complex explains the highly preserved structure.

One of the best studied homologous proteins is hemoglobin. Even at an exchange rate of up to 85% the structure and function as oxygen carrier are not affected. Mammalian α- and β-chains have a homology of about 50% like the snake venom phospholipase and its inhibitor. Although the preserved disulfide bridges point to a similar structure in this case, the function developed in different directions. Whereas one of the proteins has enzyme activity, the other shows inhibitor properties. The X-ray analysis of the vipoxin complex will elucidate the reason for the different function of two proteins with 50% homology in the primary structure.

References

1. Tchorbanov, B. and B. Aleksiev. 1981. J. Appl. Biochem. 3, 558-561.
2. Aleksiev, B. and R. Shipolini. 1973. In: Polymeri '73 - International Symposium on Polymers, Varna (Bulgaria), vol. 1, pp. 417-420.
3. Tchorbanov, B., E. Grishin, B. Aleksiev and Yu. Ovkinnikov. 1978. Toxicon 16, 37-44.
4. Aleksiev, B. and B. Tchorbanov. 1976. Toxicon 14, 474-485.
5. Rovera, G., C. Magarian and T.W. Borun. 1978. Anal. Biochem. 85, 506-518.
6. Alter, B.P., S.C. Goff, G.D. Efremov, M.E. Gravely and T.H.J. Huisman. 1980. Br. J. Haematol. 44, 527-534.
7. Hirs, C.H.W., S. Moore and W.H. Stein. 1956. J. Biol. Chem. 219, 623-642.
8. O'Donnell, I.J., B. Frangione and R.R. Porter. 1970. Biochem. J. 116, 261-268.
9. Aschauer, H., L. Vértesy, G. Nesemann and G. Braunitzer. 1983. Hoppe-Seyler's Z. Physiol. Chem. 364, 1347-1356.
10. Charlwood, P.H., R. Pitt-Rivers and H.L. Schwartz. 1970. Biochem. J. 116, 769-772.
11. Dixon, H.B.F. and R.N. Perham. 1968. Biochem. J. 109, 312-314.
12. Kratzin, H., C. Yang, J.U. Krusche and N. Hilschmann. 1980. Hoppe-Seyler's Z. Physiol. Chem. 361, 1591-1598.
13. Edman, P. and G. Begg. 1967. Eur. J. Biochem. 1, 80-91.

14. Braunitzer, G., B. Schrank, A. Ruhfus, S. Petersen and U. Petersen. 1971. Hoppe-Seyler's Z. Physiol. Chem. 352, 1730-1732.

15. Braunitzer, G., B. Schrank, A. Stangl and U. Scheithauer. 1978. Hoppe-Seyler's Z. Physiol. Chem. 359, 137-146.

16. Pfletschinger J. and G. Braunitzer. 1980. Hoppe-Seyler's Z. Physiol. Chem. 361, 925-931.

17. Zimmermann, C.L. and J.J.Pisano. 1977. Methods Enzymol. 47, 45-51.

18. Breithaupt, H., T. Omori-Satoh and J. Lang. 1975. Biochim. Biophys. Acta 403, 355-369.

19. Breithaupt, H., K. Rübsamen and E. Habermann. 1974. Eur. J. Biochem. 49, 333-345.

20. Dufton, M.J., D. Eaker and R.C. Hider.1983. Eur. J. Biochem. 137, 537-544.

21. Dufton, M.J. and R.C. Hider. 1983. Eur. J. Biochem. 137, 545-551.

22. Joubert, F.J., G.S. Townshend and D.P. Botes. 1983. Hoppe-Seyler's Z. Physiol. Chem. 364, 1717-1726.

23. Botes, D.P. and C.C. Vijoen. 1974. J. Biol. Chem. 249, 3827-3835.

24. Braunitzer, G., R. Gehring-Müller, N. Hilschmann, K. Hilse, G. Hobom, V. Rudloff and B. Wittmann-Liebold. 1961. Hoppe-Seyler's Z. Physiol. Chem. 325, 283-286.

25. Aleksiev, B., B. Tchorbanov and E. Grishin. 1979. C. R. Acad. Bulg. Sci. 32, 1227-1230.

26. Tchorbanov, B., B. Atanasov, B. Aleksiev and T. Bukolova-Orlova. 1978. 11th Int. Symposium on Chemistry of Natural Products, Golden Sands (Bulgaria), vol. 1, pp.107-110.

27. Viljoen, C.C., L. Visser and D.P. Botes. 1976. Biochim. Biophys. Acta 438, 424-436.

28. Braunitzer, G., B. Schrank, A. Stangl and C. Bauer. 1978. Hoppe-Seyler's Z. Physiol. Chem. 359, 547-558.

29. Eaker, D. 1975. Toxicon 13, 90-91.

PRIMARY STRUCTURE AND A MODEL FOR MOLECULAR ORGANIZATION OF ADRENOCORTICAL CYTOCHROME P-450$_{scc}$

V.L.Chashchin, S.A.Usanov, V.N.Lapko, T.B.Adamovich, V.M.Shkumatov, A.A.Akhrem
Institute of Bioorganic Chemistry, BSSR Academy of Sciences, Minsk, U.S.S.R.

Introduction

The conversion of cholesterol into pregnenolone is a rate-limiting step in the biosynthesis of corticosteroids in adrenal cortex (1). This reaction is catalyzed by mitochondrial multienzyme system containing an NADPH-dependent flavoprotein (ferredoxin:NADP$^+$ oxidoreductase, EC 1.18.1.2.), a 2Fe - 2S protein (adrenodoxin) and hemeprotein (cytochrome P-450$_{scc}$) specific for cholesterol side-chain cleavage (2). From investigations involving isolated cytochrome P-450$_{scc}$ (3-5) as well as membrane-bound hemeprotein (6,7), we were able to assume the following functionally important sites present in the protein molecule: (i) a catalytic center with protoheme IX as prosthetic group; (ii) site responsible for interaction with electron transport protein, adrenodoxin; (iii) hydrophobic site responsible for interaction of hemeprotein with phospholipid membrane. To elucidate the structural basis of functioning of cytochrome P-450$_{scc}$, it is necessary to understand the localization of these sites in the native hemeprotein molecule and in the polypeptide chain. The aim of the present work is cytochrome P-450$_{scc}$ primary structure investigation, studies on structure-function interrelations and proposal of the model for molecular organization of adrenocortical cytochrome P-450$_{scc}$.

Results

Limited proteolysis studies. Selective proteolytic modification of proteins proved to be a useful technique for the study of their structure and function. This approach promotes determination of compact globular sites in the protein molecule which possess a definite function. The controlled limited trypsinolysis of cytochrome P-450$_{scc}$ results in disappearance of the native hemeprotein with simultaneous appearance of two polypeptide fragments, F_1

Chemistry of Peptides and Proteins, Vol. 3

(Mr 29800) and F_2 (Mr 26600), respectively. These fragments are relatively stable to further proteolytic modification when cytochrome P-450$_{scc}$ is treated with 1:50 trypsin/hemeprotein ratio for 60 min at 20°C. However, when cytochrome P-450$_{scc}$ is incubated with trypsin for a long time or treated with high trypsin ratio, a third stable polypeptide fragment referred to as F_3 with Mr 16800 is formed with concomitant loss of the F_2 fragment .

The observed pattern of proteolytic modification indicates that (i) mild proteolytic treatment results in equimolar ratio of F_1 and F_2 fragments, the total molecular weight of which is equal to that of native enzyme; (ii)prolonged trypsin modification leads to further cleavage of the F_2 fragment with stoichiometric formation of F_3 fragment. Cytochrome P-450$_{scc}$ proteolytically modified to equimolar mixture of F_1 and F_2 or F_1 and F_3 fragments, retains the spectral and functional properties of the native hemeprotein. The aspartyl-prolyl bonds, sulfhydryl groups and surface tyrosine residues are distributed nonuniformly among fragments F_1 and F_2. These data as well as different resistance of fragments to further proteolytic modification suggests that cytochrome P-450$_{scc}$ consists of two independently folded domains linked with a short "hinge" of polypeptide chain, the domains being rigidly associated under conditions used (8,9).

Limited proteolysis of inactivated form of cytochrome P-450$_{scc}$ - cytochrome P-420 - results in appearance together with F_1 and F_2 of an additional fragment F_o (Mr 45300). Kinetic analysis of limited proteolysis of cytochrome P-420 indicates that additional, accessible to trypsin "hinge" of polypeptide chain appeared in F_1 fragment during hemeprotein inactivation.

Our further studies are aimed at elucidation of functionally important sites on cytochrome P-450$_{scc}$ domains with subsequent their localization in hemeprotein polypeptide chain.

Localization of adrenodoxin-binding domain. To identify the domain of cytochrome P-450$_{scc}$ responsible for interaction with adrenodoxin, the hemeprotein was absorbed on the immobilized adrenodoxin due to biospecific affinity, and complex has been cross-linked using bifunctional cleavable reagent. To exclude undesirable steric effects, adrenodoxin was immobilized via a long spacer arm: Sepharose-OCH$_2$-CH(OH)-CH$_2$-NH-CH$_2$-CH$_2$-NH-C(O)-CH$_2$-S-CH$_2$-CH$_2$-CH$_2$-C(NH)-NH-adrenodoxin. A general scheme of the experiment is given in Fig. 1 (10).

To avoid side reactions, the surface sulfhydryl group of F_2 domain (8,9) was first modified with N-ethylmaleimide. Subsequent treatment of cytochrome P-450$_{scc}$ with N-succinimidyl-3-(2-pyridyldithio)propionate led to modification of approximately three (3.28) lysine residues per hemeprotein molecule with concomitant introduction of three activated sulfhydryl groups.

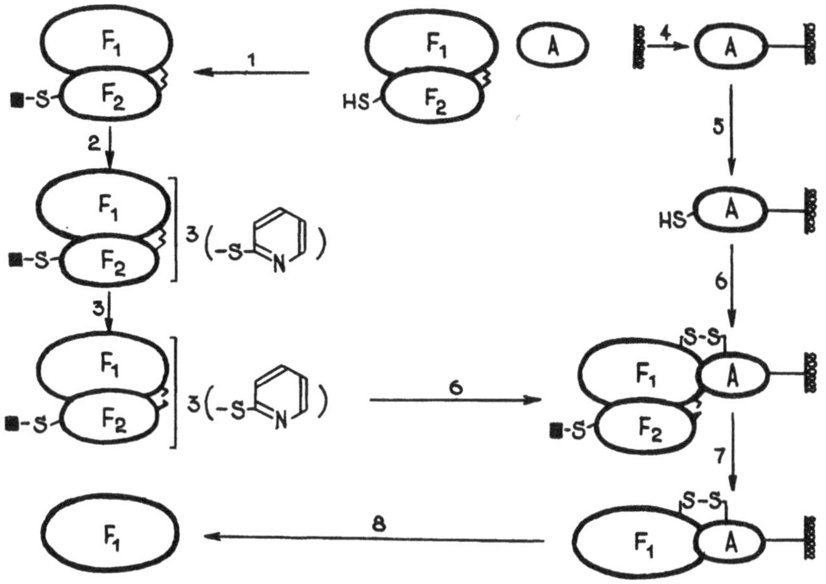

Fig. 1. General scheme of experiments on localization of adrenodoxin-binding site of cytochrome P-450scc. Chemical modification of accessible sulfhydryl group of cytochrome P-450scc with N-ethylmaleimide(1). Reaction of hemeprotein with N-succinimidyl-3-(2-pyridyldithio)propionate(2). Limited proteolysis(3). Immobilization of adrenodoxin(4). Thiolation of adrenodoxin with methyl-4-mercaptobutyrimidate(5). Cross-linking of cytochrome P-450scc with adrenodoxin(6). Removal of non-cross-linked domain(7). Elution of adrenodoxin-binding domain(8).

To identify the domain of cytochrome P-450$_{scc}$ responsible for interaction with adrenodoxin, hemeprotein cleaved to F_1 and F_2 fragments was used for cross-linking studies. This approach was chosen on the basis of previously reported fact that proteolytic modification of cytochrome P-450$_{scc}$ does not affect its ability to form functional complex with adrenodoxin (9). Since immobilized adrenodoxin contains only one free sulfhydryl group per molecule, it was additionally thiolated with methyl-4-mercaptobutyrimidate. This results in introduction of 1.22 additional sulfhydryl group per adrenodoxin molecule. Sorption of cytochrome P-450$_{scc}$ on thiolated adrenodoxin-Sepharose due to biospecific affinity was evidently accompanied by formation of cross-links between two proteins. Washing of the column with guanidinium hydrochloride allowed to elute a non-covalently bound domain of cytochrome P-450$_{scc}$. The next washing with 2-mercaptoethanol to cleave disulfide bond of the cross-linked complex resulted in elution of cytochrome P-450$_{scc}$ domain which was cross-linked with adrenodoxin. Fraction eluted after dissociation of the disulfide bond comprises 92% and 8%

of F_1 and F_2 fragments, respectively. This means that adrenodoxin preferably binds to a high-affinity binding site located on domain F_1 of the cytochrome $P\text{-}450_{scc}$ molecule.

Localization of heme group. Gel-electrophoresis of proteolytically modified cytochrome $P\text{-}450_{scc}$ with subsequent heme-peroxidase staining of the gels with tetramethylbenzidine and hydrogen peroxide, proved that heme-associated peroxidase activity is associated with fragment F_1. This fact indicates that heme group is mostly associated with the F_1 domain of cytochrome 450_{scc}.

Partial separation of the cytochrome $P\text{-}450_{scc}$ domains by gel-filtration on Sephadex G-200 with retention of heme group in domain F_1 has been achieved by incubation of the trypsin-treated hemeprotein in 50 mM phosphate buffer, pH 7.2 containing 1 M NaCI/0.3% sodium cholate/0.3% Tween 80. Gel-filtration of proteolytically modified cytochrome $P\text{-}450_{scc}$ after prolonged incubation (120 h) in the same buffer indicates that parallelly to inactivation (conversion of cytochrome P-450 to cytochrome P-420), a fraction with molecular weight 110000

appeared, in addition to fraction with molecular weight 400000. An electrophoretic analysis and spectrophotometric data indicate that fraction with molecular weight 400000 contains 80-85% of F_2 fragment, but only 10-15% of F_1. The second fraction is characterized by a reverse ratio and contains heme-bound material.Based on these results the conclusion has been drawn that heme is bound with F_1 domain of cytochrome $P\text{-}450_{scc}$(11).

Interaction with phospholipid membrane. To study orientation of cytochrome $P\text{-}450_{scc}$ in phospholipid membrane and localization of domain responsible for interaction with the membrane, the process of limited proteolysis of membrane-bound cytochrome $P\text{-}450_{scc}$ has been studied. Electrophoretic analysis indicates that treatment of submitochondrial particles with trypsin is followed by decreasing of intensity of the protein band corresponding to cytochrome $P\text{-}450_{scc}$ and concomitant appearence of protein bands corresponding to F_1 and F_2 fragments, respectively. This indicates that membrane-bound cytochrome $P\text{-}450_{scc}$ as well as isolated hemeprotein consists of two domains connected with polypeptide "hinge". The latter is accessible to trypsin and appears to be exposed.From these results we proposed that one of the domains is mostly bound with phospholipid membrane, whereas another one is exposed to the matrix.Since adrenodoxin-binding domain was shown to be in the F_1 domain and adrenodoxin is a peripheral protein, we conclude that domain F_1 is predominantly peripheral.

To prove the proposed orientation of cytochrome $P\text{-}450_{scc}$ in mitochondrial membrane, the accessibility of the surface tyrosine residues of individual domains has been studied under conditions of lactoperoxidase-dependent radioiodination. With this aim iodination of isolated and membrane-bound cytochrome $P\text{-}450_{scc}$

was studied. The radioactively labelled cytochrome P-450$_{scc}$ was isolated, subjected to limited proteolysis up to domains F_1 and F_2 and after SDS-electrophoresis the label distribution between fragments has been examined. In the case of membrane-bound hemeprotein, the radioactivity is distributed between fragments F_1 and F_2 as 80% and 20%, respectively, whereas for isolated hemeprotein this ratio is 90% and 10%. This fact indicates that being exposed into aqueous phase, F_1 domain is partially embedded into the phospholipid membrane, so that some of its surface tyrosine residues are not accessible for radioiodination as compared to isolated hemeprotein.

Further studies on topology of cytochrome P-450$_{scc}$ in membrane have been done by using immunochemical approach. Monospecific antibodies to cytochrome P-450$_{scc}$ as well as to highly purified by SDS-electrophoresis fragments F_1 and F_2 radioactivelly labelled with ^{125}I effectively interact with membrane-bound cytochrome P-450$_{scc}$. The antibodies to fragments F_1 and F_2 compete with antibodies to cytochrome P-450$_{scc}$ for binding to membrane-bound hemeprotein. This means that antigenic determinants on both domains of cytochrome P-450$_{scc}$ are accessible for antibodies and both domains have polypeptides exposed to matrix.

Based on investigations mentioned above, we can conclude that F_1 domain of cytochrome P-450$_{scc}$ represents catalytic domain responsible for catalysis, whereas the F_2 domain is responsible for protein interaction with phospholipid membrane and substrate binding.

Separation of domains. To elucidate the structural basis of cytochrome P-450$_{scc}$ dependent catalysis, it was necessary to separate cytochrome P-450$_{scc}$ fragments after proteolytic modification to further characterize them. For this purpose a thiol-disulfide exchange chromatography was chosen, since preliminary data indicate that domains differ in content of sulfhydryl groups. Chromatography on thiopropyl-Sepharose results in separation of cytochrome P-450$_{scc}$ domains: F_1 domain is eluted by guanidinium hydrochloride which causes dissociation of domain under denaturation conditions, while domain F_2 is eluted by 2-mercaptoethanol, which reduces the disulfide bond (Fig.2). Thus, the free sulfhydryl group of domain F_2 is accessible for activated sulfhydril groups of the matrix.

The structural analysis of individual fragments shows that amino acid composition of cytochrome P-450$_{scc}$ and the sum of amino acid compositions of F_1 and F_2 fragments differ insignificantly. The N-terminal amino acid sequence of fragment F_1 is identical to that of cytochrome P-450$_{scc}$ and hence this fragment represents the N-terminal sequence of hemeprotein. The C-terminal sequence of fragment F_2 coincides with that of fragment F_2, thus indicating that this domain represents C-terminal sequence of hemeprotein (11).

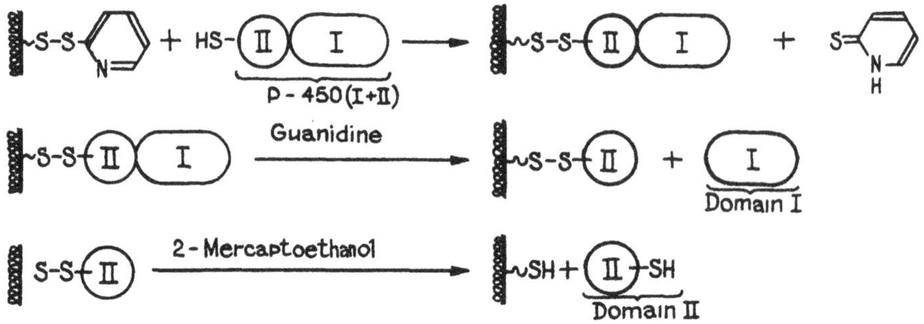

Fig. 2. Scheme for separation of cytochrome P-450scc fragments using covalent chromatography on activated thiopropyl-Sepharose.

Primary structure investigation. Two complementary approaches have been used to study the primary structure of adrenocortical cytochrome $P\text{-}450_{scc}$: (i) enzymatic (chymotrypsin, thermolysin, St. aureus protease) and chemical (cyanogen bromide) hydrolysis of the total polypeptide chain with subsequent analysis of peptides and (ii) structural analysis of highly purified fragments F_1 and F_2 after enzymatic hydrolysis. Fig. 3 shows the strategy of primary structure investigation.

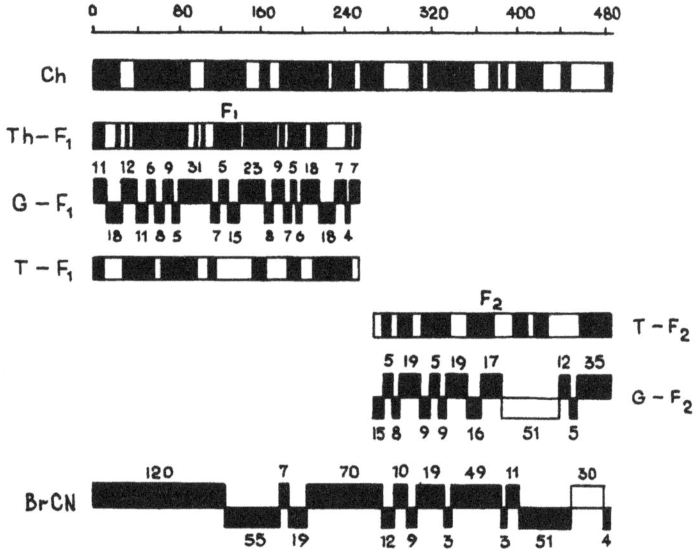

Fig. 3. Approaches used to study cytochrome P-450scc primary structure: Ch-, Th-, G-, T-, enzymatic hydrolysis with chymotrypsin, thermolysin, St. aureus protease and trypsin. In black are shown the peptides with established amino acid sequence.

The sequence analysis of peptides obtained allowed to reconstitute the complete primary structure of adrenocortical cytochrome P-450$_{scc}$, which is presented in Fig. 4 (12).

```
                                                            30
I S T K T P R P Y S E I P S P G D D G W L N L Y H F W R E K
                                                            60
G S Q R I H F R H I E N F Q K Y G P I Y R E K L G N L E S V
                                                            90
Y I I H P E N V A H L F K F E G S Y P E R Y D I P P W L A Y
                                                           120
H R Y Y Q K P I G V L F K K S G T W K K D R V V L N T E V M
                                                           150
A P E A I K N F I P L L N P V S Q D F V S L L H K R I K Q Q
                                                           180
G S G K F V G N I K E D L F H F A F E S I T N V M F G E R L
                                                           210
G M L E E T V N P E A Q K F I D A V Y K M F H T S V P L L N
                                                           240
V P P E L Y R L F R T K T W R D H V A A W D T I F N K A E K
                                                           270
Y T E I F Y Q D L R R K T E F R N Y P G I L Y C L L K S E K
                                                           300
M L L E D V K A N I T E M L A G G V N T T S M T L Q W H L Y
                                                           330
E M A R S L N V Q E M L R E E V L N A R R Q A E G D I S K M
                                                           360
L Q M V P L L K A S I K E T L R L H P I S V T L Q R Y P E S
                                                           390
D L V L Q D Y L I P A K T L V Q V A I Y A M G R D P A F F S
                                                           420
S P D K F D P T R W L S K D K D L I H F R N L G F G W G V R
                                                           450
Q C V G R R I A E L E M T L F L I H I L E N F K V E M Q H I
                                                           480
G D V D T I F N L I L T P D K P I F L V F R P F N Q D P P Q A
```

Fig. 4. Complete primary structure of adrenocortical cytochrome P-450$_{scc}$.

Recently (13), molecular cloning and nucleotide sequencing of cDNA for mRNA of cytochrome P-450$_{scc}$ was carried out. In that work authors used our data to identify the cDNA clones for structural analysis of hemeprotein. The results of primary structure determination by the two methods fully coincide, which is very important for evaluation of adequacy of the methods, since cytochrome P-450$_{scc}$ is one of the few cytochromes P-450 whose primary structure have been established by two independent methods.

Structural analysis of cytochrome P-450$_{scc}$ and determination of primary structure of hemeprotein indicates that "hinge" of polypeptide chain connecting F_1

and F_2 fragments in native **polypeptide chain is of insignificant lenght and is** limited by R (250) and N (257). The cleavage of this "hinge" results in formation of F_1 and F_2 fragments. During prolonged trypsinolysis further cleavage of polypeptide chain occurs at the C-terminal sequence of cytochrome P-450$_{scc}$ R(399) to form F_3 fragment which represents the N-terminal sequence of F_2 fragment.

Structural analysis allowed to explain the results of limited proteolysis of inactivated form of cytochrome P-450$_{scc}$,the cytochrome P-420.During inactivation,changes in N-terminus of hemeprotein take place so that new accessible to trypsin polypeptide "hinge" (R92) appears. This fact also explains why during cytochrome P-450$_{scc}$ inactivation spectral changes in the heme chromophore take place and hemeprotein loses the affinity to adrenodoxin.

Identification of putative axial ligands. In contrast to other hemeproteins which participate in reversible oxygen binding (myoglobin,hemoglobin),electron transfer (cytochrome b-type),dioxygen splitting (catalase,peroxidase),reduction of molecular oxygen (cytochrome c oxidase),cytochrome P-450 catalyzes all tipes of reactions and also the radical attack of substrate by oxo-iron species. These unique properties of cytochrome P-450 are determined by axial heme ligands which are presented with polypeptide chain amino acid radicals. A set of experiments aimed at selective chemical modification of cytochrome P-450$_{scc}$ with different sulfhydryl reagents indicate that these reagents interact with one accessible cysteine residue in native cytochrome P-450$_{scc}$.Its modification does not lead to cytochrome P-450$_{scc}$ inactivation,i.e. loss of intrinsic spectral properties inherent to the native enzyme. This,according to the present concept,indicates to intact sulfur-heme iron bond. The other cysteine residue is not accessible for reagents. Modification of this residue may be achieved only after preliminary inactivation with subsequent heme removal. It is important that both cysteine residues under definite conditions may react with thiopropyl-Sepharose,i.e. are involved in thiol-disulfide exchange.

The question may be asked: what cysteine residue is involved in the proximal ligand formation,and what is on the surface of cytochrome P-450$_{scc}$? To answer this question, the approach the general scheme of which is presented in Fig.5 was used.

As a first step,the exposed cysteine residue was irreversibly modified with N-ethylmaleimide under conditions excluding cytochrome P-450$_{scc}$ inactivation. Modification of this residue does not change spectral properties of hemeprotein, while the catalytic activity in reconstituted system is even increased.Further treatment of cytochrome P-450$_{scc}$ with SDS and removal of the heme results in regeneration of the second sulfhydryl group. When applied to thiopropyl-Sepharose, cytochrome P-450$_{scc}$ interacts with the matrix by means of thiol-disulfide exchange.

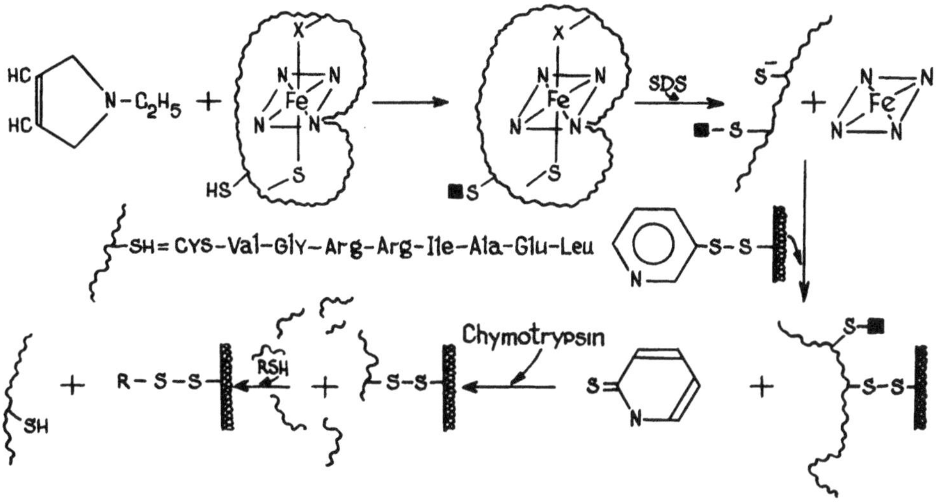

Fig. 5.The overall scheme of experiment for identification of cysteine residue involved in heme thiolation.

Subsequent proteolytic modification of immobilized cytochrome P-450$_{scc}$ with chymotrypsin results in removal of peptide material which is not connected with the carrier.The cysteine-containing peptide involved in thiol-disulfide exchange was eluted with 2-mercaptoethanol. Carboxymethylation, desalting and further sequencing allowed to identify the peptide bound with matrix as: C V G R R I A E. Thus, in contrast to theoretical conclusions about localization of the cysteine residue serving as an axial heme ligand,we were able to experimentally determine this heme-binding site, and show that if cysteine residue is indeed the fifth axial heme ligand in cytochrome P-450$_{scc}$, this residue is cysteine at 422 position in polypeptide chain of hemeprotein. There has been some controversy concerning the position of cysteine residue involved in heme ligation in polypeptide chain. Due to similarity of phisico-chemical properties of different cytochromes P-450, the primary structure and conformation of active sites were proposed to be similar for various forms of hemeproteins. To check this assumption, we compared the amino acid sequences of the known cytochromes P-450 structures near the region of cysteine residue involved in heme-binding (Fig. 6). The cytochromes P-450 from ferredoxin-dependent monooxygenases contain heme-binding cysteine least of all removed from N-terminal sequence. There is some homology between the structures presented, but it is not absolute. This might in-

P-450scc	F G W G V R Q C V G R R I A E L E M T L F	C - 422
P-450b,e	F S T G K R I C L G E G I A R N E L F L F	C - 436
P-450LM2	F S L G K R I C L G E G I A R T E L F L F	C - 436
P-450c	F G L G K R K C I G E T I G R L E V F L F	C - 461
P-450d	F G L G K R R C I G E I P A K W E V F L F	C - 456
P-450LM4	F G L G K R R C I G E T L A R W E V F L F	C - 456
P-4501	F G L G K R K C I G E T I G R S E V F L F	C - 458
P-4503	F G L G K R R C I G E I P A K W E V F L F	C - 458
P-450cam	F G H G S H L C L G Q S L A R R E I I V F	C - 355

Fig. 6. Comparison of the amino acid sequences of variuous cytochromes P-450 near heme-binding cysteine residue in C-terminal region.

dicate that either position of heme-binding cysteine is different in various forms of cytochrome P-450 or the hypothesis about conserved region near heme-binding cysteine is unlikely.

Localization of the cysteine residue of cytochrome P-450_{scc} involved in heme-binding allows to make some conclusion: (i) cytochrome P-450_{scc} contains a cysteine residue involved in heme ligation in domain F_2, which according to our data is responsible for interaction with phospholipid membrane; (ii) cysteine residue at position 264 is in the N-terminal sequence of domain F_2 polypeptide chain, exposed, accessible for sulfhydryl reagents and appear to be outside the phospholipid membrane.

In contrast to the fifth ligand, the nature of the axial trans to thiolate ligand is less defined. Based on selective chemical modification and physicochemical studies we proposed that tyrosine residue might be involved in active site formation serving as a sixth heme axial ligand (14). For more complete understanding of the functional role of this tyrosine residue, it was necessary to localize this residue in the polypeptide chain. To simplify this task, we tried to answer the question: in what domain of cytochrome P-450_{scc} putative tyrosine residue is located? To answer this question, thiol-disulfide exchange chromatography was used to separate the domains of cytochrome P-450_{scc} modified with tetranitromethane. The amino acid analysis of separated domains indicates that modification of essential tyrosine residue affects the heme-containing domain F_1. Preparation of highly purified fragment F_1 from nitrated cytochrome P-450_{scc}, proteolytic modification with St. aureus protease with subsequent chromatography of peptides on Bio-gel P-6 and further HPLC, allowed to localize the nitrotyrosine-containing peptide as: Y D I P P W L A Y H R Y Y.

Prediction of elements of the secondary structure. Theoretical analysis of polarity values indicates that polarities of the F_1 and F_2 fragments of cytochrome P-450$_{scc}$ are close to each other (38.0 and 35.5 % respectively). These data do not allow to explain the predominant interaction of F_2 domain with membrane. This however may be explained by presence in the F_2 domain of some elements of the secondary structure which are necessary for such interaction.

Calculation of secondary structure elements for cytochrome P-450$_{scc}$ performed according to the method of Ptisyn, Finkelstein (15) allowed to detect the possible α-helix and β-sheet segments in the polypeptide chain of cytochrome P-450$_{scc}$ (Fig. 7).

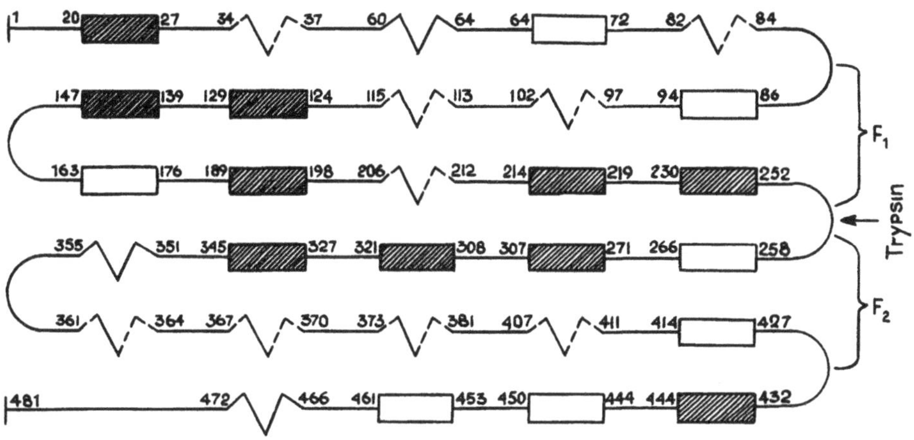

Fig. 7. Elements of secondary structure of adrenocortical cytochrome P-450scc.

- α-helix (probability 40%)
- α-helix (probability 20%)
- β-sheet (probability 40%)
- β-sheet (probability 20%)

It is evident that extended α-helices having more than three coils, which is thought to be more preferable for interaction with phospholipid membrane, are located in central- and C-terminal sequences of cytochrome P-450$_{scc}$ polypeptide chain; this does not contradict the above mentioned proposal that domain F_1 of cytochrome P-450$_{scc}$ is to a large extent exposed into aqueous phase.

Conclusion

The data presented in the present work allow to make the following conclusions.

1. Determination of the primary structure of adrenocortical cytochrome P-450 by protein chemistry methods finally proves the existence of a single functional subunit of hemeprotein.

2. Limited proteolysis of cytochrome P-450 $_{scc}$ is followed by cleavage of polypeptide chain at R(250) - N(257) region giving rise to F_1 and F_2 domains; the further cleavage of F_2 domain at R(399) results in domain F_3; cytochrome P-450 inactivation is followed by appearance of a new accessible to trypsin site at R(92).

3. The site responsible for interaction with intermediate electron transfer carrier - adrenodoxin is located in domein F_1.

4. The catalytic center of cytochrome P-450$_{scc}$ is formed by both domains, since domain F_1 is heme-containing and cysteine residue which is thought to be axial heme ligand is located in the domain F_2; such organization of cytochrome P-450 explains our failure to separate the domains with retention of the native enzyme properties.

5. Of the two cysteine residues of the F_2 domain, Cys (422) appears to be involved in heme thiolation.

6. Of the domains of cytochrome P-450$_{scc}$, the F_1 domain is predominantly exposed to matrix, while domain F_2 preferably interacts with the phospholipid membrane.

References

1. Hechter, O. , M.M. Solomon, A. Zaffaroni, G. Pinkus. 1953. Arch; Biochem. Biophys. 46. 201.
2. Simpson, E.R., G.S. Boyd. 1967. Biophys. Res. Commun. 28. 945.
3. Akhrem, A.A., V.N. Lapko, A.G. Lapko, V.M. Shkumatov, V.L. Chaschin. 1979. Acta biol. et med. germ. 38, 257.
4. Mitani, F. 1979. Mol. Cell. Biochem. 24, 21.
5. Kimura, T. 1981. Mol. Cell. Boichem. 36, 105.
6. Churchill, P.F., L.R. DeAlvare, T. Kimura. 1978. J. Biol. Chem. 253, 4924.
7. Seybert, D.W., J.R. Lancaster, J.D. Lambeth, H. Kamin. 1979. J. Biol. Chem. 254, 12088.

8. Akhrem, A.A., V.I. Vasilevsky, T.B. Adamovich, A.G. Lapko, V.M. Shkumatov, Chashchin V.L. 1980. Dev. Biochem. 13, 57.

9. Chashchin, V.L., V.I. Vasilevsky, V.M. Shkumatov, A.A. Akhrem. 1984. Biochim. Biophys. Acta. 787, 27.

10. Chashchin, V.L., I.V. Turko, A.A. Akhrem, S.A. Usanov. 1985. Biochim. Biophys. Acta. 828, 313.

11. Chashchin, V.L., V.I. Vasilevsky, V.M. Shkumatov, V.N. Lapko, T.B. Adamovich, T.M. Berikbaeva, A.A. Akhrem. 1984. Biochim. Biophys. Acta. 791, 375.

12. Chashchin, V.L., A.G. Lapko, T.B. Adamovich, V.N. Lapko, N.S. Kuprina, A.A. Akhrem. 1983. In: Abstracts of VI All-Union Symposium on Chemistry of proteins and peptides. Riga, U.S.S.R., 11.

13. Morohashi, K., Fujii-Kuriyama Y., Y. Okada, K. Sogawa, T. Hirose, S. Inayama, T. Omura. 1984. Proc. Natl. Acad. Sci. USA, 81, 4647.

14. Usanov, S.A., I.A. Pikuleva, V.L. Chashchin, A.A. Akhrem. 1984. Biochim. Biophys. Acta. 790, 259.

15. Ptitsyn, O.B., V. Finkelstein. 1983. Biopolymers. 22, 15.

STUDIES ON THE STRUCTURE OF PROTEIN HORMONES

Yu.A. Pankov, V.Yu. Butnev, V.S. Karasev, V.I. Pozdnyakov
Institute of Experimental Endocrinology and Hormone Chemistry, USSR Academy of Medical Sciences, Moscow, USSR

Introduction

Study of the relationship between biological structures and their functions has been constantly one of the most important problem at various stages of biology development. Study of the relationship between chemical structure and function of natural compounds participating in various life processes is an important part of this problem. Hormones are a large group of such compounds, and, therefore, scientists constantly paid great attention to studying their chemical structure.

A number of approaches have been used in our studies on structural and functional relationships of protein hormones. One of them is investigation of amino acid sequences in hormones isolated from endocrine glands of various species. In the course of long-term studies the primary structures of several pituitary hormones have been elucidated including those of porcine (1), bovine (2) and whale β-lipotropins (3), bovine and whale prolactins (4,5, 6,7), whale corticotropin (8,9) and growth hormone (10,11,12) and sperm-whale luteinizing hormone (13,14).

Results

Table 1 presents amino acid sequences of α- and β-subunits of sperm-whale luteinizing hormone and their comparison with the subunits of luteinizing hormones of other species. α- and β-subunits of this hormone consist of 96 and 118 amino acid residues, respectively. From comparison of primary structures of the sperm-whale LH subunits with those of human, ovine, bovine and porcine hormones the following conclusions have been drawn. The most pro-

Table. COMPARISON OF THE AMINO ACID SEQUENCES OF α- AND β-SUBUNITS OF SPERM-WHALE LH WITH THE SEQUENCES OF LH SUBUNITS OF OTHER SPECIES (only replaced or deleted amino acid residues are shown).

Species	Amino acid sequences of α-subunits of LH
	10 20 30 40 50
sperm-whale	FPNGZFTMQGCPZCKLKQNKYFSKLGAPIYQCMGCCFSRAYPTPARSKKT
human	V D E T QE PF QP L L
ovine	Q Q E PO
	60 70 80 90 96
sperm-whale	MLVPKNITSZATCCVAKAFTKATVMGNARVQNHTZCHCSTCYYHKS
human	Q V ES SYNRV GFK E A
ovine	E V E E

Species	Amino acid sequences of β-subunits of LH
	10 20 30 40 50 60 70
sperm-whale	PRGPLRPLCRPINATLAAQNZACPVCITFTTSICAGYCPSMVRVLPAALPPVPZPVCTYRQLRFASIRLP
human	S E W H I V EG VN T - Q V L Q- DV E
porcine	S Z DE R Q E I S
ovine	S Q EKE K VI M QR HE V
	80 90 100 110
sperm-whale	GCPPGVNPMVSFPVALSCHCGPCRLSSSDCGPGRAQPLACDRSPRPGL-
human	R D V R R T GPKDH T QPHL----
porcine	B T P L L
ovine	D TE HP L DIL

nounced quantitative discrepancy seems to exist with regard to LH structures of man and sperm-whale which differ in 23 and 35 amino acid residues in α- and β-subunits respectively. Only minor differences have been found in the structures of sperm-whale and porcine LH (2 residues in α-subunits and 12 in β-subunits).
A greater evolutionary conservativity of α-subunit as compared with β-subunit suggests its important role in biological action of LH. Moreover, since the α-subunit structure of LH, FSH and TSH is identical within a separate animal species, it is thought to serve as an effector or "actone" triggering a sequence of biochemical reactions that ensure biological action of all glycoprotein hormones. The specificity of hormonal action depends on peculiarities of metabolism in different organs and tissues that selectively absorb LH, FSH and TSH from the blood while information on the affinity of glycoprotein hormones to different target organs is retained in the structure of their β-subunits.
The other pituitary protein hormone which attracts attention of scientists is prolactin which possesses a wide range of biological activities. Prolactin has been isolated from pituitaries of different animal species and man. To date, the primary structures of human (15), bovine (4,5,16), ovine (17), pig (18) and rat (19) prolactins have been determined. Prolactin is currently regarded as a simple protein, its peptide chain consisting of 199 amino acid residues with three intramolecular disulphide bridges.
In recent years, along with primary structure studies, an increasing attention has been paid to the heterogeneity of prolactin preparations. Analysis of published results suggests that multiple forms of prolactin with dissimilar chemical and biological characteristics might occur in pituitaries. Some authors report molecular weight heterogeneity of lactogenic hormone (20,21,22,23), others show the existence of nonimmunoreactive forms (24-28) or observe that some forms possess enhanced biological activity (24,29). The evidence available suggests that the forms discovered are not artifacts but native prolactins with different properties.
In our studies at the final stage of isolation and purification on Sephadex G-100 prolactin preparations extracted from porcine and whale pituitaries were separated into four peaks, the 2nd, 3rd and 4th having leucine at the N-terminus and possessing

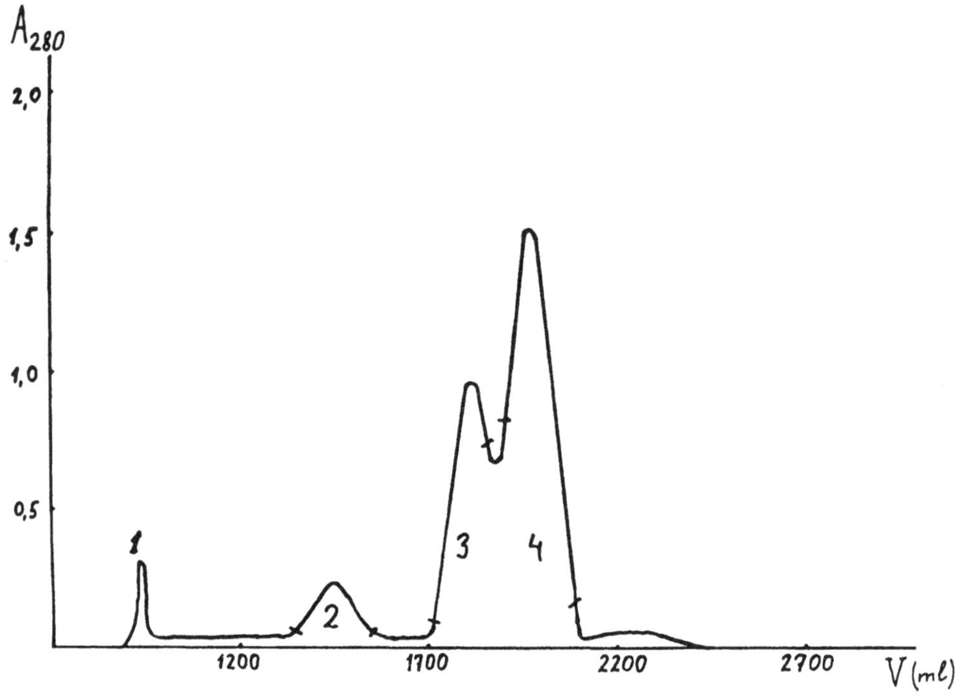

Fig. 1. Chromatography of porcine prolactin extraction on Sephadex G-100 column (4,5 x 200cm).

biological activity (Fig. 1). Our results also showed that the material of the 3rd peak had much higher lactogenic activity than that of the 2nd and 4th ones.

The electrophoretic study of the material of 3rd and 4th peaks in polyacrilamide gel with SDS revealed that each peak gave a single band which indicates the homogeneity of their molecular weight (Fig. 2). On reduction with β-mercaptoethanol the mobility of protein of the 3rd peak decreased to a smaller extent than the mobility of the 4th one.

Thus, the assumption that the 3rd peak may represent a "cleaved" form was not confirmed, for in that case on reduction of disulphide bonds by β-mercaptoethanol the major band was to be replaced by another two bands of lower molecular weights.

The mobility dependence upon log molecular weight plotted with the help of markers showed that proteins of the 4th and the 3rd

Fig.2. Electrophoresis of prolactins in polyacrilamide gel with SDS, 1 and 5-23K prolactin, 2 and 6-25K prolactin respectively after and before reduction with β-mercaptoethanol.

peaks reduced with β-mercaptoethanol had molecular weights about 23,000±2,000 and 25,000±2,000 respectively. Accordingly, the 4th peak that was the prevailing form of hormone and the 3rd peak were designated as 23K and 25K prolactins. These data do not confirm the assumption that the 3rd peak is a dimer of the 4th one. Then the only possible alternative is that these two forms had differences in the amino acid sequences or else 25K was a modified form of prolactin, the modification accounting for the increase in its molecular weight by 2000.

Amino acid analysis did not reveal any remarkable differences between 23K and 25K prolactins. To elucidate the cause of the molecular weight differences among the isolated forms, their cleavage with cyanogen bromide was performed. A porcine prolactin molecule contains 4 methionine residues and therefore is cleaved into 5 peptide fragments.

All 5 fragments were isolated and purified by gel filtration on Sephadexes, ion-exchange chromatography on carboxymethylcellulose and partition chromatography on paper. The only two fragments consisting of 36 amino acid residues and derived from the N-terminal part of the two forms of prolactin revealed essential difference in molecular weights. The peptides corresponding to amino acid sequence 29-36 of both prolactins were isolated by gel filtration from chymotriptic hydrolysates of their N-terminal cyanogen bromide fragments. The peptide from the 25K prolactin had much higher molecular weight than one could expect from its amino acid sequence and much higher than the same peptide obtained from 23K prolactin. Amino acid analysis of that peptide revealed glucosamine and galactosamine that suggested the presence of a carbohydrate moiety in the peptide. Study of this peptide with subtractive Edman procedure showed that carbohydrate moiety was attached to 25K prolactin molecule at the 31 asparagine position.

As shown above (Fig.1), the material of the 2nd peak obtained during gel filtration of prolactin extraction on Sephadex G-100 also possessed biological activity and had leucine as the N-terminal residue. An electrophoretic study showed that this fraction was heterogenous. Being homogenous upon rechromatography of Sephadex, it gave 5 bands during electrophoresis in PAAG with SDS. Upon reduction of disulphide bonds with β-mercaptoethanol three lower bands were lacking but two new bands appeared at the position at which 23K and 25K prolactins were recovered. This led us to believe that the three lower bands should be cross-reacting disulphide dimers: 23K with 23K, 25K with 23K and 25K with 25K.

As is known, the active sites of hormones are localized at the surface of the protein molecule. Presumably, post-translational modifications induce certain changes in the conformation of protein hormones, including their effector sites. Consequently, hormonal effects may be enhanced on some target organs and reduced on others which is a possible basis of functional "specialization" of different forms. In current terms most evolutionary conservative regions of the peptide chain are involved in the formation of biologically active hormonal structures of prolactin. In prolactin these regions correspond to amino acid sequences 15-47, 72-87 and 162-199. Conceivably, glycosylation of asparagine at 31 position, i.e. in one of the regions involved in the formation of the effector moiety of the molecule, is not occasional. This appears to account for the higher biological activity of 25K as compared to 23K prolactin since there are no differences between the primary structure of the two forms.

Sinha et al.(27) also isolated a highly bioactive but nonimmunoreactive form of prolactin, having higher molecular weight but no reactivity against an antiserum to the standard prolactin.
To date, quite a number of authors (25,26,28) have reported discrepancies of data on prolactin in biological fluids measured by radioimmunoassay and bioassay methods. The majority of authors share the view that there are nonimmunoreactive forms of prolactin with enhanced bioactivity. It is an interesting fact that in humans suffering from idiopathic galactorrheas without amenorrheas, which is the case of a pronounced prolactin effect serum shows a normal level of immunoreactive prolactin (31). Possibly, these pathologic states are characterized by a conversion of prolactin into a nonimmunoreactive form which is involved in the pathogenesis. It may be assumed that glycosylation makes the antigenic determinant of prolactin less accessible and affects the immune properties of the hormone.
In this respect the isolated glycosylated form of the lactogenic hormone is of special interest, as it appears to be not a mere hormone modification product, but also to have their own function in the organism, distinguished from that of prolactin itself.
Recently Suh and Frantz (22) showed that two forms of prolactin, "big" and "little", occurred in human plasma. The "little" form had the molecular weight close to that of the monomer and accounted for 80-92% of total prolactin. The molecular weight of the "big" form, 56,000, gave ground to believe that it is a prolactin dimer. This form constituted 8-20% of the overall prolactin. It was observed that the ratio of "big" to "little" forms increased during pregnancy, which was suggestive of some physiological significance of that ratio. It was also reported that the "big" prolactin in human plasma could be almost entirely transformed into monomer using β-mercaptoethanol. All this indicated that disulphide dimers did not arise during hormone extraction but were native physiological forms of prolactin present in pituitaries and secreted into blood.
Investigations of different forms of prolactin have just been initiated, and we may reasonably expect them to shed light on some important problems of hormone involvement in the regulation of metabolic processes and induction of several pathologic conditions.

Nowadays a great success has been achieved in the study of nucleic acid sequences and gene structures. More efficient approaches allow to elucidate the primary structure of nucleic acids, and amino acid sequences of many protein hormones and their precursors very quickly.

Voluminous information about the protein structure made it necessary to use modern electronic computers to devise banks of amino acid sequences for the analysis and interpretation of these findings. Adopted at this Institute is the computerized system which

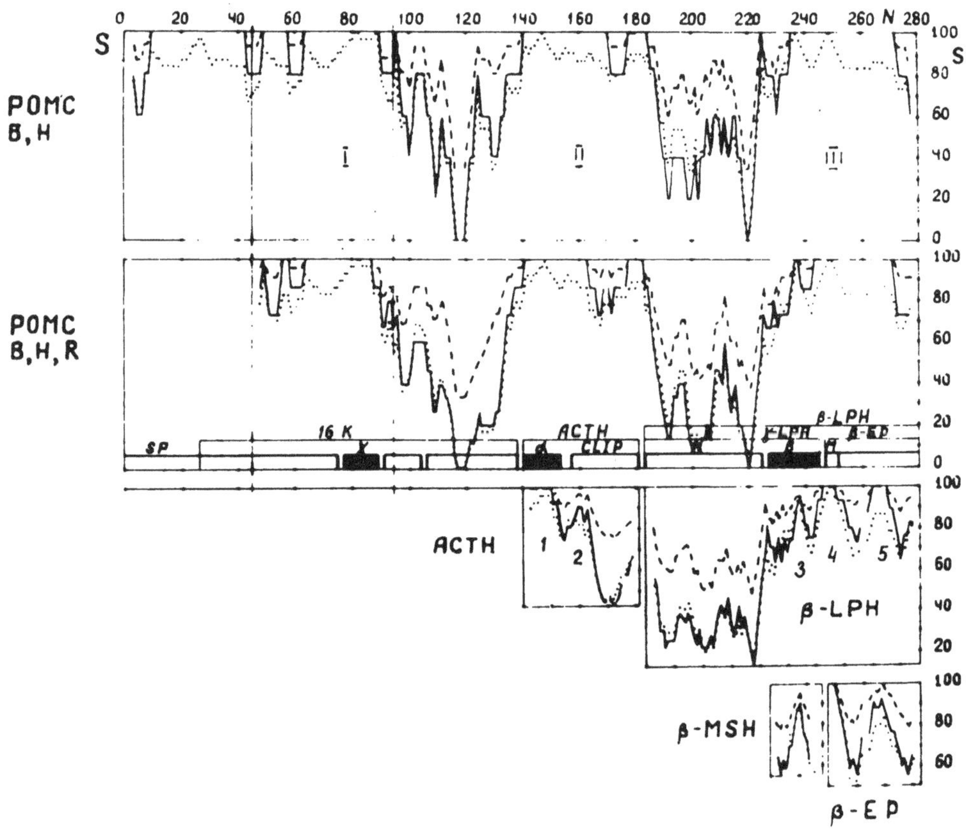

Fig.3. The profiles of local similarity within the amino acid sequences of pro-opiomelanocortin (POMC) family. Solid, broken and dotted lines indicate profiles of local similarity calculated according to the matrices of amino acid similarity 0-1, 0-3 and 0-6, respectively.

incorporates a set of programmes in FORTRAN as designed by V.I. Pozdnyakov (32-35). These programmes facilitate distributing amino acid sequences between homologous families and superfamilies, drawing maps of local similarity between two amino acid sequences to be compared, finding optimal alignments of amino acid sequences within families and superfamilies to reveal the most probable deletions and insertions, examining their evolutionary properties by plotting profiles of local similarity, analysing a number of other characteristics of the hormones under investigation. The programmes were applied to perform comparative sequence analysis in the group of precursors including pro-opiomelanocortin, pro-enkephalin and pro-dynorphin (32,33). All the data available on the structure of pro-opiomelanocortins and their fragments having been summarized, profiles of local similarity between amino acid sequences of appropriate proteins were computerized as presented in Fig.3. Analysis of these profiles has revealed coincidence between similarity maxima (evolutionary stability) and active fragments (active centres) in the structure of hormones of the pro-opiomelanocortin group. There are two such peaks in ACTH and β-endorphin molecules, three in β-lipotropin and one in β-MSH.

When analysing the local similarity maps for the amino acid sequences of pro-enkephalin and pro-dynorphin the latter have been shown to contain three known opiate peptides: β-neoendorphin, dynorphin and rymorphine which occupy the same sites as the three enkephalins within the pro-enkephalin chain. Simultaneously, a zone of amino acid sequence has been discovered, homologous to the above peptides and referred to as phorphin (Fig.4). This zone is not easily observable, and other authors failed to notice it using conventional methods. Computer analysis performed at this Institute unequivocally testifies to the high degree of similarity between this zone and both other opiate peptides within the prodynorphin structure and the corresponding section of the pro-enkephalin amino acid sequence which codes for the structure of Met-enkephalin. The data obtained indicate that this coincidence is not occasional and suggest a possible physiological potency of phorphin which is flanked by solitary basic amino acid residues within the peptide chain of the precursor.

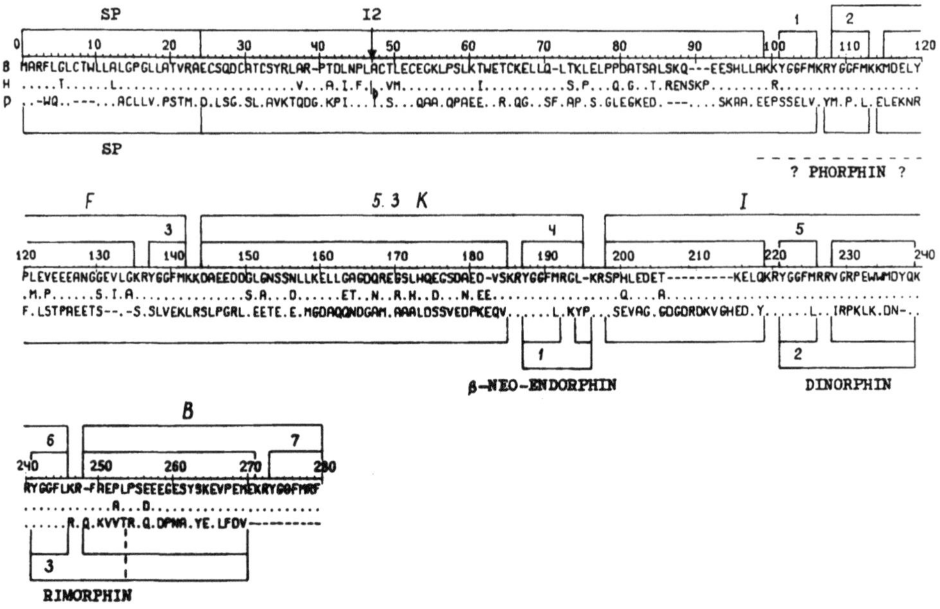

Fig.4. The alignment of the amino acid sequences in the superfamily of pro-opiomelanocortin and pro-dynorphin. The location of phorphin in porcine pro-dynorphin structure.

The results of comparison of phorphin with other peptides are presented in Fig.5. It differs from Leu-enkephalin in that a proline residue is substituted for glycine and an additional methionine residue is introduced which makes the structure more rigid and increases hydrophobic properties of the molecule. Moreover (which may be of particular interest), phorphin is somewhat similar to melanotropins, viz. γ-melanotropin, the two peptides having four common residues out of six, with one deletion. Thus, phorphin like γ-melanotropin is likely to serve as a link between corticotropins and melanotropins on the one hand and opiate peptides on

phorphin	YMGPFL	phorphin	Y-MGPFL
Leu-enkephalin	Y-GGFL	γ-melanotropin	YVMGHFRWDRF

Fig.5.

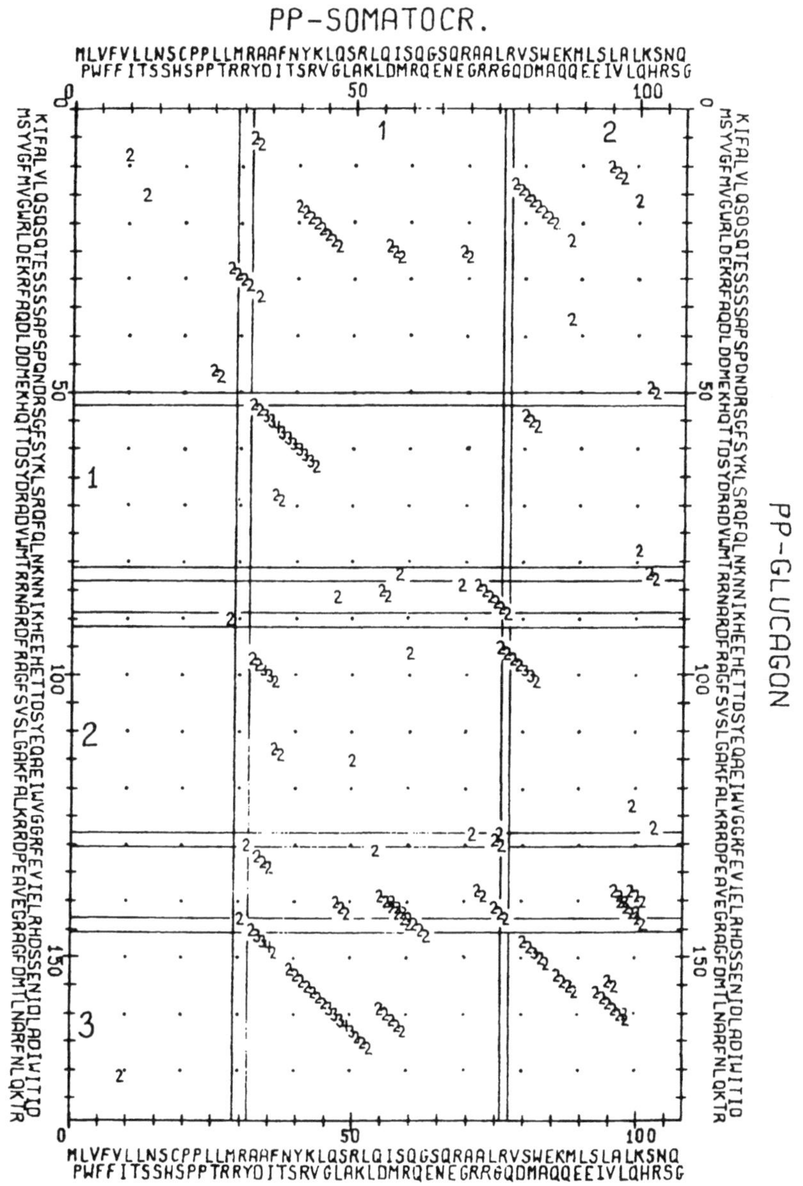

Fig.6. The map of local similarity in the amino acid sequences of pro-somatocrinin and pro-glucagon. The matrix of amino acid similarity is 0-28. The lengths of peptide fragments under comparison are 11 amino acid residues. Figures 2,3,4 and others indicate locations of similar fragments for which probabilities are between 10^{-2}–10^{-3}, 10^{-3}–10^{-4} and so on.

the other hand. This may be considered as indication of their common origin from a single precursor in the course of evolution as a result of gene duplication. In view of this finding coding corticotropins and opiates by the same gene within the pro-opiomelanocortin structure does not seem occasional and may have some biological implications. The data obtained point to feasibility of search and isolation from biological materials of phorphin-like peptides, as well as their chemical synthesis, and studies on biological activity.

Analysis of homologous sections in amino acid sequences of various peptides belonging to the pro-insulin and pro-glucagon superfamilies has brought about some interesting results (34). Fig.6 demonstrates the local similarity map for pro-somatocrinin and pro-glucagon. Somatocrinin appears to be highly homologous to the three glucagon-like sequences that occur in the pro-glucagon chain. At the same time, a peptide fragment is observable at the carboxyl end of pro-somatocrinin homologous to glucagon-like peptides, the highest degree of homology being recorded between this C-terminal peptide and the third glucagon-like sequence within the pro-glucagon structure. These findings suggest the existence of a peptide at the carboxyl end of pro-somatocrinin which is released from its protein precursor together with somatocrinin and like two other peptide fragments of pro-glucagon is supposed to perform some biological function to be elucidated.

Interesting data have been obtained by application of computer analysis to the assessment of local similarity between two variants of the precursor of mouse epidermal growth factor (35) (Fig.7). The precursor proves to be very long as compared to the growth factor that contains 53 amino acid residues. This suggests that its structural units are likely to recur though identical fragments are lacking in this molecule opposite to those of pro-enkephalin and pro-dynorphin. Scott J. et al.(36) have reported a family of 8 fragments similar to epidermal growth factor which in fact closes this recurrent structure. We have termed them as egfins (from EGF - epidermal growth factor). However, the symmetry of tracks on the local similarity map we have computed is indicative of the presence of another family within the precursor in addition to Scott's one that we have termed as coegfins. The two

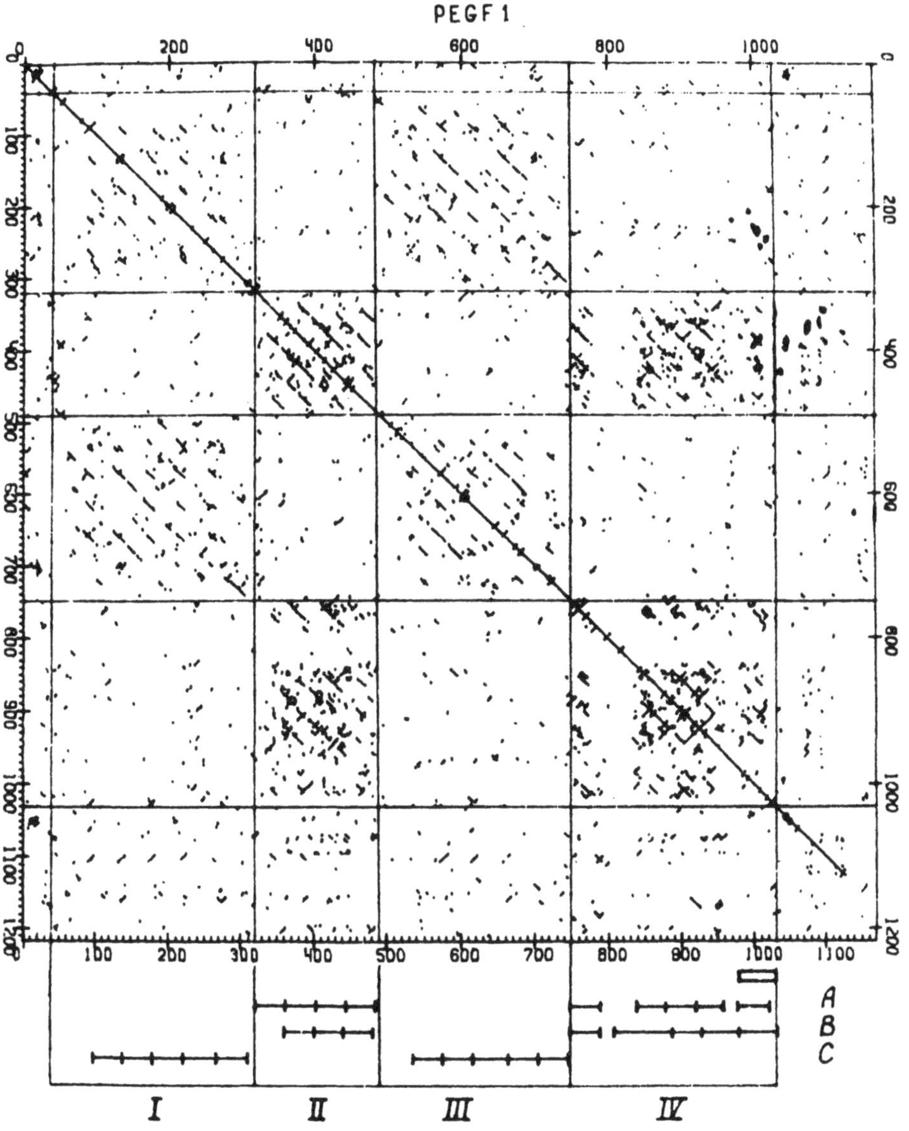

Fig.7. The map of local similarity in the amino acid sequences of epidermal growth factor precursors (PEGF). The matrix of amino acid similarity is 0-28. The lengths of compared fragments are 11 amino acid residues. Dots indicate the location of peptides for which the values of similarity are higher than 150. Rectangle indicates locations of EGF.

families are different though consist of approximately the same number of similar sections. The map demonstrates that both families contain two interpenetrative recurrent groups that alternate beginning from the amino-terminus of the precursor in such a way that the first group of family (coegfins) is followed by the first group of Scott's one (egfins), next to it comes the second group of the former family (coegfins) and finally the second group of the latter (egfins).
A current concept of molecular endocrinology suggests that peptide hormone precursors are polyproteins or polyhormones with their active centres coinciding with homologous fragments of the structure. Bearing this in mind there are reasons to believe that peptides corresponding to the two families with recurrent structural patterns and formed after selective proteolytic cleavage of the growth factor precursor (which does not necessarily affect pairs of positively charged amino acids) can possess some biological activity that remains unknown. Fig. 8 represents the structure of these recurrent fragments having N-terminal sequences of the **LFWTD** or **IYWTD** types that may be responsible for the biological activity of corresponding peptides that have been termed as coegfins.
Thus, recent progress in science induces a radical turn in the strategy of search and investigation of biologically active peptides. Formerly, the sequence of events included a discovery of a new biological function as an initial step followed by preparation of raw tissue extracts, their purification, obtaining of

Fig.8. The alignment for coegfins .

pure biologically active product and as a final step elucidation of its chemical structure. Nowadays, information about the chemical structures of proteins or peptides seems to be obtained in the first place, prompting research on their biological function.

References

1. Pankov, Yu.A., N.A. Yudaev. 1972. Biokhimiya (in Russian) 37, 991-1004.
2. Pankov, Yu.A. 1973. Voprosy Medicinskoy Khimii (in Russian) 19, 330-332.
3. Pankov, Yu.A., O.P. Bochkova. 1985. Biokhimiya (in Russian) 50, 543-553.
4. Yudaev, N.A., Yu.A. Pankov, G.P. Elizarova, Z. Han, O.P. Nikolaeva. 1975. Bioorganicheskaya Khimiya (in Russian) 1, 97-112.
5. Pankov, Yu.A., G.P. Elizarova, O.P. Nikolaeva. 1976. Voprosy Medicinskoy Khimii (in Russian) 22, 360-363.
6. Elizarova, G.P., L.I. Kolesnichenko, O.P. Bochkova, Yu.A. Pankov. 1982. Biokhimiya (in Russian) 47, 515-527.
7. Karaseva, L.I., Yu.A. Pankov. 1985. Biokhimiya (in Russian) 50, 1528-1535.
8. Pankov, Yu.A., O.P. Nikolaeva, G.P. Elizarova. 1976. Bioorganicheskaya Khimiya (in Russian) 2, 855-856.
9. Pankov, Yu.A., O.P. Nikolaeva, G.P. Elizarova. 1977. Biokhimiya (in Russian) 42, 2044-2050.
10. Pankov, Yu.A., A.A. Bulatov, T.A. Osipova, Yu.M. Keda, A.L. Sinitsina. 1976. Biokhimiya (in Russian) 41, 2047-2055.
11. Yudaev, N.A., Yu.A. Pankov, A.A. Bulatov, T.A. Osipova. 1982. Biokhimiya (in Russian) 47, 1059-1069.
12. Pankov, Yu.A., A.A. Bulatov, T.A. Osipova. 1982. Int.J.Peptide Protein Res. 20, 396-399.
13. Pankov, Yu.A., V.S. Karasev. 1984. Biokhimiya (in Russian) 49, 111-126.
14. Pankov, Yu.A., V.S. Karasev. 1984. Biokhimiya (in Russian) 49, 1004-1018.

15. Shome, B., A. Parlow. 1977. J.Clin.Endocr.Metab. 45, 1112-1115.

16. Wallis, M. 1974. FEBS Letters 44, 205-208.

17. Li, C.H., J. Dixon, T.-B. Lo, K.D. Schmidt, Y.A. Pankov. 1970. Arch.Biochem.and Biophys. 141, 705-737.

18. Li, C.H. 1976. Int.J.Peptide Protein Res. 8, 205-224.

19. Parlow, A., B. Shome. 1976. Federat.Proc. 35, 219.

20. Lawson, D.M. 1980. Proc.Soc.Exptl.Biol.and Med. 165, 364-367.

21. Swearingen, K.C. 1971. Endocrinology 89, 1380-1388.

22. Suh, H.K., A.G. Frantz. 1974. J.Clin.Endocr.Metab.39, 928-935.

23. Wallis, M., M. Daniels, Sh.A. Ellis. 1980. Biochem.J. 189, 605-614.

24. Sinha, Y.N., S.R. Baxter. 1979. Biochem.and Biophys.Res. Communs. 86, 325-330.

25. Asawaroengchai, H., S.M. Russell, C.S. Nicoll. 1978. Endocrinology 102, 407-414.

26. Leung, F.C., S.M. Russell, C.S. Nicoll. 1978. Endocrinology 103, 1619-1628.

27. Sinha, Y.N., S.R. Baxter. 1979. J.Endocrinol. 81, 299-314.

28. Asawaroengchai, H., C.S. Nicoll. 1977. J.Endocrinol. 73, 301-308.

29. Pankov, Yu.A., G.P. Elizarova. 1971. Problemy Endokrinologii (in Russian) 17, 92-95.

30. Butnev, V.Yu., Yu.A. Pankov. 1984. Biokhimiya (in Russian) 49, 1828-1839.

31. Kleinberg, D.L., G.L. Noel, A.G. Frantz. 1977. New Engl.J.Med. 296, 589-600.

32. Pozdnyakov, V.I., Yu.A. Pankov. 1983. Biokhimiya (in Russian) 48, 1939-1953.

33. Yudaev, N.A., Yu.A. Pankov, V.N. Babichev, V.P. Fedotov, Yu.P. Shvachkin. 1984. Voprosy Medicinskoy Khimii (in Russian) 30, 8-15.

34. Pozdnyakov, V.I., Yu.A. Pankov. 1985. Biokhimiya (in Russian) 50, 32-46.

35. Pozdnyakov, V.I., Yu.A. Pankov. 1985. Doklady Akademii Nauk SSSR 282, 215-219.

36. Scott, J., M. Urdea, M. Quiroga, R. Sanchez-Pescador, N. Fong, M. Selby, W.J. Rutter, G.I. Bell. 1983. Science 221, 236-240.

CURRENT PROBLEMS OF SPACE STRUCTURE DESCRIPTION FOR OLIGOPEPTIDE MOLECULES IN SOLUTION

G.V.Nikiforovich, B.Vesterman, I.Sekacis,
J.Bētinš, L.Podinš, B.S.Kataev, J.Balodis

Institute of Organic Synthesis,
Latvian SSR Academy of Sciences, Riga 226006, USSR

Introduction

Interpretation of results of physico-chemical experiments is considerably complicated because all steric parameters A^{exp} experimentally measured for a given molecule must be regarded as mean values obtained for a statistical conformer ensemble existing in solution. Unfortunately, most experimental studies on the space structure of peptides implicitly suggest the presence of a single molecular conformation in solution which must be satisfactorily described by the available experimental data. At the same time, the current methods of energy calculations applied to oligopeptides allow to distinguish, with a fair degree of reliability, between sterically allowed and high-energy conformations (1). However, several limitations inherent in the method (entropy contribution to the conformational free energy, solvent effects, etc. are neglected) make the energy estimates unsuitable for assessment of conformer statistical weights using the Boltzmann distribution.
Therefore, neither experimental nor computational methods per se can solve the problem of space structure description for oligopeptide molecules in solution.
To cope with the situation, we propose combined use of computational and experimental data after the following scheme:

$$\sum_{i=1}^{N} w_i A_{ki}^{calc} \longrightarrow A_k^{exp},$$

where N is the number of sterically allowed conformers found by computations, A_{ki}^{calc} are structural parameter A_k values calculated for each structure, w_i are conformer weights. The proposed approach involves the selection of weights so that the weighted sum

Chemistry of Peptides and Proteins, Vol. 3

of estimated values A_{ki}^{calc} would provide the best fit with A_k^{exp} for all structural characteristics being measured. The merit of the present approach was demonstrated in the case of conformationally rigid (tuftsin cycloanalogue Thr-Lys-Pro-Arg) and conformationally labile (spin-labelled angiotensin) molecules using CD and NMR spectroscopy.

Methods

Synthetic procedure. Spin-labelled angiotensin (SL-AT) was prepared by reacting an angiotensin amide (AT) with a 1,3 fold excess of 2,2,5,5-tetramethyl-pyrroline-1-oxyl-3-carboxylic acid N-hydroxysuccinimide ester (Aldrich). The product was purified by HPLC on a

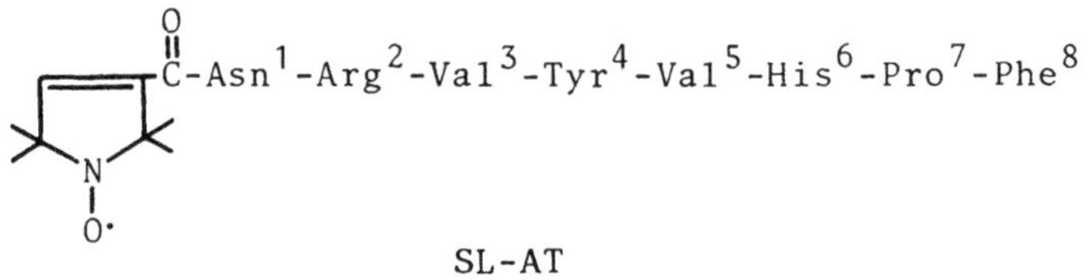

SL-AT

Zorbax C_8 column (Du Pont) using 40% CH_3CN + 60% 0,1 M NH_4OAc for elution. SL-AT yield amounts to 70%. For NMR vicinal coupling constants measurements and fluorescence spectroscopy the nitroxyl radical in SL-AT was reduced by a 50-fold excess of sodium ascorbate, the reduction efficiency being monitored by the intensity of ESR signal.

Cyclotuftsin was synthesized in the Laboratory of Peptide Chemistry (2).

NMR and ESR spectroscopy. NMR spectra were recorded with a WM-360 Bruker-Physik instrument at 360 MHz. For determination of relaxation times solution concentrations did not exceed 1 mM, for coupling constant measurements - 10 mM, temperature 30°C. Spin-lattice relaxation times T_1 were determined using the pulse sequence $180°-\tau-90°-T_\infty$ (3) with calculations according to the equation $I(\tau) = A + B \cdot \exp(-\tau/T_1)$ with concurent assessment of T_1 accuracy carried out on a Wang-2200 VP minicomputer.

ESR spectra were recorded at 10°C to 40°C with a RE1306 instrument (USSR). The rotational correlation time of spin label τ_s was determined as in (4).

Fluorescence spectroscopy. Corrected fluorescence excitation and emission spectra were recorded with a Hitachi-850 spectrofluorimeter. Internal filter effects were corrected. Solutions were prepared with bidistillated water, their concentration did not exceed 40 μM. Intramolecular distances between the centers of aromatic rings of Tyr and Phe (r_{14}) were determined by measuring the efficiency of singlet-singlet energy transfer T between these chromophores (5). T values were estimated by comparing fluorescence excitation spectra of the peptide in study with that of Tyr or Ile^8-AT. The shift in the excitation spectrum of the Tyr chromophore in the oligopeptide with respect to the free amino acid is determined and taken into account.

Energy calculations. Semiempirical conformational analysis was performed for rigid valence geometry using a system of atom-atomic potential functions (6). The proline residue was considered in the trans-conformation (7,8). The value of dielectric constant ε was chosen equal to 80 (SL-AT) or 2.0 (cyclotuftsin, for explanations see (9)). Cycle closure was obtained by using penalty functions of the type described in (6). Combinations of local energy minima in the peptide backbones of type B, R and L corresponding to φ, ψ angle signs -,+; -,- and +,+, respectively, were the "starting points" for intramolecular energy minimization; in some cases, a more detailed classification of the minima was applied (see Table 1).

CD spectrum calculations. The CD spectra were calculated after (10) with parameters being borrowed from (10-12). Only $n\pi^*$ and $\pi\pi^*$ transitions of the peptide chromophores were taken into account with the starting Gaussian bond parameters (position and half-width of a peak) being $\lambda_0 = 215$ nm, $\Delta\lambda_0 = 13.4$ nm; $\lambda_0 = 187$ nm, $\Delta\lambda_0 = 14$ nm, respectively; for tertiary amide $\lambda_0 = 198$ nm, $\Delta\lambda_0 = 14$ nm.

Results and Discussion

Cyclotuftsin: a conformationally rigid molecule. Energy calculations for the analogue Thr-Lys-Pro-Arg (cyclic Lys–Arg) were conducted after the scheme described in (7). The final set contains 14 low-energy conformation types ($\Delta U < 7.8$ kcal/mol). This set can be further subdivided into 5 groups with respect to the signs of dihedral angles in the backbone, responsible for the mutual location of peptide

Table 1. A Set of Low-Energy Backbone Structures Found for a Tuftsin Cycloanalogue *.

No	Backbone structure type	Backbone angles (degrees)									
		Thr[1]		Lys[2]			Pro[3]		Arg[4]		
		ψ	ω	φ	ψ	ω	ψ	ω	φ	ψ	ω
1	B B_2R B_2	170	180	-149	86	-171	-54	161	-121	80	-155
	R B_2R B_2	-60	180	-149	85	-170	-55	160	-121	77	-155
	B B_3R B_2	169	179	-151	134	-178	-39	169	-131	70	-164
2	B L_1R B_2	170	179	59	85	-171	-54	161	-121	77	-155
	R L_1R B_2	-60	179	59	85	-170	-55	160	-121	77	-155
	B L_1R B_3	170	180	61	93	160	-29	173	-105	17	173
	R L_1R B_3	-60	-179	59	94	162	-29	173	-106	16	173
3	B B_2R R_2	148	180	-130	109	-174	-41	167	-115	-56	168
	B B_4R R_1	170	180	-90	146	144	-50	-173	-97	-63	161
	B B_3R R_1	170	180	-150	147	144	-50	-173	-97	-63	160
	R B_4R R_1	-60	180	-90	146	144	-50	-173	-97	-63	161
	R B_3R R_1	-60	180	-150	147	144	-50	-173	-97	-63	160
4	B L_2R R_1	170	-179	63	148	144	-50	-173	-97	-61	160
5	R B_3B L_1	-60	177	-155	157	177	100	-156	50	75	-165

* Indicated are the values of local minima of the potential peptide backbone map, whose centers correspond to the following angle pairs φ and ψ: B_1 (-80, 75); B_2 (-150, 150); B_3 (-150, 45); B_4 (-75, 140); R_1 (-75, -35); R_2 (-160, -60); L_1 (55, 45); L_2 (65, 185).

chromophores (see Table 1). CD spectra were calculated for each low-energy conformation. The estimated spectra of conformations grouped together in Table 1 appear to be qualitatively similar, Fig.1a. Selection of weights for the five groups in Table 1 was performed by minimization of the function

$$[\theta^{exp}(\lambda) - \sum_{i=1}^{5} w_i \theta_i^{calc}(\lambda)]^2$$

for λ ranging from 185 to 245 nm, where θ^{exp} stands for the experimental CD spectrum of the molecule in aqueous solution given in (7); θ_i^{calc} corresponds either to arithmetical means of spectra (Fig.1b)

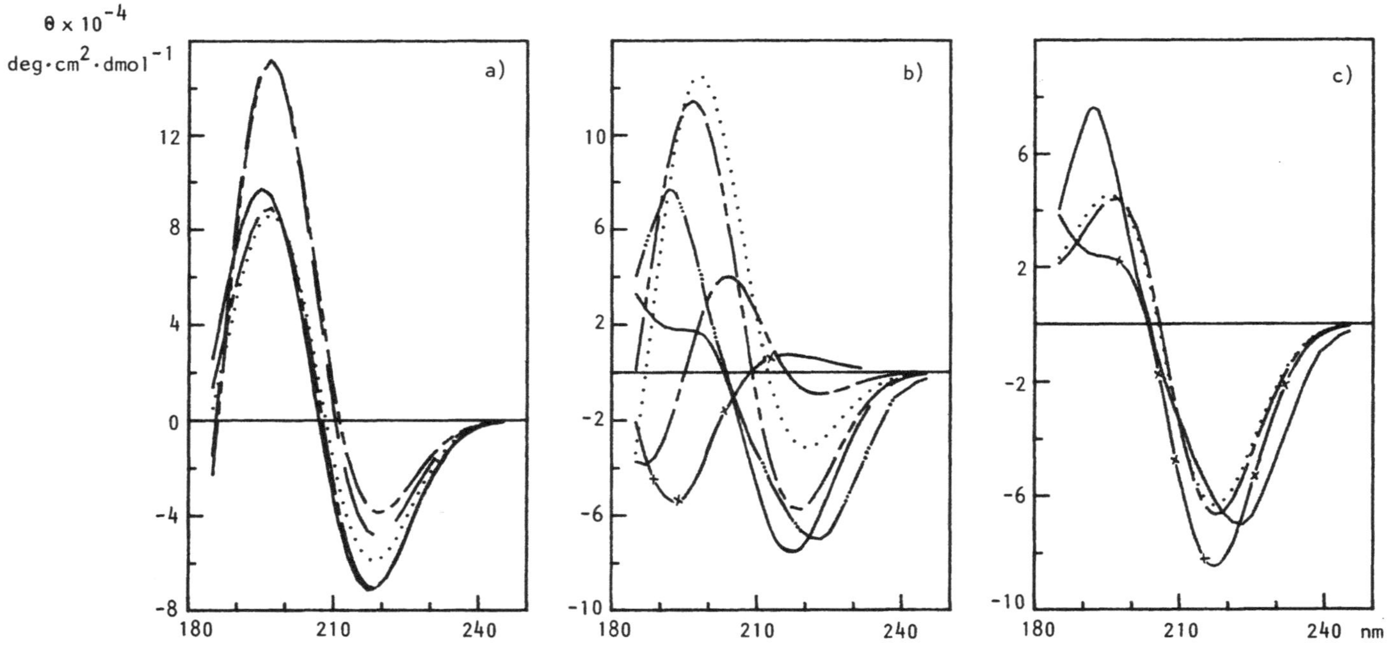

Figure 1. CD spectra of tuftsin cycloanalogue: (a) calculated spectra for BBRR conformations; (b) (—···—) experimental spectrum in water and calculated spectra averaged for BBRB (———), BLRB (—-—), BBRR (—-—-—), BLRR (·····), RBBL (—x—) conformations; (c) (———) experimental spectrum in water and overall calculated CD spectra (—-—) and (·····) using selected set of weights; (—x—) using Boltzmann weight estimates.

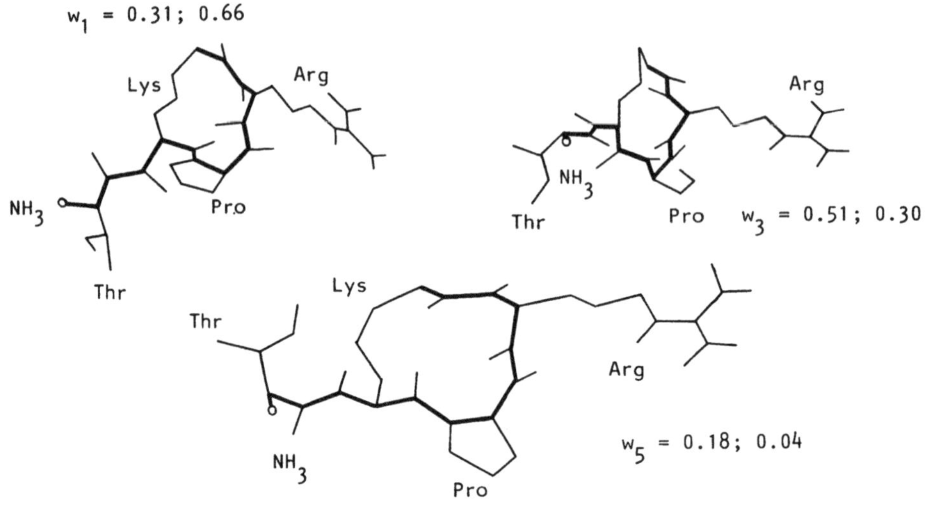

Figure 2. Three sterically allowed cyclotuftsin conformations in solution.

or to estimated spectra of the most stable conformation within each group. The appropriate selection procedure for the two cases in question yields the following sets of weights for the five groups: 0.31; 0.00; 0.51; 0.00; 0.18 and 0.66; 0.00; 0.30; 0.00; 0.04. The corresponding estimated overall CD spectra are presented in Fig.1c. They appear practically identical and are generally in good agreement with the experimental curve, whereas the estimated spectrum obtained with the aid of calculated Boltzmann weights is clearly different from the experimental one. Hence, application of the above procedure leads to the suggestion that a high degree of similarity with experimental CD spectroscopy data is the result of dynamic equilibrium existing between three types of peptide backbone conformations in the aqueous solution (Fig.2).

Spin-labelled angiotensin: a conformationally labile molecule. Computation of SL-AT conformations was based on a similar procedure performed for AT that has revealed 10 sterically allowed peptide backbone structures (13). Introduction of the spin label (valence geometry was borrowed from (14)) increases the number of possible variants for SL-AT. Following the conformational energy minimization, 12 low energy peptide backbone structures ($\Delta U < 10$ kcal/mole) remain, Table 2.

Table 2. A Set of Low-Energy Backbone Structures Obtained by Semiempirical Conformation Analysis for SL-AT *.

No	Backbone structure type	Backbone angles (degrees)														
		Asn[1]		Arg[2]		Val[3]		Tyr[4]		Val[5]		His[6]		Pro[7]	Phe[8]	
		φ	ψ	φ	ψ	φ	ψ	φ	ψ	φ	ψ	φ	ψ	ψ	φ	ψ
1	BBLRRBBB	-138	151	-124	147	61	92	-98	-46	-120	-53	-127	154	150	-114	142
2	BRBRRBBB	-81	99	-132	-57	-113	70	-147	-56	-96	-41	-119	151	164	-150	151
3	BBRRRBBB	-147	115	-157	143	-72	-27	-71	-28	-83	-57	-148	165	130	-138	142
4	BBLRRBRB	-145	146	-80	156	55	39	-161	-57	-102	-36	-145	151	-27	-140	144
5	BRBRRBRB	-78	96	-129	-56	-116	70	-145	-55	-96	-39	-119	151	-39	-148	150
6	BBBRRBBB	-76	92	-144	131	-116	66	-139	-55	-94	-41	-125	152	144	-158	151
7	BBBRRBRB	-150	149	-149	125	-120	153	-62	-30	-72	-49	-155	150	-45	-147	150
8	BBRRBBBB	-152	140	-92	103	-94	-38	-142	-50	-125	140	-62	139	147	-146	146
9	RBLRRRBB	-89	-41	-128	150	60	91	-96	-44	-122	-53	-130	154	150	-114	140
10	RRBRRBBB	-84	-54	-172	-55	-124	105	-146	-56	-109	-46	-113	153	156	-146	146
11	RBLRRBRB	-89	-51	-86	149	58	36	-163	-55	-101	-38	-146	153	-26	-142	144
12	RBRRBBBB	-71	-41	-124	98	-89	-37	-144	-51	-128	135	-66	139	146	-147	147

* Calculations suggest two possible orientations for the label relative to AT molecule, only one of them being generally preferred. Label orientation slightly affects the backbone conformation.

Table 3. Structural Parameters of SL-AT Molecules *.

Parameter	Experiment	1	2	3	4	5	6	7	8	9	10	11	12
		Parameter values for calculated structures											
$1/T_{1M}(H_3^{\alpha})$	14.47 2.03	2.52 0.11	3.86 0.17	3.29 0.31	2.37 0.03	3.88 0.17	3.89 0.19	2.05 0.01	3.69 0.24	9.48 0.17	22.94 0.91	13.16 0.61	54.58 1.79
$1/T_{1M}(H_4^{\delta})^a$	15.58 1.16	9.84 0.93	4.25 1.17	4.75 0.56	12.86 1.79	3.83 1.18	3.22 0.09	0.27 0.01	0.63 0.04	43.32 6.35	268.4 30.2	29.01 5.01	1.73 0.04
$1/T_{1M}(H_4^{\varepsilon})^a$	19.01 0.55	50.85 6.55	5.04 3.36	2.46 0.29	6.70 1.11	3.56 2.32	2.51 0.07	0.14 0.01	0.28 0.03	48.30 6.85	88.64 11.85	14.37 2.52	0.72 0.05
$1/T_{1M}(H_5^{\alpha})$	6.47 0.82	0.66 0.03	0.85 0.06	14.66 3.33	27.92 2.27	0.79 0.08	1.65 0.08	0.32 0.01	1.49 0.08	6.27 0.34	85.50 3.26	3.71 0.25	1.29 0.02
$1/T_{1M}(H_6^{\delta})$	5.95 1.43	0.29 0.01	0.75 0.03	36.87 18.79	89.46 41.24	1.20 0.05	19.01 1.73	0.72 0.04	2.98 0.54	0.66 0.02	10.64 2.28	3.61 0.30	1.34 0.52
$1/T_{1M}(H_6^{\varepsilon})$	8.26 0.84	0.42 0.02	4.08 0.29	14.93 10.26	420.1 170.2	3.12 0.54	92.90 9.01	3.82 0.36	3.15 1.58	0.57 0.02	136.1 41.0	6.53 1.55	0.66 0.21
$1/T_{1M}(H_7^{\beta_1})^b$	5.85 0.73	0.30 0.01	0.32 0.01	0.57 0.08	1.16 0.05	0.31 0.02	1.26 0.07	6.57 0.16	76.68 10.11	0.81 0.04	1.75 0.21	0.94 0.04	1.76 0.10
$1/T_{1M}(H_7^{\beta_2})$		0.39 0.01	0.32 0.01	0.58 0.07	1.19 0.03	0.31 0.01	0.92 0.04	3.88 0.13	92.01 8.85	1.47 0.09	2.07 0.21	1.16 0.07	3.11 0.20
$1/T_{1M}(H_7^{\delta_1})$	3.85 0.70	0.21 0.01	0.36 0.02	1.91 0.44	4.71 0.24	0.35 0.02	2.54 0.18	1.68 0.05	11.46 1.82	0.77 0.02	5.41 1.02	1.09 0.01	2.13 0.18
$1/T_{1M}(H_7^{\delta_2})$		0.26 0.01	0.33 0.01	1.59 0.36	3.88 0.14	0.31 0.01	1.50 0.08	1.22 0.01	15.09 2.24	1.31 0.05	5.36 0.82	1.03 0.02	3.97 0.25
$1/T_{1M}(H_8^{\delta})^a$	5.95 0.47	1.30 0.04	2.68 0.14	3.62 0.38	1.54 0.22	0.49 0.01	21.76 2.52	47.83 4.99	4.44 1.91	1.61 0.03	9.32 1.72	0.81 0.06	0.24 0.02
$T_{48}{}^c$	0.50 0.03	0.95 0.01	0.32 0.04	0.38 0.04	0.13 0.01	0.14 0.01	0.55 0.02	0.20 0.01	0.22 0.04	0.96 0.01	0.60 0.01	0.11 0.01	0.21 0.03

J Asn[1]	7.3	8.2	5.7	9.5	7.3	5.7	6.7	6.2	8.7	6.4	5.3	4.9	4.9
	0.5	0.8	0.6	0.9	0.8	0.8	0.8	0.8	0.8	0.8	0.8	0.8	0.7
J Arg[2]	7.1	6.3	10.0	6.9	6.3	8.4	9.5	7.5	8.1	10.1	5.4	5.2	8.8
	0.3	1.1	0.8	0.7	0.6	1.0	0.8	0.7	0.8	0.8	0.7	0.7	0.8
J Val[3]	8.0	8.6	8.4	6.8	8.7	8.3	10.8	10.8	9.5	8.6	10.6	8.7	7.9
	0.3	0.7	0.8	0.9	0.7	0.8	0.8	0.8	0.8	0.7	0.8	0.7	0.8
J Tyr[4]	7.6	8.9	7.9	5.4	6.6	8.3	9.8	3.8	9.1	9.0	9.5	6.2	7.6
	0.3	0.7	0.7	0.9	0.7	0.8	0.7	0.6	0.8	0.7	0.7	0.7	0.8
J Val[5]	8.2	10.8	8.2	6.5	9.5	8.1	8.5	5.8	9.9	10.8	10.5	9.5	9.4
	0.3	0.8	0.8	0.8	0.8	0.7	0.8	0.7	0.8	0.8	0.8	0.8	0.8
J His[6]	7.0	10.7	10.6	9.2	8.9	10.8	10.4	7.5	2.6	10.3	10.1	8.0	4.5
	0.3	0.8	0.8	0.7	0.7	0.8	0.8	0.7	0.5	0.8	0.8	0.8	0.7
J Phe[8]	7.5	10.6	9.0	9.2	7.8	8.9	6.6	9.2	8.5	10.6	9.6	8.0	8.4
	0.3	0.7	0.8	1.1	0.9	0.8	0.7	0.8	0.9	0.8	0.7	0.9	0.8

* $1/T_{1M}$ was measured in s^{-1}, vicinal coupling constants J in Hz; subscripts correspond to amino acid number; the table cites the means (standard deviations placed below them) of parameters being measured.

[a] T_{1M} values for two equivalent protons were averaged after the fast exchange limit $0.5/T^{a}_{1M} + 0.5/T^{b}_{1M}$.

[b] A more precise assignment of Pro^{7} signals is absent; the values attributable to $C^{\beta_1}H$ and $C^{\delta_2}H$ were chosen as T_{1M}, as being most consistent with experimental data.

[c] Critical Förster distance for Phe-Tyr $R_o = 12.1$ Å.

Statistical samples of SL-AT conformations consisting of 15,000 structures were constructed in the vicinity of each of these minima using Monte-Carlo techniques. This enabled to calculate the mean values and standard deviations for all experimentally measured parameters for each of the 12 structures: $1/T_{1M}$ for 9 protons, the efficiency of singlet-singlet energy transfer T and 7 vicinal coupling constants $J(HNC^{\alpha}H)$. Value T was calculated after the formula from (3), the coupling constants as in (15) and $1/T_{1M}$ after the Solomon-Blombergen equation (16).
To assess the contribution of intramolecular mobility to the dipole-dipole relaxation the spin label rotational correlation time τ_s was used instead of τ_c. Experimental and calculated means and standard deviations for all the structural characteristics being studied are summarized in Table 3. The standard deviations of J values also include errors (≈0.5 Hz) occurring due to uncertainty of the parameters of the Carplus-Bystrov curve (15).
The fitting between calculated mean parameters A_k^{calc} (with standard deviations D_k^{calc}) and experimental values A_k^{exp} was carried out by selection the conformer weights w_i with minimization the object function represented by the sum of Student's coefficients for comparison of Gaussian distributions:

$$\sum_{k=1}^{17}\left\{(A_k^{exp} - \sum_{i=1}^{12} w_i A_{ki}^{calc})^2 / [(D_k^{exp})^2 + \sum_{i=1}^{12}(w_i D_{ki}^{calc})^2]\right\}$$

that the value $\left| A_k^{exp} - \sum_{i=1}^{12} w_i A_{ki}^{calc} \right|$ for each k would not exceed the corresponding 95% confidence limits. The minimization procedure resulted in several sets of weights fulfilling the aforesaid requirements (set I-IV) (see Table 4, 5). Set IV is "minimal", however if we take into account that the limited nature of the constructed statistical conformation samples must lead to underestimated D_k^{calc} values and propose less strict requirements, the sets of weights V, VI (Table 4) consisting of 5 structures can also be considered as satisfactory (see Fig.3). A further reduction in the number of SL-AT structures under consideration leads to drastic discrepancies with experimental data. Therefore, it can be claimed that: (i) none of the SL-AT structures taken alone satisfies the entire set of experimental data; (ii) some molecule structures, especially conformation 3 and 8 from Table 2 are "indispensable"

Table 4. Selected Sets of Conformer Weights for SL-AT.

Set number	Weight values							
	1	3	6	7	8	9	11	12
I	0.04	0.21	0.08	0.05	0.07	0.25	0.12	0.18
II	-	0.26	0.06	0.05	0.07	0.30	0.07	0.19
III	0.03	0.27	0.09	0.04	0.07	0.27	-	0.23
IV	-	0.26	0.08	0.05	0.07	0.32	-	0.22
V	-	0.31	0.12	-	0.09	0.29	-	0.19
VI	-	0.45	-	0.05	0.06	0.30	-	0.14

Table 5. Results of Matching the Computational and Experimental Data.

Parameter	Measured	Calculated for sets of weights from Table 4			
		I	II	III	IV
SL-$C^{\alpha}H$ Val^3, Å	9.8±0.2	9.8±0.1	9.6±0.1	9.6±0.1	9.6±0.1
SL-$C^{\delta}H$ Tyr^4, Å	9.7±0.1	9.7±0.2	9.6±0.2	9.9±0.2	9.7±0.2
SL-$C^{\epsilon}H$ Tyr^4, Å	9.4±0.1	9.6±0.2	9.6±0.2	9.7±0.2	9.7±0.2
SL-$C^{\alpha}H$ Val^5, Å	11.3±0.2	11.5±0.2	11.3±0.2	11.3±0.2	11.3±0.2
SL-$C^{\delta}H$ His^6, Å	11.4±0.4	10.4±0.6	10.2±0.7	10.1±0.7	10.2±0.7
SL-$C^{\epsilon}H$ His^6, Å	10.8±0.2	10.1±0.3	10.3±0.4	10.0±0.3	10.2±0.4
SL-$C^{\beta}H$ Pro^7, Å	11.5±0.2	11.1±0.2	11.2±0.2	11.2±0.2	11.3±0.2
SL-$C^{\delta}H$ Pro^7, Å	12.3±0.3	12.9±0.1	12.9±0.1	12.8±0.1	12.9±0.1
SL-$C^{\delta}H$ Phe^8, Å	11.4±0.2	11.5±0.1	11.5±0.1	11.5±0.1	11.4±0.1
r_{48} Å	12.1±0.2	12.2±0.1	12.1±0.1	11.9±0.1	11.9±0.1
J_1 Hz	7.3±0.5	6.3±0.3	6.4±0.3	6.5±0.3	6.4±0.3
J_2 Hz	7.1±0.3	7.5±0.3	7.7±0.3	7.8±0.3	7.9±0.4
J_3 Hz	8.0±0.3	7.8±0.3	7.6±0.3	7.6±0.3	7.6±0.3
J_4 Hz	7.6±0.3	6.9±0.3	6.8±0.3	7.0±0.3	6.9±0.3
J_5 Hz	8.2±0.3	8.3±0.3	8.1±0.3	8.1±0.3	8.1±0.3
J_6 Hz	7.0±0.3	7.4±0.3	7.4±0.3	7.4±0.3	7.4±0.3
J_8 Hz	7.5±0.3	8.3±0.3	8.4±0.4	8.4±0.4	8.4±0.4

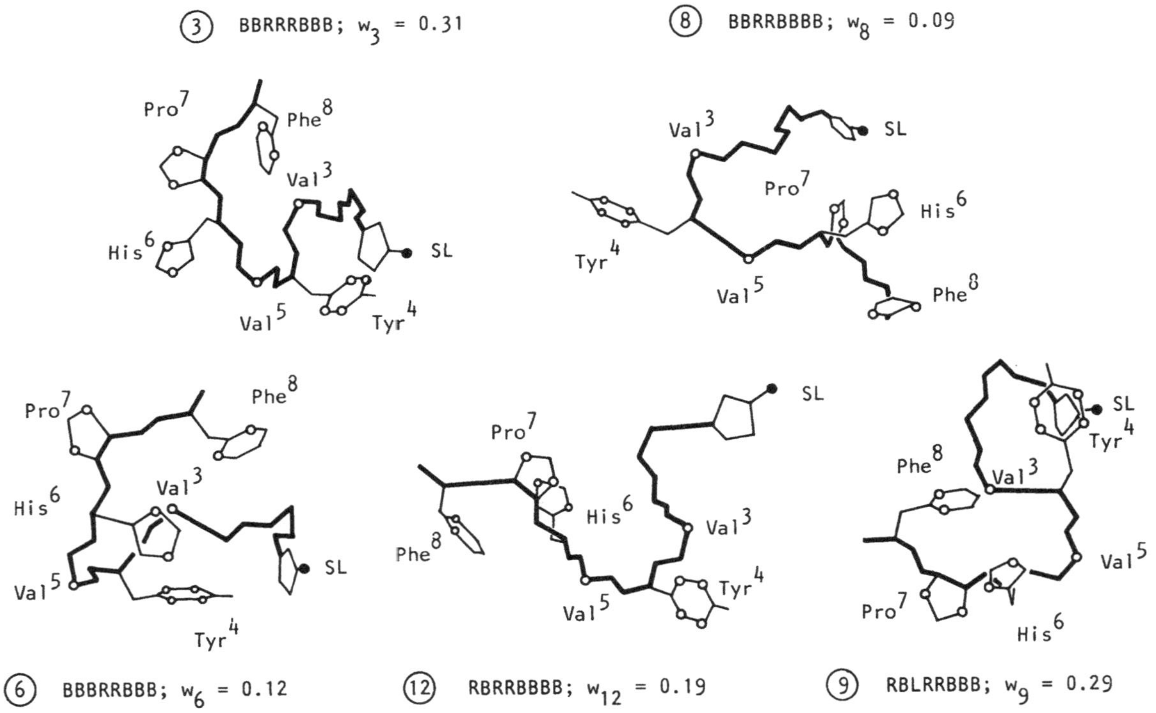

Figure 3. SL-AT structures constituting set V in Table 4. Blank circles designate those protons in the molecule which relaxation is induced by the spin label electron shown as a filled circle.

for reaching agreement with experimental data. It should be pointed out that the results obtained can be equally applied to AT, because, as has been demonstrated, SL-AT shows the same affinity and specificity towards specific antiAT antibodies, as AT.

Conclusions

Agreement with experimental data supports the applicability of the approach proposed for evaluation of conformer weights both for conformationally rigid (cyclotuftsin) and conformationally labile (spin-labelled angiotensin) compounds. In both cases statistically reliable qualitative conclusions are made about the possibility of excluding from the set of the structures "unnecessary" for reaching agreement between experimental and computational data as well as the "indispensability" of certain structures within such set. A complete and,hopefully,the most correct solution of the problem of space structure description for oligopeptide molecule in solution must be represented by the determination of the possible limits of conformer weight variation. The same apparently applies to conformationally mobile molecules of non-peptide structure.

References

1. Nikiforovich, G.V., S.G. Galaktionov, G. Chipens. 1983. Conformations of Peptide Bioregulators. Meditsina, Moscow.
2. Chipens, G.I., N.I. Veretennikova, G.V. Nikiforovich, Z.A. Atare. 1981. In: Peptides-1980 (K.Brunfeldt ed.). Scriptor, Copenhagen, p. 445.
3. Dwek, R.A. 1973. Nuclear Magnetic Resonance (NMR) in Biochemistry. Claredon Press, Oxford.
4. Kuznetsov, A.N., A.M.Wasserman, A.I. Volkov, N.N. Korst. 1971. Chem. Phys. Lett. 12, 103.
5. Förster, Th. 1965. In: Modern Quantum Chemistry (O. Sinanoglu, ed.). Academic Press, N.Y. - London, part III, p. 93.
6. Momany, F.A., R.F. McGuire, A.W. Burgess, H.A. Scheraga. 1975. J. Phys. Chem. 79, 2361.

7. Nikiforovich, G.V., I.T. Liepina, I.P. Sekacis, E.E. Liepins, B.S. Katayev, N.I. Veretennikova, G.I. Chipens. 1984. Int. J. Pept. Prot. Res. 23, 271.

8. Deslauriers R., R. Walter, J.C.P. Smith. 1973. Biochem. Biophys. Res. Commun. 53, 244.

9. Nikiforovich, G.V., S.A. Rozenblit, G.I. Chipens. 1982. In: Chemistry of Peptides and Proteins (W.Voelter, E. Wünsch, J. Ovchinnikov and V. Ivanov, eds.). Walter de Gruyter & Co., Berlin - N.Y., vol. 1, p. 407.

10. Madison, V. 1973. Biopolymers 12, 1837.

11. Madison, V., K.D. Kopple. 1980. J. Am. Chem. Soc. 102, 4855.

12. Bayley, P.M., E.B. Nielsen, J.A. Schellman. 1969. J. Phys. Chem. 73, 228.

13. Nikiforovich, G.V., M.D. Shenderovich, S.G. Galaktionov. 1976. Bioorg. Khimia 2, 1268.

14. Lajzerowicz, J. 1976. In: Spin Labelling. Theory and Applications (L.J. Berliner ed.). Academic Press, N.Y. - San Francisco - London, p. 264.

15. Bystrov, V.F., V.T. Ivanov, S.L. Portnova, T.A. Balashova, Yu.A. Ovchinnikov. 1973. Tetrahedron 28, 873.

16. Solomon, I. 1955. Phys. Rev. 99, 559.

PREDICTION AND SYNTHESIS OF SOME IMMUNOLOGICALLY ACTIVE FRAGMENTS OF INTERFERON MOLECULES

G. Chipens, R. Vegners, O. Papsuevich, N. Ievina, G. Rosenthal, T. Petrova, I. Rituma, G. Afanasyeva, A. Krikis

Institute of Organic Synthesis, Latvian SSR Academy of Sciences, Riga, USSR

Along with the hormone and central nervous system the immune system is one of the most intricate regulatory systems in the living organism, as indicated by the enormous number of compounds involved in immunoregulation. They include hormones from the thymus and other endocrine glands, interleukins, monokines, immunoglobulins, major histocompatibility complex proteins, etc. Most of them exert regulatory functions via immunocompetent cells, which participate in all cellular and humoral immune responses leading to the elimination of genetically foreign material of exo- and endogenic origin.

According to the biochemical universality concept, one might assume that the different groups of immunoregulators listed above rely in their action on related biochemical mechanisms. This must be reflected by common principles of their biogenesis and similar structural elements of their active sites. Peptides and proteins are known to exert their biological action not only as full-length molecules via functionally active sites located on the molecule surface but also as precursors of low-molecular peptides being formed in limited proteolysis reactions, e.g., proopiomelanocortin. It should be pointed out that in either case the active sites of molecules are frequently characterized by quasicyclic or cyclic structures which are stabilized by ionic, hydrogen or disulphide bridges and comprise, on an average, 6 to 8 amino acid residues (1, 2). The profile of the quasicycle/cycle is complementary to the active site ("pocket") of the cell receptor (enzyme or some other receptive molecule), which accounts for the high selectivity of interaction.

Additional evidence in support of this assumption has been obtain-

Chemistry of Peptides and Proteins, Vol. 3

Figure 1.

The peptide chain fragment of IgG Eu domain C_H3

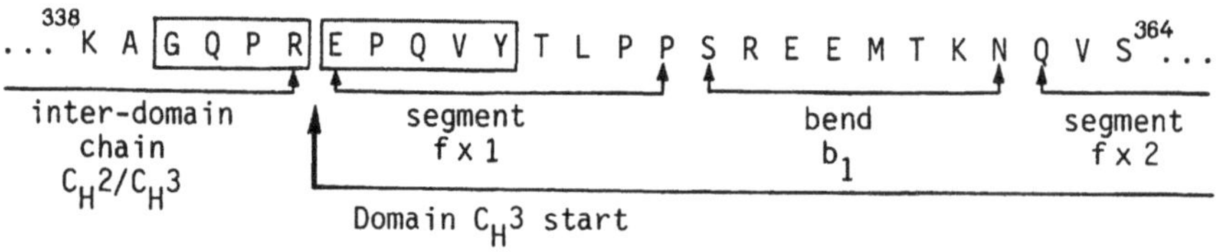

The structures of rigin ($IgG_{340-344}$) and immunopoietin G ($IgG_{345-349}$) are boxed.

ed by examining the possible pathways of biogenesis of the immunomodulatory peptide designated rigin (3) (Fig.1). This approach has also led to the discovery of immunopoietins - a novel class of immunologically active peptides represented by immunoglobulin fragments (Fig.1) (4,5). Rigin is a potent macrophage activator, whereas immunopoietins that share structural similarity with peptides containing the active sites of thymopoietin and splenin:

Immunopoietin G	E P Q V Y
Immunopoietin M	R P D V Y
Immunopoietin E	A A P E V Y
Immunopoietin D	Q P L G V Y
Thymopentin	R K D V Y
Splenopentin	R K E V Y

affect T-lymphocytes. The linkage (juxtaposition) of rigin and immunopoietin G in the IgG molecule (Fig.1) suggests that the amino acid sequences in precursors of immunologically active peptides may code for a certain order in which the reactions of lymphocyte and macrophage activation occur, indicating the crucial role of such linkage in the regulation of immune responses. Since the formation of low-molecular immunoregulatory peptides presumably occurs on the surface membranes of immunocompetent cells during the process of their cooperation, the peptides can be easily transferred to the neighbouring cells resembling the flow of neurotransmitters in synapses, as discussed more elaborately in (4). It has been suggested that immunoregulatory peptides can be formed via this route also from other proteins participating in immune reactions. Such a possibility in relation to interferons (IFN) is

analysed in this paper.
IFN are known to stimulate two main groups of reactions responsible for inhibition of viral proliferation within cells and for regulation of the immune system. IFN fragments, if immunologically active, can be expected to possess enhanced specificity and selectivity of action, lower toxicity and less marked species-specificity as compared to the full-length IFN molecule. They must be particularly useful for elucidation of structure-function organization of IFN molecules. In an attempt to locate potentially active sites in IFN molecules, we performed analysis of their primary structures. Some peptides containing amino acid sequences of the established putative active sites were synthesized and assayed for immunological activity.

Primary structure analysis relies on the assumed equifunctionality of some of amino acids as deduced from the signature and equivocation principles of information theory (1). According to these principles, the biological activity of a compound is determined by a set of properties (signature), similar signatures being shared in certain circumstances by compounds with different structures. For instance, the side chains of Arg, Lys, Asn, Gln can behave equifunctionally upon hydrogen bonding with receptor groups. The Val, Ile, Pro residues, being responsible for the general molecular architecture of a peptide, bend its backbone in a similar fashion. Val and Ile exert their action due to the branching at the C^{β} atom, while Pro can do so owing to the rigidly fixed structure of the pyrrolidine cycle. Gly can also be conformationally equifunctional to Val, Ile and Pro since it unlike the sterically hindered residues acts as a hinge and provides a similar bend (1). The occurrence rates of mutual amino acid substitutions in the course of evolution support the equifunctionality of the side chains differing as to their chemical structure. For instance, based on the matrix of relative substitution frequencies at an evolutionary distance of 256 accepted point mutations per 100 residues, natural amino acids can be arranged in the following sequence (6):

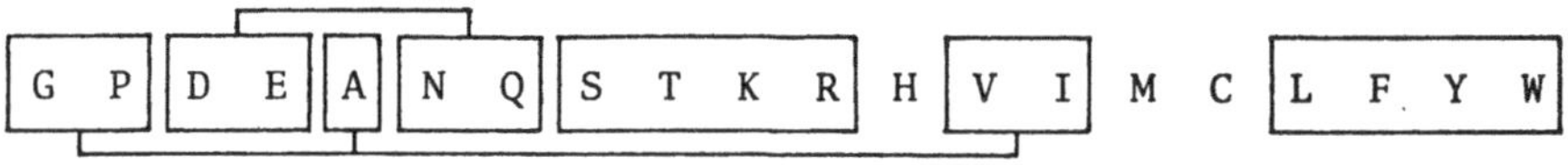

Figure 2. SIMILAR STRUCTURAL FEATURES SHARED BY VARIOUS INTERFERON MOLECULES.

The primary structure of interferon molecules with arbitrary division into distinct fragments

```
                 (1)      |      (2)      |     (3)     |       (4)        |      (5)       |  (6)  |83
IFN-αA      CDLPQTHSLGSRRTLMLLAQMRKISLFSCLKDRHDFGFPQE EF-GNQFQKAETIPVLHEMIQQIFNLFSTKLSSAAWDETLLDK
IFN-β1   MSYNLLGFLQRSSNFQCQKLLWQLNGRLEYCLKDRMNFDIPEE IKQLQQFQKEDAALTIYEMLQNIFAIFRQDSSSTGWNETIVEN 86
IFN-γ           CYCQDPVVKEAENLKKYFNAGHSDVADNGTLFLGIL KNWKEESDRKIMQSQIVSFYFKLFKNFKDDQSIQKSVETIKED 79

  |    (7)     |     (8)     |     (9)     |     (10)     |     (11)     |    (12)
  FYTELYQQLNDLEACVIQGVGVTETPLMKEDSILAVRKYFQRITLYLKEKKYSPCAWEVVRAEIMRSFSLSTNLQESLRSKE
  LLANVYHQINHLKTVLEEKLEKEDFTRGKLMSSLHLKRYYGRILHYLKAKEYSHCAWTIVRVEILRNFYFINRLTGYLRN
  MNVK-FFNSNKKKRDDFEKLTNYSVTDLNVQRKAIHELIQVMAELSPAAKTGKRKRSQMLFRGRRASQ
```

Sketchy depiction of the interferon molecules

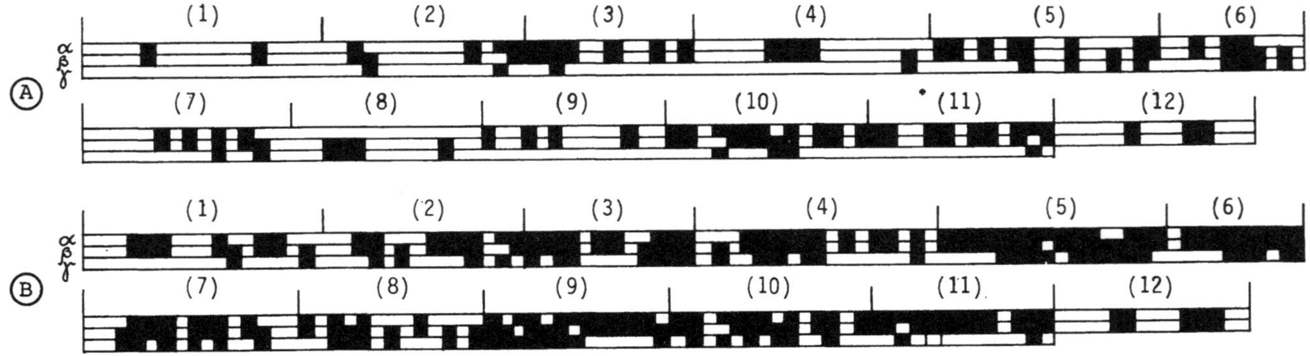

The like amino acid residues are blackened taking into account:

A - solely the identical amino acids, B - both the identical and equifunctional amino acids together.

The amino acids located in close proximity in this sequence have been frequently involved in mutual substitutions during evolution. Framed amino acids and those connected by lines are regarded as potentially equifunctional in this study.

A comparative analysis of the primary structures of IFN-α, -ß and -γ is presented in Fig.2. The IFN sequences are arranged one below the other, the N-terminal amino acid residues of IFN-ß and -γ being removed with respect to IFN-α along the chain to provide maximum coincidence. For the same reason, potential deletions in position 44 (IFN-α) and 88 (IFN-γ) were introduced. The identical and presumably equifunctional amino acid residues are shaded. The analysis reveals marked homology between IFN-α and -ß, as already reported in the literature (7); IFN-γ shows greater structural similarity only in the region containing the amino acid residues 58-128. Comparative analysis allows to divide the primary structures of IFN into twelve arbitrary regions (Fig.2). The boundaries between the regions were selected taking into account the occurrence density of the identical and equifunctional amino acid residues and the location of disulphide bridges, the prevailing molecular size of hormones, tachy- and pachykinins, and other biologically active oligopeptides seldom exceeding more than 10-15 amino acid residues. The selected IFN-α fragments with the numbers of their terminal amino acid residues indicated are as follows:

(1)	1-15	(5)	58-73	(9)	112-124
(2)	16-29	(6)	74-83	(10)	125-138
(3)	30-41	(7)	84-97	(11)	139-151
(4)	42-57	(8)	98-111	(12)	152-165

Amino acid sequence conservation over the course of evolution (as exemplified by fragments (9) and (10) in Fig.3) and the extent of structural similarity are postulated to be a measure of functional importance of each particular region. For instance, out of the 23 IFN-α structures available the sequence RITLYL carries no mutations, and fragment YFQR has only two presumably equifunctional substitutions of the Gln ⟷ Arg type, whereas the amino acid Thr following the RITLYL sequence can be replaced for Met, Lys or Ile as a result of point mutations.

Since the primary structure of IFN-γ differs considerably from that of IFN-α and -ß, comparative analysis is presented in two versions (Fig.4) with either including IFN-γ for comparison or omitting it.

Figure 3. Amino acid sequence conservatism of regions (9) and (10) from various interferons.

	Region (9)	Region (10)
Variants (boxed)	N: K, Y; E: V, A; S: F; A: T; R: K; Q: R	T: M, K, I; E: K, G; K: R; P: A, S
Consensus	N E D S I L A V R K Y F Q	R I T L Y L T E K K Y S P C

IFN-α	Region (9)	Region (10)
A	Ⓚ E D S I L A V R K Y F Q	R I T L Y L Ⓚ E K K Y S P C
B	Ⓨ E D S I L A V R K Y F Q	R I T L Y L T E K K Y S S C
C	N E D S I L A V R K Y F Q	R I T L Y L Ⓘ E Ⓡ K Y S P C
C_1	N E D S I L A V R K Y F Q	R I T L Y L T E K K Y S P C
D	N Ⓥ D S I L A V Ⓚ K Y F Ⓡ	R I T L Y L T E K K Y S P C
E	N Ⓥ D S I L A V R K Y F Q	R I T L Y L T Ⓚ K K Y S P C
F	N Ⓥ D S I L A V Ⓚ K Y F Q	R I T L Y L T E K K Y S P C
G	N Ⓥ D S I L Ⓣ V R K Y F Q	R I T L Y L T E K K Y S P C
H	N E D S I L A V R K Y F Q	R I T L Y L Ⓜ E K K Y S P C
I	N E D S I L A V R K Y F Q	R I T L Y L T E K K Y S P C
J	N E D Ⓕ I L A V R K Y F Q	R I T L Y L Ⓜ E K K Y S P C
K	N E D S I L A V R K Y F Q	R I T L Y L T E K K Y S P C
L	N E D S I L A V R K Y F Q	R I T L Y L Ⓘ E Ⓡ K Y S P C
N	N E D S I L A V R K Y F Q	R I T L Y L Ⓜ Ⓖ K K Y S Ⓐ C
$\lambda 2C_1$	N E D S I L A V R K Y F Q	R I T L Y L T E K K Y S P C
λ2h	N E D S I L A V R K Y F Q	R I T L Y L Ⓜ E K K Y S P C
1	N Ⓐ D S I L A V Ⓚ K Y F Ⓡ	R I T L Y L T E K K Y S P C
2	Ⓚ E D S I L A V R K Y F Q	R I T L Y L Ⓚ E K K Y S P C
4b	N Ⓥ D S I L A V R K Y F Q	R I T L Y L T E K K Y S P C
5	N Ⓥ D S I L Ⓣ V R K Y F Q	R I T L Y L T E K K Y S P C
6	N E D S I L A V R K Y F Q	R I T L Y L T E K K Y S P C
7	N E D Ⓕ I L A V R K Y F Q	R I T L Y L Ⓜ E K K Y S P C
8	Ⓨ E D S I L A V R K Y F Q	R I T L Y L T E K K Y S Ⓢ C
10	N E D S I L A V R K Y F Q	R I T L Y L Ⓘ E Ⓡ K Y S P C
IFN-β_1	Ⓚ Ⓛ Ⓜ S Ⓢ L Ⓗ Ⓛ Ⓚ Ⓡ Y Ⓨ Ⓖ	R I Ⓛ Ⓗ Y L Ⓚ Ⓐ K Ⓔ Y S Ⓗ C
IFN-γ	N Ⓥ Ⓠ Ⓡ Ⓚ Ⓐ Ⓘ Ⓗ Ⓔ Ⓛ Ⓘ Ⓠ Ⓥ	Ⓜ Ⓐ Ⓔ L Ⓢ Ⓟ Ⓐ Ⓐ K Ⓣ Ⓖ Ⓚ Ⓡ Ⓚ

Figure 4. Mutual relatedness of the primary structures of interferon fragments.

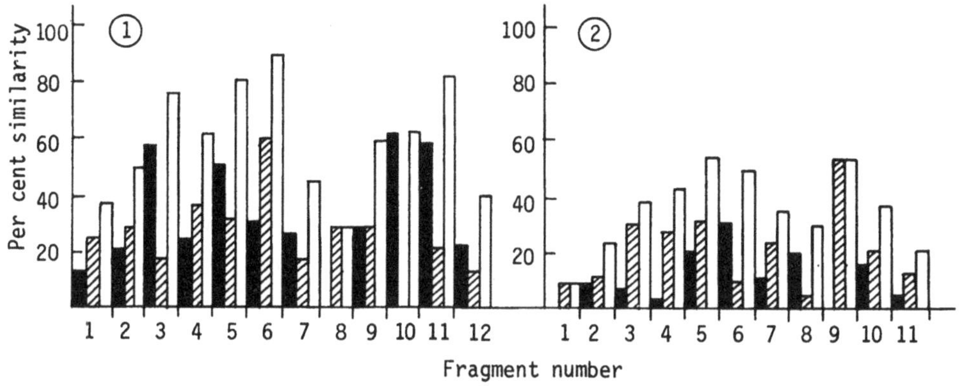

1 - comparison of the primary structures of IFN-α and IFN-ß;

2 - comparison of the primary structures of IFN-α, IFN-ß and IFN-γ.

■ - identical amino acids; ▨ - equifunctional amino acids;

□ - the sum of identical and equifunctional amino acids.

The structural similarity is most pronounced in fragments (5)-(6) and (9)-(11). As fragments (5) and (6), in compliance with the existing space structure models for IFN, constitute the α-helical "backbones" of the molecule, whereas fragments (9)-(11), on the contrary, predominantly occur in the functionally more active ß-structure, including the reverse turns (potential active centres) and the sites attacked by proteolytic enzymes, our attention has been primary focussed on the C-terminal part of IFN molecules. Fragment (10) has the highest content of identical amino acids when IFN-α and -ß structures are compared (Fig.4). It contains the pentapeptide RITLY (that shares common structural features and organization principles with the immunopoietins, thymopentin and splenopentin) and the reverse turn KKYS with a pair of basic amino acids (marker of proteolytic enzyme action). Conversely, fragment (9) is characterized by a low identical amino acid content, but abounds in potentially equifunctional amino acids.

According to our model of consecutive limited proteolysis and quasicyclization reactions (CLPQ reactions) (4,5), enzyme hydrolysis of the peptide chain leads to the formation of free amino and carboxyl groups that may serve as intramolecular quasicyclization centres (interaction with other amino acid residues in the chain by way of ionic, hydrogen and other bonds). Such interactions may lead to new steric structures (potentially active sites) responsible for their interaction with cell receptors (a new biological function emerges) or with new enzyme systems (cleaving off a chain fragment - a potential bioregulator). In the latter case new quasicyclization centres are formed, which may trigger a new cycle of reactions. The points ar which CLPQ reactions start generally lie on the surface of the molecule, in the interdomain regions or in the C-terminal part of the peptide chain, where the free C-terminal carboxyl group usually constitutes one of the quasicyclization sites. A pair of basic amino acids or a basic amino acid in combination with aminodicarboxylic acid amide or acidic amino acid can often serve as markers for proteolytic cleavage sites located within the peptide chain (4,5). In the case of IFN molecules, anomalous peptide chain breakage induced by trypsin in the post-tyrosine position deserves special attention. For example, native IFN molecules are attacked by trypsin in regions (<u>10</u>) and (<u>8</u>) (Tyr^{129} in IFN-α and Tyr^{101} in IFN-γ) (7). This allows to conclude that the regions in question are located on the IFN surface and are involved in loop or quasicycle formation, whose profile corresponds to that of the active site of the enzyme (or that of the cell receptor). Consequently, region (<u>10</u>) contains several peptide bonds that can serve as targets for trypsin type enzymes. The cleavage of bonds 131-132, 129-130 results in fragments with a free carboxyl group in the C-terminal amino acid (Lys^{131} or Tyr^{129}), which on interaction with the side chains of Arg^{125}, Gln^{124}, Lys^{121}, Arg^{120}, etc. by way of ionic and/or hydrogen bonds provides the formation of specific steric (quasicyclic) structures. The latter can act as new recognition sites for receptors or enzymes. It must be noted, though, that little is known about the enzymes participating in peptide processing and their substrate specificity. It has been only recently established that a family of kallikrein genes codes for at least 25-30 structurally related, highly specific enzymes of the trypsin group participating in the processing of many growth factors and T-lymphocyte transforming factors (8). Among the peptides being formed as a result of region (<u>10</u>) processing, the first to be mentioned is the sequence RITLY whose counterparts are also contained in the IFN-ß molecule (Fig.2). Owing to the potential equifunctionality of Val, Ile and Pro residues, considerable structural homology is noted for immunopoietin M, thymopentin, splenopentin and the IFN fragment (10)

Thymopoietin	31	Q	R	K	D	V	Y	V	E	37
Splenin	31	Q	R	K	E	V	Y	V	E	37
Immunoglobulin M	450	H	R	P	D	V	Y	L	L	457
Interferon-αA	124	Q	R	I	T	L	Y	L	K	131
Interferon-ß	158	N	R	L	T	G	Y	L	R	165
Structural elements of thymopentin essential for immunological activity			1 R	○	D	○	Y 5			

The prevalent features responsible for the biological activity of thymopentin are the amino acid residues in position 1, 3 and 5, whereas the amino acids in positions 2 and 4 can be significantly varied (9). Thus, the structural features of RITLY allow to expect that the peptide would interact with T-lymphocytes.

To test this assumption RITLY was synthesized by solid-phase technique (R_f= 0.46 in the Py-AcOH-H_2O-AcOEt system, 20 : 60 : 11 : 120; $[\alpha]_D$ = -39.2^o (c 0.5; 1 M AcOH)) as well as RITLYLR (R_f = 0.66 in the same system, $[\alpha]_D$ = -50.0^o (c 0.5; 1 M AcOH)), because of the uncertainty as to the site of enzyme attack.

For biological assessment of synthesized compounds in relation to cell-mediated immunity, we chose the delayed-type hypersensitivity (DTH) assay. Mice sensitized with sheep erythrocytes were challenged 7 days later with the antigen and the extent of swelling was measured after 20 hours. IFN is known to inhibit various reactions of cellular immunity, the DTH included (10). It can be seen from Fig.5 that the pentapeptide RITLY and its elongated analogue also suppress the DTH. These results demonstrate important functional involvement of the amino acid in position 3 of the thymopentin type peptides. A substitution of the acidic amino acid in position 3 for a hydroxyl-containing one (Thr) reverses the effect of these compounds (inhibition instead of stimulation of DTH). A notable alteration in the spectrum of thymopentin action is brought about even when Asp is replaced by Glu (11).

To test the validity of our hypothesis postulating linkage of active sites in immunologically active proteins affecting T-lymphocytes (T-sites) and macrophages (A-sites), such as rigin and immu-

Figure 5.

The effect of IFN-$\alpha A_{125-129}$, its analogue and thymopentin on the delayed - type hypersensitivity in mice.

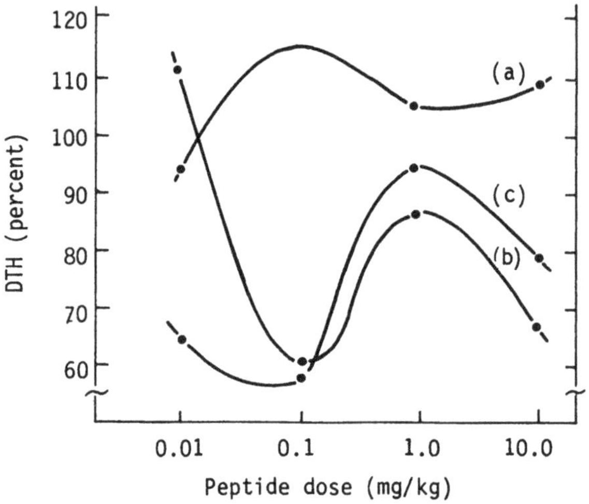

The control DTH without a peptide - 100%.

(a) - thymopentin (Arg-Lys-Asp-Val-Tyr);

(b) - IFN-$\alpha A_{125-129}$ (Arg-Ile-Thr-Leu-Tyr);

(c) - IFN-$\alpha A_{125-129}$ extended analogue (Arg-Ile-Thr-Leu-Tyr-Leu-Arg).

nopoietin G (Fig.1), it was interesting to synthesize amino acid sequences preceding RITLY and fragment (9). For this purpose, a $2\xi 2\pi$ sequence was chosen, where π stands for polar charge-carrying amino acid or aminodicarboxylic acid amide and ξ is hydrophobic, aromatic or branched aliphatic amino acid. The IFN-$\alpha A_{122-125}$ fragment (YFQR) represents such a sequence. In some immunological tests it resembles IFN. For instance, YFQR stimulates phagocytosis by mouse macrophages, as judged by the level of active phagocytes and by their ability to engulf bacteria (Fig.6), which is also characteristic of native interferon molecules (12). That the peptide presumably affects the binding ability of Fc-receptors in killer cells is also supported by antibody-dependent cell-mediated erythrocyte lysis assays. This effect is more distinctly pronounced when lymphocytes are obtained from patients with altered cytotoxic potential (Fig.7).

Figure 6.

The effect of IFN-$\alpha A_{122-125}$ and tuftsin
on phagocytosis by mouse peritoneal macrophages.

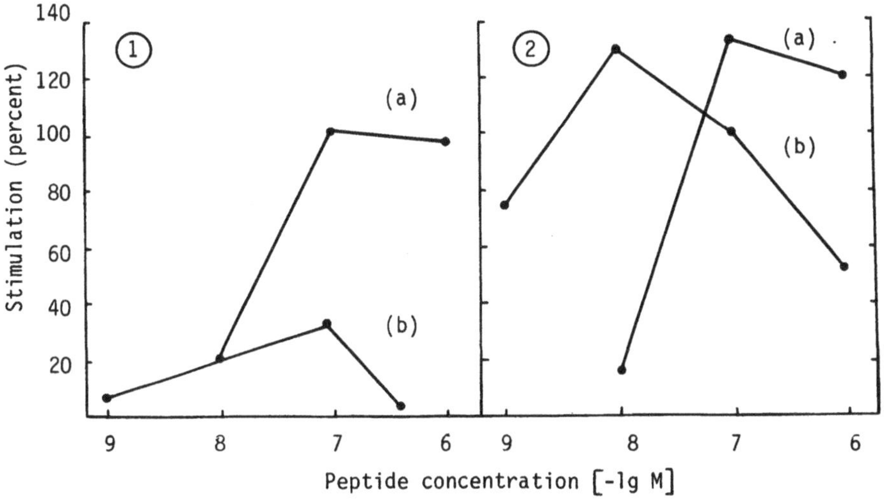

1 - phagocytosis index; 2 - bacterial ingestion.
(a) - IFN-$\alpha A_{122-125}$ (Tyr-Phe-Gln-Arg); (b) - tuftsin (Thr-Lys-Pro-Arg).

It is interesting to note that the tetrapeptide YFQR also activates CNS neurones resulting in enhanced learning ability and long-term memory in rats (13).

In order to verify the amino acid sequence of fragment (9) stimulatory to immunocompetent cells, synthesis of other peptides in this region is definitely required. However, the results obtained support the assumed linkage of A- and T-sites in IFN. Comparative primary structure analysis conducted for other proteins (various classes of immunoglobulins, histocompatibility antigens, bacterial toxins) indicates that this phenomenon is of general occurrence. Amino acid sequences composed of 2 hydrophobic and 2 polar amino acid residues ($2\xi 2\pi$)or vice versa ($2\pi 2\xi$) frequently occur in the active sites of biologically active peptides (hormones, tachy-, pachykinins, etc.). It is believed that the sequences of the $2\pi 2\xi$ type in IFN: RKYF (IFN-$\alpha_{120-123}$), KRYY or FINR (IFN-$\beta_{123-126}$, IFN-$\beta_{156-159}$) are involved in the binding of IFN molecules to immunocompetent cells, because these sequences share some homology with the binding region of the cholera toxin, which interferes

Figure 7.

The effect of the tetrapeptide Tyr-Phe-Gln-Arg on the antibody-dependent cytotoxicity of K-cells.

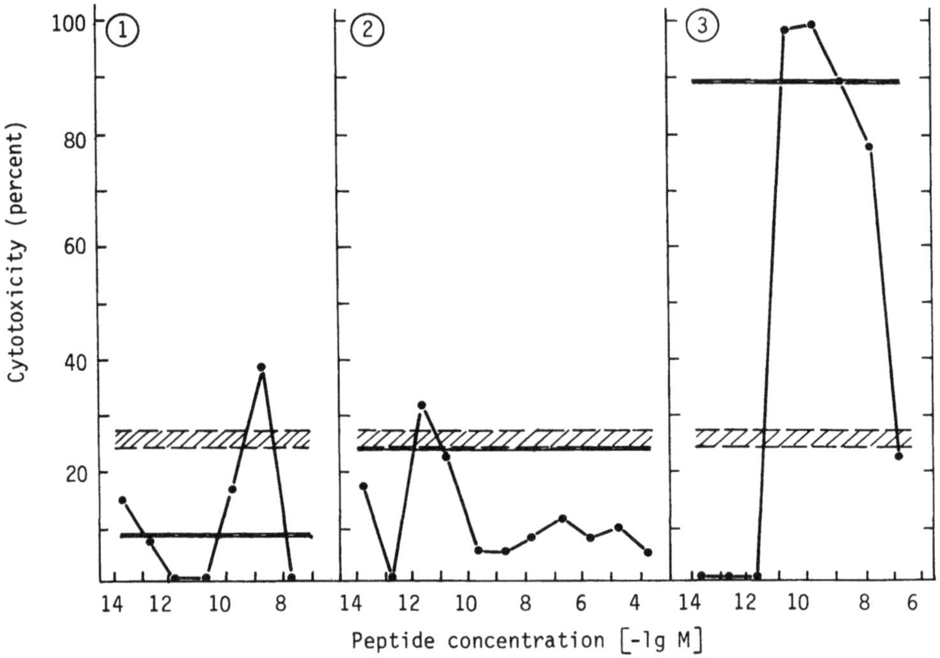

Lymphocytes are isolated from patients with diagnosis of
1 - chronic hepatitis; 2 - viral encephalitis; 3 - mammary carcinoma.
[hatched bar] - mean cytotoxicity in healthy controls,
[solid bar] - cytotoxicity in the particular patient without the peptide.

with IFN binding to cell receptors (14). If so, it should be also present in the primary structure of IFN-γ. Analysis of the latter revealed that the sequence KKYF is located in the N-terminal part of the molecule (IFN-γ_{15-18}) and is apparently derived by convergent evolution. Comparison of potential binding regions of all IFN types shows a certain similarity of their structures:

IFN-α	^{118}A	V	R	K	Y	F	Q	R	I	T^{127}
IFN-ß	^{121}H	L	K	R	Y	Y	G	R	I	L^{130}
IFN-γ	^{13}N	L	K	K	Y	F	N	A	G	H^{22}

The hydrophobic amino acid residue (L or V, see the first column) is followed by a more or less typical 2π2ξ2π sequence terminating in a hydrophobic amino acid. In IFN-γ the A-site is also linked

Figure 8.

Conjectural immunoregulatory sites linked in interferon molecules.

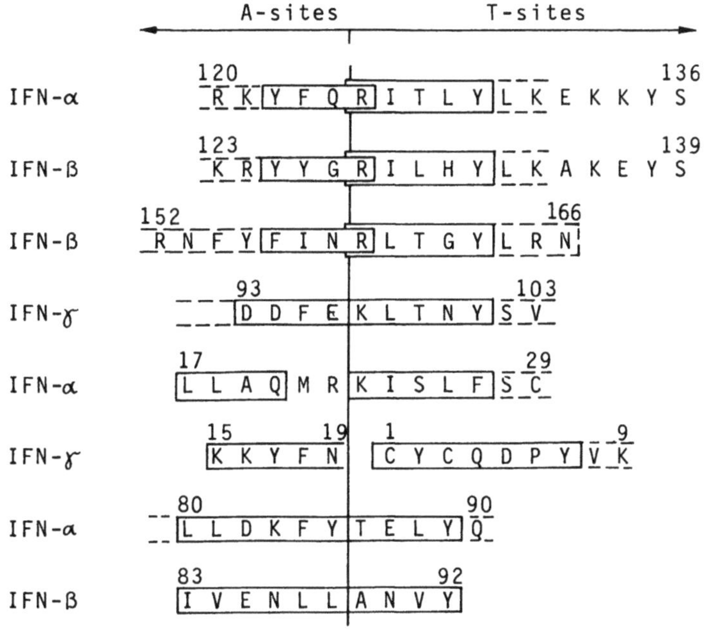

with a tyrosine-containing amino acid sequence which is a potential T-active region CYCQDPYVK (IFN-γ_{1-9}) (Fig.8). The IFN-γ_{4-9} sequence contains an amino acid sequence equifunctional to thymopentin or splenopentin: QDPYV (IFN), RKDVY (thymopentin), RKEVY (splenopentin). As many as 20 similar conjugated amino acid sequences have been found in IFN-α and -ß (Fig.8). It should be pointed out that the putative A- and T-active sites are linked in all cases.

In summary, the present study reports results of systems analysis conducted for IFN-α, -ß and -γ structures with special emphasis on the similarity, conservatism, and spatial arrangement of their hypothetical immunologically active fragments. More than 20 potentially active sites were identified sharing structural similarity with the active sites of thymus hormones and immunoglobulins. Linked sequences sometimes showing overlapping were revealed in these regions. Some IFN fragments identified on the basis of analysis data were synthesized and examined. They were found to possess detectable immunological activity.

References

1. Chipens, G.I., A.Y. Kriķis, L.K. Polevaya. 1979. In: Biological and Biochemical Information Transfer in Recognition (J.G. Vassileva-Popova and E.V. Jensen, eds.). Plenum Press, New York, p. 23.

2. Chipens, G.,G. Nikiforovich, N. Veretennikova, I. Vosekalna, A. Sosnov, J. Ancans, N. Mishlyakova, E. Liepins, I. Sekacis, M. Breslav. 1979. In: Peptides: Structure and Biological Function, Proc. Am. Pept. Symp., 6th (E. Gross and J. Meienhofer, eds.). Pierce Chemical Company, Rockford, USA, p. 567.

3. Veretennikova, N.I., G.I. Chipens, G.V. Nikiforovich, Y.R. Betinsh. 1981. Int. J. Peptide Protein Res. 17, 430.

4. Chipens, G.I. 1984. Latv. PSR Zinātņu akadēmijas vēstis No 12, p. 23 (in Russian).

5. Chipens, G.I. 1985. Surv. Immunol. Res. (in press)

6. Shulz, G.E., R.H. Schirmer. 1979. In: Principles of Protein Structure. Springer-Verlag, New York, Heidelberg, Berlin, p. 14.

7. Levy, W.P.. M. Rubinstein, J.E. Shively, U. Del Valle, C.Y. Lai, J.A.Moschera, L.D. Brink, L.D. Gerber, S. Stein, S. Pestka. 1981. Proc. Natl. Acad. Sci. U.S.A. 78, 6186.

8. Mason, A.J., B.A. Evans, D.R. Cox, J. Shine, R.I. Richards. 1983. Nature 303, 300.

9. Goldstein, G., T. Audhya, G.A. Heavner. 1981. In: Peptides: Synthesis - Structure - Function, Proc. Am. Pept. Symp., 7th (D.H. Rich and E. Gross, eds.). Pierce Chemical Company, Rockford, USA, p. 535.

10. Hooks, J.J., B. Detrick-Hooks. 1982. Mol. Aspects Medicine 5, 183.

11. Audhya, T.,M.P. Scheid, G. Goldstein. 1984. Proc. Natl. Acad. Sci. U.S.A. 81, 2847.

12. Friedman, R.M., S.N. Vogel. 1983. Adv. Immunology 34, 97.

13. Papsuevich, O., G. Chipens, V.D. Bakharev, T.A. Petrova. 1985. Khim. farm. zh. 19, 35 (in Russian).

14. Lai, C.Y. 1977. J. Biol. Chem. 252, 7249.

ANTIGENIC DETERMINANTS OF PROTEINS AND PEPTIDES

S.V. Komissarenko

Department of Molecular Immunology, Palladin Institute of Biochemistry, Kiev, 252030 USSR

Proteins and peptides are the most important and numerous among antigens (Ags). Estimation of their antigenic structure is highly important for understanding the immunocompetent cells' receptors organization, these cells cooperation in the immune response and for artificial Ags design. Immune system effector molecules (mainly antibodies) can also serve as useful tools in protein chemistry for affinity purification and quantification of molecules as well as for their function, primary structure and conformation analysis. Almost any bioorganic molecule or its part can be an antigenic determinant (AD) or epitope - the site of Ag recognized by the immune system and to which the immune effector molecules can be elicited. Important to note that AD analysis is impossible without taking into account the corresponding antigen recognizing regions. Only during the interaction of these complementary structures one can determine the properties of such interaction (e.g. boundaries, forces, role of particular residues, etc). What are the protein and peptide structures recognized by antibodies (Abs), B and T cells of the immune system?

Antigenic Determinants (Epitopes) Recognized by Antibodies and by B Cell Immunoglobulin Receptors

The immune recognition of Ags by Abs (the latter belong to immunoglobulins) and by immunoglobulin receptors of B cells is identical and is the most comprehensively studied field in immunology. This recognition occurs due to summation of weak forces of interacting complementary conformations of an epitope on the surface of Ag and of immunoglobulin binding site for Ag. Immunochemical analysis of protein and peptide ADs by Abs is usually done through the

Chemistry of Peptides and Proteins, Vol. 3

study of epitope quantity, epitope size and localization, amino acids (or non-protein components) as well as some local conformation contributions into the formation of a molecule antigenic structure. Synthetic and natural peptides, protein fragments, amino acid and protein bonds modifications are used for this purpose as well as Ag structure comparison between homologous proteins with known amino acid sequence and conformation. The quantity of epitopes is usually calculated from the stoichiometry of Ag binding to polyclonal Abs active sites or from the Ag interaction with monoclonal or monospecific Abs of various specificities. On analysing the stoichiometry of Ag-Ab interaction it is necessary to remember about possible steric hindrance for simultaneous binding to an Ag of Abs of different specificities. For some Ags this difficulty can be overcome using F_{ab} or F_v fragments of Abs. Immunochemical analysis of apamin - bee venom neurotoxin consisting of 18 amino acids was done in our laboratory (1). Radioimmunoassay of anti-apamin Abs with ^{3}H- or ^{125}I-labelled apamin and a set of apamin derivatives with modified Cys^{1}, Lys^{4}, Arg^{13}, Arg^{14}, His^{18} revealed that $Arg^{13,14}$ were crucial for apamin antigenic integrity. An example of monoclonal antibodies (MAbs) usage is a study of bacteriorhodopsin molecule topography (2). Immunochemical analysis of MAbs of different anti-bacteriorhodopsin specificities with this protein fragments and modified peptides helped to localize at least five epitopes exposed from the purple membrane.

In 1975 M.Atassi published "the complete immunochemical anatomy" of sperm whale myoglobin showing that this myoglobin had five epitopes consisting of six to eight consecutive amino acids, 15-22, 56-62, 94-99, 113-119 and 145-151, on the protein surface (3). The author claimed that all anti-myoglobin Abs could be exhausted on immobilized synthetic peptides corresponding to the found epitopes and that only found sites could be myoglobin epitopes. Sperm whale myoglobin has no disulfide bonds. On the contrary, hen egg-white lysozyme (HEL) has four ones which stabilize its conformation. Reduction or oxidation of HEL S-S bonds denatures the protein and completely changes HEL antigenicity - Abs to native HEL don't cross react with denatured protein. Several groups of investigators have shown that HEL epitopes recognized by Abs to the native protein were assembled or discontinuous (4, 5). There were three such epitopes in HEL formed by residues 5,7,13,14,125; 33,34,113,114,116 and 62,87,89,93,96,97 standing apart in amino acid sequence but put together in HEL native conformation. Such assembled epitopes were called conformational i.e. dependent on protein conformation. "Surface-simulation synthesis" was used with success to create artificial epitopes bringing together amino acids standing apart in protein (6). These artificial peptides inhibited Ab binding to native HEL as did peptides of continuous epitopes of proteins with topographic AD, e.g. sperm whale myoglobin or cytochrome c (7). So, some 5-6 years ago the main principles of protein antigenicity seemed to be established. Proteins were considered to have two types of ADs: "conformational" or assembled as in HEL (4) and "topographic" formed by a segment of continuous amino acid sequence on the protein surface as in myoglobin and in cytochrome c (5,7).

However, recent experiments using highly sensitive radioimmuno and immunoenzyme techniques, advances in peptide synthesis and mainly hybridoma technology providing monoclonal Abs gave an invaluable impact for immunochemistry of proteins. They produced a lot of new data which necessitated the reappraisal of former Ag structure principles (8). Thus, "the complete immunochemical anatomy" of myoglobin (5) appeared not to be complete. Other epitopes were found (9) among which there were not only continuous but also assembled ones (10). Various clones secreting MAbs to HEL were obtained (11,12). These MAbs reacted with HEL surface epitopes which were different from those previously described (5). Interesting data were obtained by Fujio et al. who showed that polyclonal Abs recognize HEL fragment 38-54 and on this fragment three overlapping, 38-45, 44-48 and 46-54, continuous and non cross-reactive epitopes (13). The abovementioned and numerous other data make possible to conclude that all protein surface sterically available for immune recognition "consists of a complex array of overlapping potential antigenic determinants, in aggregate these approach a continuum" (8), which are formed by amino acid residues (sometimes by non amino acid components in proteins) either continuous in a polypeptide chain or conformationally close assembled. These two groups of epitopes, continuous and discontinuous (or assembled) compose topographic ADs. As the basis of immune recognition is an interaction of complementary conformations of AD and of an active site of Ag-binding molecule and as any site of any Ag has certain spacial structure or consists of a set of conformers it is evident that all epitopes, protein epitopes as well, have to be and are in fact conformational regardless how they are formed. Topographic epitopes of protein surface differ from sequential or linear epitopes of short peptides because the latter usually have no stable conformation and their structure consists of statistic set of conformers. It is possible to obtain the immune response, Abs as well, to short peptides(14). In this case peptide binding to Ab or lymphocyte receptor active site may "freeze" one of the peptide conformations.

Recent understanding that all protein surface include multiple overlapping potential epitopes (8) conforms well to the concept that in principle the immune system can recognize any conformation accessible to Ag-binding structures. Nevertheless, experimental data show that protein surface sites are not antigenically equal during the immune recognition. Usually only a limited number of Ab specificities are formed against so called immunodominant sites of an Ag in the immune response to proteins. What is the nature of immunodominance, how many immunodominant sites are there on a protein and where are they? What is the relationship between immunodominant sites, epitopes and immunogenicity? Many authors answer these questions quite in an opposite way. Atassi et al. claimed that quantity and localization of ADs are independent of immunized species and are the intrinsic properties of an Ag (15). The other point of view is that immunodominance depends on the structural differences between donor and host homologous proteins used as an Ag (7). In our laboratory we have studied rabbit and BALB/c mouse humoral immune response to horse myocardium cytochrome c. Earlier it was shown that horse cytochrome c had three ADs localized in places of donor-host differences in protein structure (7). This protein of 104 amino acids is a convenient model for immunochemical analysis because of known primary and tertiary structures of horse as well as of different species cytochromes c and because its conformation is very stable in evolution. For cytochrome c analysis we used its synthetic peptides,

1-13,2-13,14-22,23-104,46-56,57-77,61-77,61-69 and 92-104, which were a generous gift of Prof.R.Evstigneeva from Moscow Institute of Fine Chemical Technology. Nonapeptide 14-22 had a heme. By means of fluorescence quenching and of immune complexes molecular weight analysis it was found that native horse cytochrome c had four immunodominant sites (16). Monospecific Abs to individual epitopes were purified by isoelectric focusing (17). Ionic interaction was essential for these Abs binding to cytochrome c and the binding caused conformational changes in Abs' active site. With the aid of radioimmuno and enzymeimmunosorbent assays with synthetic peptides cytochrome c immunodominant sites were localized in the sequences: 12-16, 46-50, 64-69, 92-104 (18). Heme was absolutely needed for all but one, 46-50, epitopes recognition. Cytochrome polymerization by glutaraldehyde, the procedure often used to enhance antigenicity of the protein, changed the Ag structure and masked the epitope 64-69. Four immunodominant sites of horse cytochrome c were found to be common for both rabbit and BALB/c mouse immune recognition.

The data obtained permitted to suggest that immunodominance does not depend on structural differences between donor and host homologous proteins. An example in proof is autoimmune response to self-proteins. Provided this suggestion is valid, why are Abs formed in many cases against those Ag sites where donor-host protein differences occur? This is the possible explanation. Evolutionary changes in homologous protein structures are usually localized in regions not essential for molecule conformation stability and that is why they are of the least importance for this protein biologic activity. Evidently, in most cases these regions represent the groups of high atomic mobility on the protein surface that can be of exceptional importance for immunodominant site formation. In general, there seems to be no principal differences for antigenecity between "flexible" (or mobile) and "rigid" sites if the immune system of the host possesses Ag-binding structures (receptors) for both types of these sites. However, high mobility regions have an advantage because they can tune to or adopt to Ag-recognizing conformations thus augmenting the quantity and affinity of Ag-Ab or Ag-receptor interactions. If it is the case, under other equal conditions the quantity of immunodominant sites formed from mobile regions will be larger than that of rigid regions and will have preference for antigenicity at the cell population level. Now more experimental data appeared in proof which show that anti-peptide Abs against highly mobile regions but not against rigid regions react with native protein (19). So, Ab formation found in experiments against Ag sites where donor-host protein differences occur can be a coincidence of these sites with high atomic mobility sites because the latter and protein structure differences are mainly localized in "evolutionary unstable" regions. Some authors propose to quantify protein ADs from the number of hydrophilic clusters in the protein amino acid sequence suggesting that hydrophilic regions exposed on Ag surface represent immunodominant regions (20).

Antigenic Determinants Recognized by T Lymphocytes

Effective Ab biosynthesis against complex Ags as most proteins or as hapten-carrier conjugates requires interaction of Ab producing B cells with T cells which provide positive regulatory signal for Ab production (21). Immune regulation even in one type of immune response, in Ab biosynthesis, occurs through the multilevel circuits of interacting subsets of lymphocytes, mainly T cells (22,23). How T cells recognize an Ag? In early studies it was suggested that T lymphocytes recognition of Ag was much less specific than that of B cells (24) and that T cells couldn't recognize native protein topographic epitopes but interact with continuous and sequential determinants of denatured protein (25). Numerous later experiments refuted coarse specificity of T cell receptor and confirmed the data that various T cell subsets interacted with different protein determinants and that at least the majority of these determinants were localized on the fragments of denatured protein Ag having been processed by antigen presenting cells (APCs) or on continuous segments of immunogenic peptides (26). Thus, guinea pig T cells could recognize even one amino acid substitution in fibrinopeptide B or angiotensin II (27). They could also recognize the differences between conformations of fibrinoctapeptide and its inverted sequence analog because these two peptides were not cross-reactive in T cell activation (28).

From numerous regulatory T cell subsets T helper (Th) and T suppressor (Ts) cells interacting with B cells seems to have the major influence, correspondingly positive and negative, on Ab synthesis. It appeared that there were determinants on proteins usually separate and non-overlapping which specifically activated Th or Ts lymphocytes. Thus, HEL suppressor determinant is on the NH_2-end of the polypeptide chain and its activity depends on the presence of Phe^3, important to note, - in the certain major histocompatibility complex (MHC) haplotype mouse (29). HEL injection into mice with $H-2^b$ haplotype gives stable suppression of Ab formation to HEL. Such type of suppressor determinant is a constant or a permanent one. But it is sufficient to cleave off tripeptide from HEL NH_2-end to make the rest of HEL molecule immunogenic. The second type of suppressor determinant that can be called inducible is found on β-galactosidase. At the beginning of the immune response to β-galactosidase, after this Ag administration, there is Ab formation to β-galactosidase due to Th cells activation. Later, Ts cells are activated and Ab synthesis is inhibited (30). This suppressor determinant being induced in the course of the immune response was localized in the region of β-galactosidase fragment 27-37. Another type of so called hidden or covered suppressor determinant can be found only artificialy by experimental cleavage and separation of active peptides as, for example, peptide 44-89 of myelin basic protein. On this protein administration into guinea pigs experimental allergic encephalomyelitis usually develops. Though it is mainly cell mediated immunity not Ab production which provokes this disease, the immune response to myelin basic protein in guinea pigs and in rats can be blocked by preliminary peptide 45-89 injection into animals due to Ts cells activation (31). T helper determinants are localized in the regions of protein molecules distinct from those for suppressor determinants. Their nature can also be different. Thus, though epitope for Ts cell is found on NH_2-end of HEL (29), there are several helper determinants on HEL molecule which activate various haplotype Th cells in a different way. Mouse $H-2^a$ Th cells are stimulated by an epitope assembled from at least three regions of HEL polypeptide chain. $H-2^b$ Th cells recognize the determinant on peptide 74-94. Residues 113 and 114 definitely take part in helper determinant formation for $H-2^d$ Th cells (32). Nevertheless, Th and

Ts cells interaction with Ag is not sufficient for T cell activation. Moreover, most T cells recognition of Ag fundamentally differs from B cells recognition. It is genetically determined and happens only if Ag is processed by APCs and is presented to T cells in the form of immunogenic complex of Ag determinants with proteins which are the MHC genes products.

Antigen Processing by Antigen Presenting Cells and Immunocompetent Cells Cooperation in Antibody Biosynthesis

Ab biosynthesis occurs in B lymphocytes and in their differentiated progenitors - plasma cells. B cells can directly bind Ag epitopes through B cell immunoglobulin receptors but for most Ags, so called thymus-dependent Ags, this Ag-receptor binding is not sufficient for B cells to be stimulated for Ab production. To proliferate and/or to differentiate they need additional activation signals from T cells. The latter in turn have to be activated by other factors (as IL-1) and have to recognize specific T cell determinant on an Ag in the immunogenic complex formed by APC. APCs can be cells of different origin but mainly macrophages serve as a source of APCs. They hydrolyze and/or denature protein Ags and present them to T cells on APC surface in association with their own Ia proteins. Ia proteins in mouse have three polypeptide chains, α, β and γ, with molecular mass 34, 28 and 31 kd correspondingly (33). α and β chains are polymorphic but γ or i (invariant) chain is practically identical within the animal species. Ia proteins belong to two groups of I-A and I-E molecules encoded by A_α, A_β and E_α, E_β genes that are class II genes of the MHC or immune response (Ir) genes. On the cell surface membrane Ia proteins have only single α and single β chain because before Ia expression on the membrane the intracellular complex $[\alpha \cdot \beta \cdot \gamma]$ dissociates in the cell liberating γ chain (34). Interesting to suggest that this γ chain dissociation from $[\alpha \cdot \beta \cdot \gamma]$ complex can be linked to Ag processing and occurs when Ag binds to Ia protein. The formation of a complex of Ia protein with denatured Ag or with Ag fragment can be also a process that regulates further Ag degradation thus determinning the structure and properties of the immunogenic complex presented by APCs to T cells.

The detailed analysis of Ia molecule role in single T cell clones activation was done by R.Schwartz group (35). COOH-end CNBr fragment 81-104 of pigeon cytochrome c and synthetic fragment 81-103 of tobacco hornworm moth cytochrome c were used as Ags. B10.A and B10.A(5R) strains of mice give good immune response to fragment 81-103 from moth cytochrome c, but in a different way respond to fragment 81-104 from pigeon cytochrome c: B10.A mouse is a high responder and B10.A(5R) is a low responder to this Ag. Ia protein α chain of these mouse strains is encoded in the I-E subregion and β chain in the I-A subregion of the MHC. So, in B10.A the Ia molecule is designated $A/E^k_\beta:E^k_\alpha$ and in B10.A(5R) - $A/E^b_\beta:E^k_\alpha$. Though these two Ia proteins differ only by several amino acid residues in their β chain this difference plays a crucial role in Ag presentation by APCs (35). Thus, T cell specific clones from non-responder to pigeon cytochrome c fragment 81-104 B10.A(5R) strain were activated when high responder B10.A APCs were used to present this Ag. The authors suggest that the state of immune responsiveness in the system studied is determined by the structure of Ia molecule controlled by Ir genes. They propose also the existence of a special site on an Ag, called agretope, which binds to a site on Ia protein, called desetope, on formation of [Ag·Ia protein] immunogenic complex (35). If it is so, Ag-Ia protein or agretope-desetope interaction can induce conformational changes giving a new determinant for T cell receptor recognition, the determinant existing only on [Ag·Ia protein] complex. A comprehensive study in what form Ag is presented by APCs to T cells was done on sperm whale myoglobin (26). The same epitope of this protein but in different forms - on the native and modified myoglobins and on the peptide fragment 132-153 as well as after Ag processing by APCs - was presented to a single specific T cell clone. Inhibitors of macrophage processing of Ag - lysosomal protease competitive inhibitor leupeptin, NH_4Cl and chloroquine that raise lysosomal pH and cationic ionophore monensin that inhibits lysosomal and Golgi apparatus functions - were added to APCs to study the requirements for myoglobin processing. It was found that Ag presentation by APCs treated with inhibitors was impaired if native myoglobin served as Ag. Apomyoglobin presentation was inhibited to a lesser extent. Lysosomal inhibitors had no effect on S-methylmyoglobin and peptide 132-153 presentation to T cells. These and other data demonstrate that proteolysis and protein fragmentation are not obligatory, as it was shown for larger proteins (36), for the presentation of protein Ags of rather low molecular size as sperm whale myoglobin. It is sufficient to denature myoglobin native conformation to expose its hydrophobic sites on the surface of the molecule that can be necessary for Ag-Ia protein interaction (26). Ag recognition by immunocompetent cells is obligatory but not sufficient requirement for the most types of the immune response for which the cooperation of different immunocompetent cells is needed. Such cooperation is MHC restricted in the sense that it is controlled and/or regulated by the products encoded in the MHC (22,37). Regulatory Th cells interact with those B cells and APCs that have the same and of definite structure Ia proteins. That is, syngeneic T, B cell and APC cooperation occurs not because of the identical MHC genes products on their surface but due to the T cell receptor recognition of the same Ia proteins on B cells and on APCs (38). MHC restriction analysis revealed the existance of distinct B and T cell subpopulations interacting for Ab production: Lyb 5^+ and Lyb 5^- B cells (22,39) having different properties in immune response and responding in a different way to T cell factors and various Th cells which are or aren't MHC restricted in their cooperation with Lyb 5^+ and Lyb 5^- B cells (40-43) and which can either induce or augment Ab biosynthesis as well as can regulate Ab isotype selection.

Conclusions

Though potentially there is a multitude of overlapping epitopes on the protein surface (8) polyclonal antibodies are mainly synthesized and hybridoma clones are mainly directed against a limited quantity of protein surface regions, immunodominant sites, due to the antigen intrinsic molecular properties. Such as: local high atomic mobility or density, special molecular geometry, the presence of hydrophilic and/or hydrophobic clusters on the Ag. These special antigen properties create the conditions for immunodominant sites formation and these sites preferential binding to Ag-recognizing structures of the immune system: B and T cell receptors, Ia proteins of APCs, Ag-specific factors, antibodies.

From the other side, antigen immunodominance is controlled by and is realized through the regulatory mechanisms of the host immune system: Ag processing and Ag presentation to different immunocompetent cells, these cells cooperation in the immune response. Various populations of immunocompetent cells: antigen presenting cells, B cells and different subsets of T cells recognize distinct antigenic determinants on proteins. Evidently, the type of regulatory T cell to be activated (and the type of immune response) depends on antigenic determinant presentation or orientation in the immunogenic complex.

References

1. Komissarenko, S.V., S.V.Vasilenko, E.G.Elyakova, E.A.Surina, A.I.Miroshnikov. 1981. Mol.Immunol. 18, 533.

2. Ovchinnikov, Yu.A., N.G.Abdulaev, R.G.Vasilov, I.Yu.Vturina, A.B.Kuryatov, A.V.Kiselev. 1985. FEBS Letters. 179, 343.

3. Atassi, M.Z. 1975. Immunochem. 12, 423.

4. Arnon, R. 1977. In: Immunochemistry of Enzymes and their Antibodies (M.R.J. Salton, ed.). Wiley, p. 1.

5. Atassi, M.Z., C.-L.Lee. 1978. Biochem.J. 171, 429.

6. Atassi, M.Z., Sh.Sakata. 1982. Biochem.J. 201, 669.

7. Urbanski, G.J., E.Margoliash. 1977. In: Immunochemistry of Enzymes and their Antibodies (M.R.J. Salton, ed.). Wiley, p.203.

8. Benjamin, D.C., J.A.Berzofsky, I.J.East, F.R.N.Gurd, Ch.Hannum, S.J.Leach, E.Margoliash, J.G.Michael, A.Miller, E.M.Prager, M.Reichlin, E.E.Sercarz, S.J.Smith-Gill, P.E.Todd, A.C.Wilson. 1984. Ann.Rev.Immunol. 2, 67.

9. East, I.J., P.E.Todd, S.J.Leach. 1980. Mol.Immunol. 17, 519.

10. Berzofsky, J.A., G.K.Buckenmeyer, G.Hicks, F.R.N.Gurd, R.Feldmann, J.Minna. 1982. J.Biol.Chem. 257, 3189.

11. Kobayashi, T., H.Fujio, K.Kondo, Y.Dohi, A.Hirayama, Y.Takagaki, G.Kosaki, T.Amano. 1982. Mol.Immunol. 19, 619.

12. Smith-Gill, S.J., A.C.Wilson, M.Potter, E.M.Prager, R.Feldmann, C.R.Mainhart. 1982. J.Immunol. 128, 314.

13. Takagaki Y., A.Hirayama, H.Fujio, T.Amano. 1982. Arch.Biochem. and Biophys. 214, 750.

14. Lerner, R.A. 1982. Nature. 299, 592.

15. Twining, S.S., H.Lehmann, M.Z.Atassi. 1980. Biochem.J. 191, 681.

16. Komissarenko, S.V., M.V.Skok, G.A.Vasilieva, R.P.Evstigneeva. 1982. Doklady USSR Acad.Sci (in Russian). 264, 752.

17. Skok, M.V., P.V.Denysiuk, S.V.Komissarenko, G.A.Vasilieva, R.P.Evstigneeva. 1982. Doklady Ukr.Acad.Sci (in Russian). 10, 72.

18. Skok, M.V., S.V.Komissarenko, G.A.Vasilieva, N.E.Fedorova, R.P.Evstigneeva. 1983. Doklady USSR Acad.Sci (in Russian). 270, 1259.

19. Tainer, J.A., E.D.Getzoff, H.Alexander, R.A.Houghten, A.J. Olson, R.A.Lerner. 1984. Nature. 312, 127.

20. Hopp, T.P., K.R.Woods. 1981. Proc.Natl.Acad.Sci.USA. 78, 3824.

21. Rajewsky, K., V.Schirrmacher, S.Nase, N.K.Jerne. 1969. J.Exp. Med. 129, 1131.

22. Singer, A., R.J.Hodes. 1983. Ann.Rev.Immunol. 1, 211.

23. Green, D.R., P.M.Flood, R.K.Gershon. 1983. Ann.Rev.Immunol. 1, 439.

24. Hoffman, M., J.W.Kappler. 1972. J.Immunol. 108, 261.

25. Ishizaka, K., T.Kichmoto, G.Delespresse, T.King. 1974. J. Immunol. 113, 70.

26. Streicher, H.Z., I.J.Berkower, M.Busch, F.R.N.Gurd, J.A. Berzofsky. 1984. Proc.Natl.Acad.Sci.USA. 81, 6831.

27. Thomas, D., K.-H. Hsieh, J.Schauster, G.Wilner. 1981. J.Exp. Med. 153, 583.

28. Thomas, D., M.Hoffman, G.Wilner. 1982. J.Exp.Med. 156, 289.

29. Goodman, J.W., E.E.Sercarz. 1983. Ann.Rev.Immunol. 1, 465.

30. Krzych, U., A.V.Fowler, E.E.Sercarz. 1983. In: Protein Conformation as an Immunological Signal (F.Celada, V.Schumaker, E.E. Sercarz, eds). Plenum.

31. Swanborg, R., J.Holda, J.Killen. 1982. In: Immunobiology of Proteins and Peptides - II (M.Z.Atassi, ed.). Plenum. p. 3.

32. Katz, M.E., A.Miller, U.Krzych, L.Wicker, R.Maizels, J.Clarke, N.Shastri, A.Oki, E.E.Sercarz. 1983. In: Ir Gene: Past, Present and Future (C.Pierce, S.Cohen, J.Kapp, B.Schwartz, D.Schreffler, eds.). Humana, p. 311.

33. Benacerraf, B., ed. 1981. In: New Initiatives in Immunology. NIH publication 81-2215, 281 p.

34. Brodsky, F. 1984. Immunol.Today. 5, 350.

35. Heber-Katz E., D.Hansburg, R.H.Schwartz. 1983. J.Mol.Cell. Immunol. 1, 3.

36. Shimonkevitz, R., J.Kappler, P.Marrack, H.Grey. 1983. J.Exp. Med. 158, 303.

37. Katz, D.H., T.Hamaoka, M.E.Dorf, P.H.Maurer, B.Benacerraf. 1973. J.Exp.Med. 138, 734.

38. Sprent, J. 1978. J.Exp.Med. 147, 1838.

39. Singer, A., P.J.Morrissey, K.S.Hathcock, A.Ahmed, I.Scher, R.J.Hodes. 1981. J.Exp.Med. 154, 501.

40. Erb, P., M.Feldmann. 1975. J.Exp.Med. 142, 460.

41. Singer, A., K.S.Hathcock, R.J.Hodes. 1979. J.Exp.Med. 149, 1208.

42. Tada, T., T.Takemori, K.Okumura, M.Nonaka, T.Tokuhisa. 1978. J.Exp.Med. 147, 446.

43. Asano, Y., A.Singer, R.J.Hodes. 1983. J.Immunol. 130, 67.

GENETIC ENGINEERING AS A TOOL FOR THE IDENTIFICATION OF FUNCTIONAL DOMAINS ON PROTEINS: THE MAJOR HISTOCOMPATIBILITY COMPLEX ANTIGENS

B. Arnold, U. Horstmann, S. Kvist*, G.J. Hämmerling
Institute for Immunology and Genetics, German Cancer Research Center, D-6900 Heidelberg, Germany; *Swiss Institute for Experimental Cancer Research, CH-1066 Epalinges s./Lausanne, Switzerland

Introduction

One way for the elucidation of functional domains or sites on proteins is the chemical approach. This can involve cleavage of proteins, synthesis of the protein or peptides (with or without defined modifications), and functional analysis of the resulting fragments. An alternative way is genetic engineering of genes encoding the respective proteins. With this approach well defined modifications are introduced into genes at the DNA level. The manipulated genes can be introduced and expressed in eukaryotic cells and the modified proteins analysed in suitable functional assay systems.
As an example we describe here the genetic modification of major histocompatibility complex (MHC) class I antigens, designated H-2 in the mouse and HLA in the human. Of the many class I antigens, H-2, Qa and Tl, the most important ones are the classical H-2 antigens K, D, and L. These antigens are crucial for a variety of cellular interactions in the immune system. They are not only the major targets in transplant rejections but they also serve as guidance structures for the recognition of foreign antigens. Thus, all cytotoxic T lymphocytes (CTL) recognize foreign antigens on neoplastically transformed or virus infected cells only in association with the H-2 antigen on the target cell (1). CTL do not recognize antigen in the absence

Chemistry of Peptides and Proteins, Vol. 3

of H-2. This phenomenon has been designated H-2 restriction. Little is known about the underlying molecular mechanism. The MHC system is highly polymorphic. For each of the H-2 loci there exists a large number of alleles (about 10 - 50) encoding cell surface glycoprotein of approximate molecular weight 45.000, which is noncovalently associated with another polypeptide ß2 microglobulin (Fig. 1).
The H-2 genes consist of several exons and introns (2) (see Fig. 2). The exons correspond to distinct domains on the proteins (3), such as $\alpha 1$, $\alpha 2$ and $\alpha 3$ domains which are located on the outside of the cell membrane, the membrane spanning domain M by which the H-2 molecule is anchored in the plasma membrane, and the cytoplasmic regions C (Fig. 2). A schematic view of the domain structure of H-2 is presented in Fig. 1.

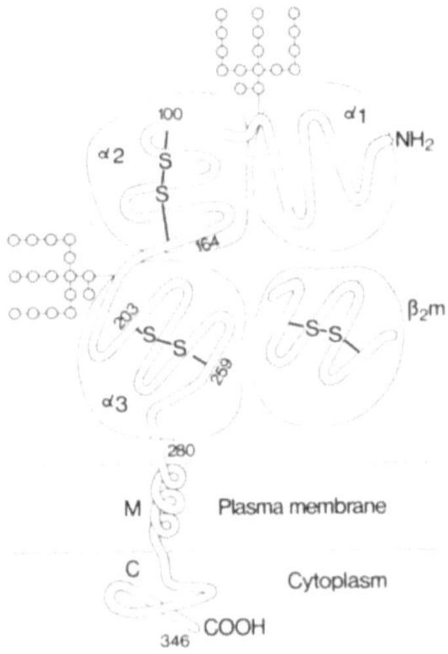

Fig. 1. Domain structure of major histocompatibility class I antigens.

Since H-2 molecules are composed of distinct protein domains the question was raised which of these domains are involved in immune recognition by antibodies and T lymphocytes. This question was approached by exchange of exons between different alleles of H-2 genes, expression of these hybrid genes in suitable cells and subsequent analysis with antibodies and CTL.

Results

1. H-2 hybrid genes. For the present study H-$2K^d$ and H-$2K^k$ genes have been cloned and their nucleotide sequence determined (4,5). Fig. 2 depicts the gene organization of the parental wild type genes and the hybrid genes generated by exchange of exons between K^d and K^k. Construction of these hybrid genes by the deletion subcloning technique has been described in detail elsewhere (6,7). The hybrid genes were introduced into fibroblasts, such as 3T3 or 1T22-6 or into tumor cells using the calcium phosphate method (8) as previously described (6).

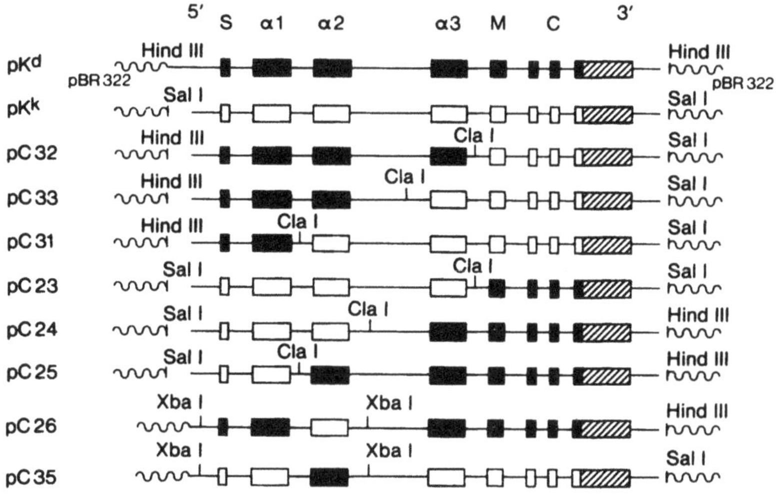

Fig. 2. Organization of H-$2K^k$ adH-$2K^d$ wild type ad hybrid genes.

Cotransfection of the neophosphotransferase gene allowed selection of transfected cells in medium containing the drug G418 G(9). Expression of transfected H-2 genes in G418 resistant colonies was determined with monoclonal antibodies (mAb), either by immunoprecipitation or by analysis with a fluorescence activated cell sorter (FACS).

2. Localization of H-2 determinants recognized by monoclonal antibodies. In a first set of experiments the binding of 29 distinct monoclonal anti-K^d and anti-K^k antibodies (10,11 and unpublished) to the hybrid antigens was analysed. Three different patterns of antibody binding were obtained. Typical examples are depicted in Table 1 and Fig. 3. The first group of mAb consisting of 8 mAb such as 34-1-2 and 20-8-4 react only with hybrid molecules when the $\alpha 1$ domain is of the K^d allele.

Table 1.
Binding of K^d mAb to K^d/K^k Hybrid Antigens.

K^d/K^k Hybrid				Antibody		
$\alpha 1$	$\alpha 2$	$\alpha 3$	/ MC	34-1-2	31-3-4	K9-18
d	d	d	/ d	314	327	273
d	d	d	/ k	313	337	321
d	d	k	/ k	395	400	0
d	k	k	/ k	175	10	0
k	d	d	/ d	0	0	254
k	k	d	/ d	0	4	374
Domain required for binding:				α_1	α_{1+2}	α_3

The binding of monoclonal anti-K^d antibodies to cells transfected with hybrid genes was determined with a fluorescence-activated cell sorter. Numbers are mean fluorescence values, the background is subtracted.

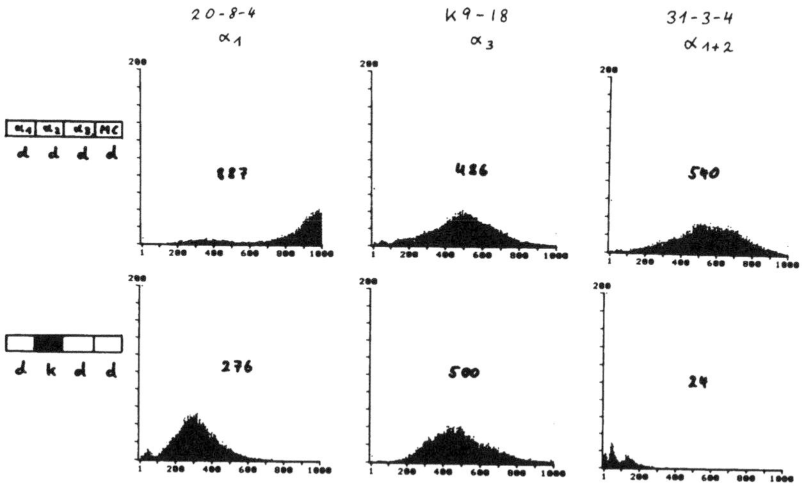

Fig. 3. FACS analysis of binding of monoclonal H-2 antibody to H-2 hybrid antigens.
Legend: see Table 1.

However, it can also be seen that the binding is reduced when the neighboring α2 domain is of a different allele. The second group (18 out of 29 mAb) with 31-3-4 as a representative binds only if α1 and α2 are of the same allele. In contrast, the third group (3 out of 29 mAb), with K9-18 as a representative, binds to the α3 domain independent of the neighboring domains. These findings are summarized in Table 3.

In conclusion, the data show that most epitopes recognized by mAb are conformational determinants formed by interaction of α1 and α2 domains and that α3, membrane and cytoplasmic domains have no influence. These conclusions are also supported by other investigators (12,13).

3. Domains recognized by alloreactive and H-2 restricted CTL.

K^d and K^k specific alloreactive killer cells (CTL) were generated by stimulation of lymphocytes from appropriate mouse strains (7) (anti-K^d, B10 anti-B10.GD; anti-K^k, B10 anti-B10.A(4R)). Influenza specific and K^d or K^k restricted CTL were obtained by infection of DBA/2 (H-2^d) or C3H(H-2^k) mice with influenza A virus as described (6). The CTL were tested on cell lines

Table 2 . Reactivity of K^d/K^k hybrid antigens with alloreactive and K^d or K^k restricted influenza specific CTL .

					% specific lysis with CTL			
Gene	α_1	α_2	α_3	/ MC	anti-K^d	anti-K^d/Flu	anti-K^k	anti-K^k/Flu
K^d	d	d	d	d	65/50	60/30	10/ 8	5/ 5
C32	d	d	d /	k	61/52	62/53	20/15	5/ 5
C33	d	d /	k	k	54/45	50/35	10/ 5	5/ 5
C31	d /	k	k	k	18/ 5	5/ 5	10/15	5/ 5
C35	k /	d /	k	k	2/ 2	nd	10/5	nd
K^k	k	k	k	k	20/10	8/ 5	72/60	40/35
C23	k	k	k /	d	5/ 5	nd	38/40	nd
C24	k	k /	d	d	25/10	nd	65/55	nd
C25	k /	d	d	d	25/10	nd	25/15	nd
C26	d /	k /	d	d	20/10	nd	5/10	nd

Reactivity of CTL against cells transfected with hybrid genes was determined in a standard chromium release assay. Effector : target ratio: 80:1/40:1.

transfected with H-2 hybrid genes. The results summarized in Table 2 show that K^k specific CTL recognize hybrid antigens only if both, $\alpha 1$ and $\alpha 2$ domains, are of the K^k allele. Likewise K^d restricted influenza specific CTL recognize also only influenza together with hybrid antigens which carry both, $\alpha 1$ and $\alpha 2$ domains, of the K^d allele. Respective results were obtained with K^k specific alloreactive CTL and K^k restricted influenza specific CTL, and with TNP specific K^k restricted CTL (not shown). Other investigators have obtained similar results (12-15). No contribution of $\alpha 3$, M or C domains could be observed.

However, all these experiments have been performed with bulk cultures of CTL, that is with mixed CTL populations. To determine whether there are some few CTL clones which recognize a particular domain independent of the allelic type of the other domains, experiments were designed utilizing K^k and K^d alloreactive CTL clones generated in a limiting dilution system (7). The results can be summarized as follows (Table 3). Of 76 K^k specific CTL clones tested, 75 recognized the hybrid antigens only when $\alpha 1$ and $\alpha 2$ are K^k, one clone recognizes hybrid C31 which has $\alpha 1$ of K^d and $\alpha 2$ of K^k. Among the 64 K^d specific CTL clones 5 react with $\alpha 1$. Thus, as in the case of antibodies, there are only few CTL clones which can recognize $\alpha 1$ independently. No CTL clones reactive with $\alpha 3$ were found, which is in contrast to the antibody studies.

Table 3.
Summary: Frequency of mAb and CTL clones with specificity for distinct H-$2K^d$ and K^k domains.

H-2 domain recognized	Antibodies (29 tested)	CTL clones (140 tested)
$\alpha 1$	8[a]	5
$\alpha 1 + \alpha 2$	18	135
$\alpha 3$	3	0

a: decreased binding to hybrid antigen as compared to binding to wild type H-2.

Summary and Concluding Remarks

For the assignment of functional properties to individual H-2 protein domains, H-2 hybrid genes were constructed by exchange of exons derived from different alleles. These hybrid genes were introduced and expressed in cell lines and analysed with monoclonal antibodies and cytotoxic T lymphocytes. The results show that the vast majority of antibodies, alloreactive and H-2 restricted CTL react with determinants on the first two external domains, $\alpha 1$ and $\alpha 2$, of the H-2 molecule. Only few mAb or CTL clones were found able to recognize $\alpha 1$ or $\alpha 2$ independently of the neighboring domain. These findings demonstrate that the H-2 determinants recognized by the immune system are conformational determinants generated by interaction of $\alpha 1$ and $\alpha 2$ domains. These conclusions are also supported by work of other investigators (12-15). No significant contribution of $\alpha 3$, membrane or cytoplasmic domain to these determinants was found. In further studies we deleted the exon M encoding the membrane spanning region of the H-2 molecule. As a result the H-2 molecule can no longer be anchored in the plasma membrane and is therefore secreted by the cell (unpublished). These examples demonstrate how by genetic engineering techniques functional properties can be assigned to descrete regions of a molecule of biological importance.

Acknowledgements

This work was supported by Deutsche Forschungsgemeinschaft (Ar 152/1-2) and the Swiss National Science Foundation.

References

1. Zinkernagel, R.M., P.C. Doherty. 1979. Adv. Immunol. 27, 51.

2. Steinmetz, M., K.W. Moore, J.G. Frelinger, T.B. Sher, F.W. Shen, E.A. Boyse, L. Hood. 1981. Cell 25, 683.

3. Nathenson, S.G., H. Uehasa, B.M. Ewenstein, T.J. Kindt, J.E. Coligan. 1981. Ann. Rev. Biochem. 50, 1025.

4. Kvist, S., L. Roberts, B. Dobberstein. 1983. EMBO J. 2, 245.

5. Arnold, B., A. Archibald, H.G. Burgert, S. Kvist. 1984. Nucl. Acids Res. 12, 9473.

6. Arnold, B., H.-G. Burgert, U. Hamann, G. Hämmerling, U. Kees, S. Kvist. 1984. Cell 38, 79.

7. Arnold, B., U. Horstmann, W. Kuon, H.-G. Burgert, G.J. Hämmerling, S. Kvist: Proc. Nat. Acad. Sci. USA (in press)

8. Wigler, M., A. Pellicer, S. Silverstein, R. Axel, G. Urlaub, L. Chasin. 1979. Proc. Nat. Acad. Sci. USA 76, 1373.

9. Southern, P.J., P. Berg. 1982. J. Mol. Appl. Genet. 1, 327.

10. Lemke, H., Hämmerling, G.J. 1981. In: Monoclonal Antibodies and T Cell Hybridomas (G.J. Hämmerling, U. Hämmerling and J.F. Kearney, eds.). Elsevier North Holland, p. 102.

11. Sachs, D.H., N. Mayer and K. Ozato. 1981. In: Monoclonal Antibodies and T Cell Hybridomas (G.J. Hämmerling, U. Hämmerling and J.F. Kearney, eds.). Elsevier North Holland, p. 95.

12. Ozato, K., G.A. Evans, B. Shykind, D. Margulies, J.G. Seidman. 1983. Proc. Nat. Acad. Sci. USA 80, 2040.

13. Allen, H., D. Wraith, P. Pala, B. Askonas, R.A. Flavell, 1984. Nature 309, 279.

14. Stronowski, I., A. Orn, R.S. Goodenow, M. McMillan, J. Forman, P.R. Brayton, J. Frelinger, L. Hood. 1984. Immunogenetics 20, 141.

15. Reiss, C.S., G.A. Evans, D.H. Margulies, J.G. Seidman, S.J. Burakoff. 1983. Proc. Nat. Acad. Sci. USA 80, 2709.

PROTEIN CHEMISTRY OF HALORHODOPSIN BASED ON ITS CYSTEINE RESIDUE

E.S.Schegk, J. Tittor, F. Lottspeich, and D.Oesterhelt

Max-Planck-Institute of Biochemistry, D-8033 Martinsried, FRG

1. Introduction

In the past few years it has been shown by several groups that the cytoplasmic membrane of Halobacterium halobium contains in addition to bacteriorhodopsin (BR)*) a second retinal-binding protein called halorhodopsin (HR) (for review see (1)) transporting upon illumination chloride ions into the cell's interior (2,3,4).

Amino acid analysis of HR revealed that this protein contains one cysteine residue. After addition of the thiolspecific reagent $HgCl_2$ to cell vesicles containing HR, no more chloride transport activity could be detected, indicating cysteine to play a functional role in the transport process. Furthermore the photochemical reactions of HR are strongly affected by $HgCl_2$ (5). The localisation of the cysteine residue within the primary structure of the protein is not only of importance for functional but also for structural reasons.

*) Abbreviations used in the text: BR bacteriorhodopsin, HR halorhodopsin, HPLC high performance liquid chromatography, SDS-PAGE sodium dodecylsulfate polyacrylamide gelelectrophoresis, MOPS morpholinopropanesulfonic acid, EDTA ethylenediamine tetraacetate sodium salt, DABIA 4'dimethylaminoazobenzene 4-iodoacetic acid, DTNB 5,5'dithiobis(2-nitrobenzoic acid), NTCB 2-nitro 5-thiocyanobenzoic acid, CNBr cyanogene bromide, Tris tris(hydroxymethyl)aminomethane

Chemistry of Peptides and Proteins, Vol. 3

First, a reliable identification of sequences of HR is necessary, because the yield of peptides of this very hydrophobic protein (purity not better than 95%) could be very low compared to peptides from contaminating proteins. Second, the thiol group of the cysteine can be used for a specific covalent coupling of HR to a solid matrix and facilitates the isolation of peptides after cleavage of the protein by several methods. Third, the strong influence of the modified thiol group on the photochemical properties of HR indicates the importance of the cysteine for the chromophore structure within the protein.
In this report we describe experiments which provide evidence for answers to the question of localisation of the cysteine residue and therefore allow some more insight to the structure of HR.

2. Materials and Methods

All chemicals used were either from Merck, Darmstadt or from Sigma. The HPLC system used was from Knauer with a widepore column from Baker. HR was isolated according to the method of Steiner and Oesterhelt (6). Purity was checked by the absorption-ratio at 280 nm/578 nm (less than 1.6).

Titration of the Cysteine Content of HR with DTNB

The standard procedure (7) was modified for hydrophobic proteins and with respect to the fact that after denaturation of a retinal-protein the absorption of free retinal disturbs the photometric test. For this purpose, BR, which contains no cysteine, was taken as a reference.
Throughout this paper buffer consists of 1M NaCl, 1% octylglucoside (w/v), 10 mM MOPS, pH=7.0. HR, 6.5 nmol in 150 µl buffer,

was denatured with 150 µl 20% SDS (w/v), 1mM EDTA within a few minutes. After addition of 400 µl 50 mM MOPS, pH=7,0, 1mM EDTA, a baseline was stored in the spectrophotometer (Kontron UVIKON 810). The reaction was started upon addition of 100 µl 1mM 5,5'dithiobis(2-nitrobenzoic acid) (DTNB) in MOPS buffer to both cuvettes and the difference spectra recorded at intervals of few minutes.

For titration of the protein in the native form, the same test was carried out without SDS and the native state was checked with the help of the absorption band of HR at 578 nm.

A control with bovine serum albumin under the same conditions showed, that the described modifications are applicable without affecting accuracy of the analysis.

Carboxymethylation of HR and Modification with DABIA

Synthesis of 4-N,N-dimethylaminoazobenzene - 4'iodoacetamide (DABIA) and its reaction with HR was carried out as described in ref.(8).

For determination of the cysteine content by carboxymethylation, 3 µl of 0.5 M mercaptoethanol and 100 µl of 20% SDS were added to 7,4 nmol HR in 100 µl buffer. After 2 min., 50 µl of 100 mM EDTA, 250 µl of 1.5 M Tris-HCl (pH=8.6) and 500 µl of 4.4 mM iodoacetic acid in dimethylformamide were added and the solution stirred for 2h in the dark under nitrogen. The reaction was stopped with 400 µl of trifluoroacetic acid and the mixture first dialyzed once against 500 ml 50% acetic acid for 14h at 4^{0}C and then twice against 500 ml water. Amino acid analysis of the precipitate was carried out following standard procedures (9) after removal of residual SDS by ion-pair extraction (10).

The preparative coupling of DABIA to HR was performed in a similar way. For details see ref. (11).

Cleavage of HR with NTCB at the Cysteine Site

200 nmol HR in 4 ml buffer were denatured by the addition of 6 µl of 50 mM mercaptoethanol in 1M Tris-HCl (pH=8.0) and 2 ml of 20% SDS in 100 mM EDTA. An additional of 5 ml Tris-HCl (pH=8.0) and 1.11 mg solid 2-nitro 5-thiocyanobenzoic acid (NTCB) were added under nitrogen and incubated after readjustment of pH at 37°C for 1h (12). The reaction was stopped with acetic acid and the

mixture first dialyzed against 50% acetic acid and then against water. The precipitate was incubated in 10% SDS, 0.1 M sodium borate (pH=9.0) and an aliquot subjected to SDS-PAGE.

Cyanogen Bromide Cleavage and HPLC

HR or modified HR was dissolved in 98% formic acid and then diluted with water to 70% under nitrogen-flushing. CNBr dissolved in 70% formic acid, was added in a 200fold excess over methionine (final concentration 5%). After incubation for 5h in the dark at room temperature the whole mixture was lyophilized and the peptides were extracted with 10% aqueous formic acid.

The peptides were separated on a Bakerbond C_{18} Widepore column with a gradient (5% to 50%) of acetonitrile in 10% formic acid (flow rate 1ml/min). Sequencing was performed with a liquid phase sequencer (modified Edman-type).

Coupling of HR to a Iodoacetamido-propyl-glass Support

Aminopropylglass, type Fractosil^{R} (Merck, Darmstadt, FRG) was used as the starting material for the synthesis of this thiol-specific solid phase support (13) . The glass support was reacted with iodoacetic acid and dicyclohexylcarbodiimide and afterwards 40 µmol thiolactive groups/g were detected by amino acid analysis following glutathiontreatment. The coupling of HR to the column was similar to the above described modification with DABIA except that the pH was lowered to 8.25, mercaptoethanol was reduced to a 5-fold excess and mixing was done by gentle shaking the reaction vessel. The reaction was stopped with pure mercaptoethanol and the glass support extensively washed with SDS-buffer, 10% formic acid, 98% formic acid and finally water.

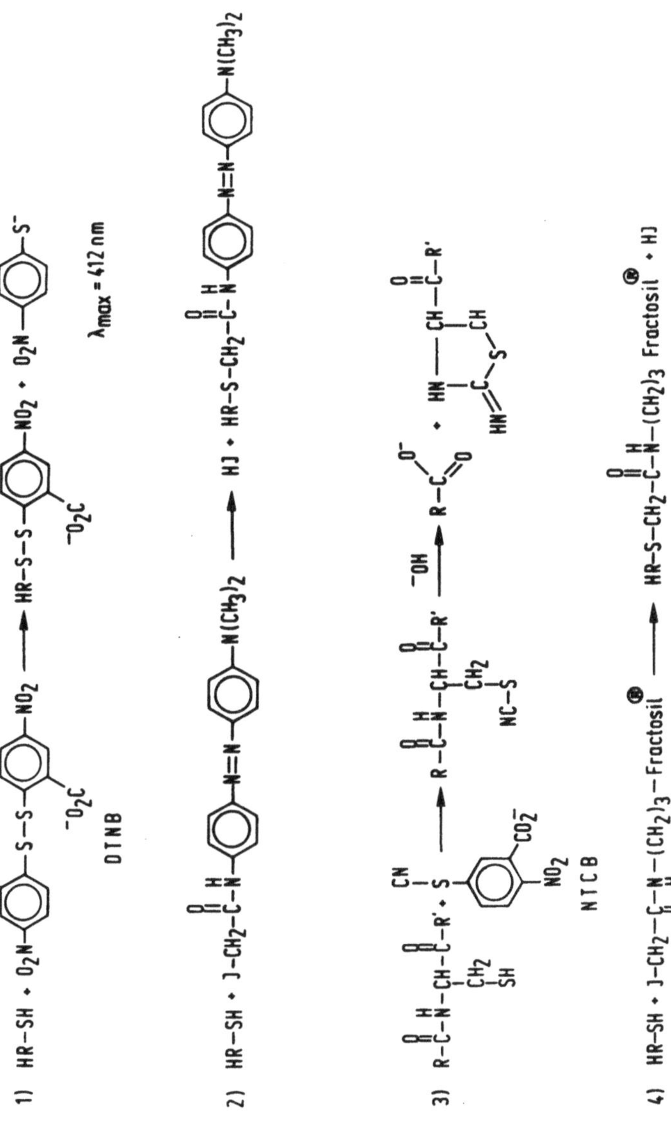

Figure 1: Reactions of the cysteine residue in HR described in this work.

V8-Protease digestion was performed at 37°C for 24h in 0.1% SDS, 0.1 M ammonium bicarbonate (pH=7.8) with shaking (14).
Glass support samples for sequencing were directly applied to a gas phase sequencer.

3. Results

Experiments with the Uncleaved Protein

In order to confirm the occurrence of a cysteine residue in HR, we titrated the protein after denaturation in SDS with Ellman's reagent, DTNB (see reaction 1 in Fig.1). As it is shown in Figure 2 one free thiol group per molecule of HR was found, but was only detectable in the denatured and not in the native protein.

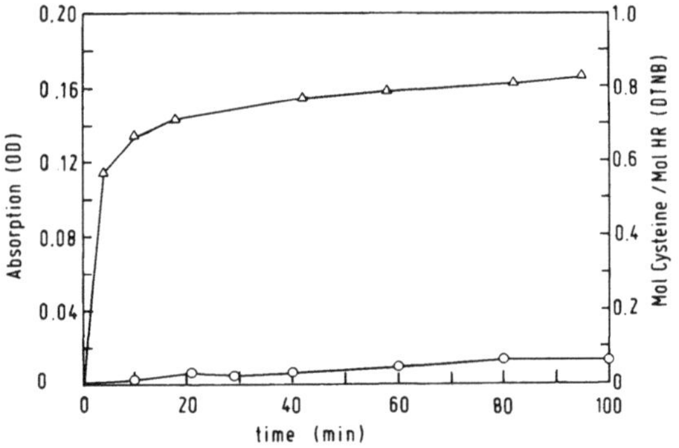

Figure 2: Titration of denatured and native HR with DTNB
With denatured HR 0.85 mol cysteine per mol of protein was found, whereas native HR has no detectable thiol group.

The amino acid composition is given in Table 1 together with the values reported in the literature. The cysteine content was quantitated by carboxymethylation and the amount of carboxymethylcysteine was found to be 1.1 mol per mol HR. The comparision of various amino acid analyses in Table I shows considerable variations of up to 30% for some amino acids and is not easy to explain except for those mentioned in the legend.

Table I:

Amino Acid Composition of HR

Amino Acid	**a**	**b**	**c**	**d**
Asp	12	12	13	12
Thr	18	17	18	13
Ser	16[1]	21	19	13
Glu	13	12	10	13
Pro	11	6	7	11
Gly	23	25	22	25
Ala	36	33	31	29
Cys	1	1	1*	0
Val	22	22	27	21
Met	7[2]	6	9	9
Ile	14	11	12	15
Leu	33	33	30	36
Tyr	7	5	6	11
Phe	10	8	9	13
Lys	2	2	2	7
His	2	2	2	0
Arg	9	9	12	7
Trp	7*	7	7*	8
Total residues	243	235	236	248
Calculated MW	25kD	24kD	25kD	26kD

columm **a**: HR after carboxymethylation, this work, values were obtained with Phe as reference amino acid, 10 residues, yielding a molecular weight of 25kD
columm **b**: from ref. (5)
columm **c**: from ref. (16)
column **d**: BR values from amino acid sequence (17)

*): values for Trp and Cys are assumed to be 7 and 1, if not experimentally determined
1): not corrected for destruction during hydrolysis
2): often underdetermined because of oxidation to the corresponding methionine sulfone

Compared with BR which contains neither cysteine nor histidine, the content of serine and threonine is higher in HR and the content in tyrosine and lysine is lower. The average hydrophobicity of both proteins is about the same (0.3 according to (15)).
The purple azo dye DABIA was reacted with a sample of purified native HR (reaction 2 in Fig.1) and the modified protein subjected to a SDS-PAGE. In lane c of Figure 3 a single DABIA-stained band is seen, which could be stained afterwards with Coomassie Blue (lane b).

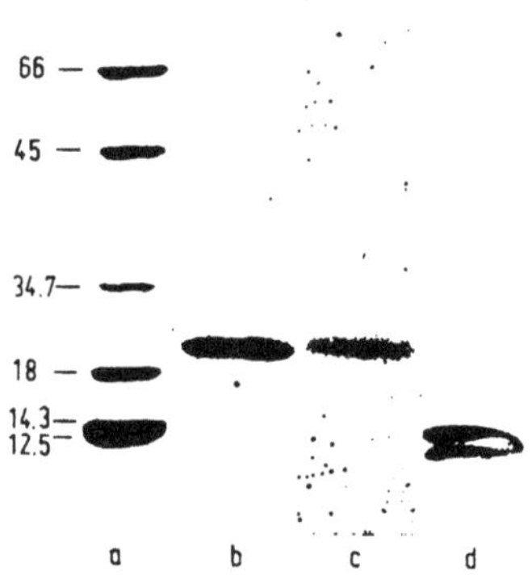

lane a: marker proteins in Coomassie-stain

lane b: HR modified with DABIA in Coomassie-stain

lane c: as b, but before stain

lane d: HR, cleaved by NTCB

Figure 3: SDS-PAGE of HR, modified with DABIA (c) or cleaved with NTCB (d).

In lane d of this Figure the two peptides with apparent molecular weights of 12 and 13 kD respectivly are shown which were obtained after cysteine-specific cleavage with NTCB (reaction 3 in Fig. 2). This result clearly positions the cysteine residue in the middle of the polypeptide chain of HR.

Cleavage and Analysis of the Peptides from HR
DABIA-modified HR was cleaved with CNBr and the peptides obtained separated by HPLC. A corresponding chromatogram is shown in Figure 4. Simultaneous detection at 535 nm and 280 nm allowed the identification of one coloured peptide which was characterized by amino acid analysis (Asp 1,Thr 1,Ser 2,Glu 3,Gly 4,Ala 2,Val 1,Ile 1,Leu 2,Arg 1) but could not be sequenced.

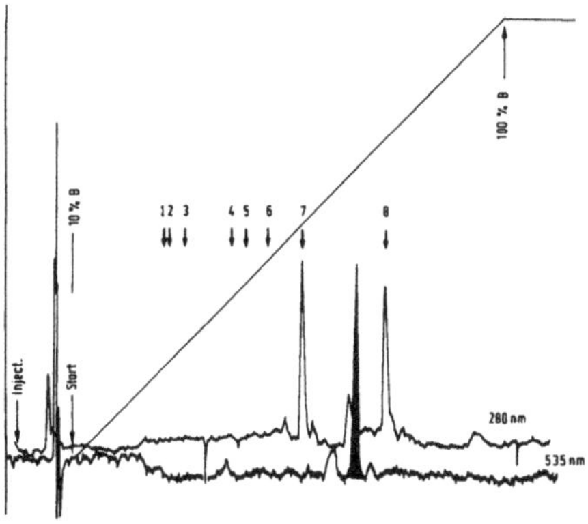

Figure 4: Purification of the DABIA-modified cysteine peptide (filled in black) after CNBr cleavage by HPLC on a Bakerbond C_{18} Widepore column. Solvent A: 10% formic acid in water. solvent B: 10% formic acid and 50% acetonitrile in water, gradient elution was from 10% B to 100% B within 45 min (1ml/min.). Arrows indicate the positions in the chromatogram, where peptides 1 - 8 according to (15) elute. For explanation see Discussion.

In addition to this DABIA-modified peptide, two more peptides appeared, being identical to peptides number 7 and 8 from CNBr cleavage with denatured isolated HR (15). Note that only trace amounts of peptides 1 to 6 were detected in this preparation as will be discussed later.
In order to circumvent the problems of the strongly hydrophobic DABIA residue which made necessary the use of 10% formic acid instead of 0.19% triflouroacetic acid in the water/acetonitrile system and which prevented successful runs in the liquid phase sequencer, we used ^{14}C-iodoacetic acid as a label for a cysteine-containing peptide. Isolation of this modified peptide was possible and yielded an amino acid composition comparable with the DABIA modified peptide. Unfortunatly, sequencing was not possible because Edman-degradation was blocked. During the isolation of the CNBr peptides, however, we could increase the sequence-information of the polypeptide stretch

around peptide 2 which starts with

P A G . . .

The newly CNBr-found peptide starts with

I E M(O_2) P A G . . .

i.e. it contains an uncleaved, but oxidized methionine. This clearly indicates that in native HR part of the methionine residues becomes oxidized irreversibly during the isolation procedure and therefore are not susceptible to CNBr cleavage.

HR Coupled to a Solid Phase Support

HR can be specifically coupled to iodoacetamidopropyl-Fractosil (reaction 4 in Fig.1) in about 20% yield as shown by amino acid analysis of the support-bound protein and the yield of carboxymethylcysteine.

Treatment with chymotrypsin, trypsin and even CNBr yielded a heterogeneous population of support bound peptides with only minor differences in the amino acid composition compared with untreated HR coupled to the glass support demonstrating neglegible cleavage of the polypeptide chain under these conditions. Because the same amount of unoxidized methionine was found, the low extent of CNBr cleavage in this case cannot be explained by oxidation but must have different reasons. As expected from this result no unique sequences could be detected from the support-bound residual material.

Only V8-protease digestion lead to the sequence:

N A L L . . .

which appeared to be part of the initial sequence detected with intact HR coupled to this glass support. Although HR is described to have a blocked N-terminus (15), we found about 25% of the molecules unblocked. Edman-degradation yielded the sequence

A V R E N A L L . . .

The overlapping sequences demonstrate cleavage of HR by V8-protease after a glutamic acid residue near the N-terminus. Since the cleavage was carried out in ammonium bicarbonate buffer only glutamic acid residues are expected to be sites of cleavage. However, as seen already with trypsin and chymotrypsin, the glass support prevents cleavage of most of the 13 possible sites in HR by V8-protease.

4. Discussion

Although isolation of cysteine peptides labeled with two different compounds and binding of HR to a glass support was possible we could not obtain partial sequences including the cysteine residue. Nevertheless, some conclusions about its localization in HR can be drawn. From the experiments with DTNB we conclude that HR has a single free thiol group in that part of its structure which is not accessible to DTNB in the native protein.

The stoichiometry was confirmed by amino acid analysis after carboxymethylation of the denatured protein. Modification of cysteine with a thiol-specific azo-dye stains exclusivly the 25 kD band proving that this cysteine residue resides exclusivly in HR. The cleavage of HR with NTCB leads to two peptides of similar molecular weight, so that the cysteine must be located in the middle part of the primary structure. This experiment is a further hint of the existence of only one cysteine per HR. Unfortunately, the procedure modifies the cysteine at the N-terminus of the second peptide preventing its Edman-degradation (12) (reaction 3 in Fig. 1).

DABIA could be used as a coloured label for the cysteine peptide, but it increased its hydrophobicity. This might be the cause for washing out in the liquid phase sequencer.

CNBr cleavage of the native isolated HR confirmed the existence of peptides 7 and 8 obtained from HR isolated in the denatured form (15). However, in our preparations, peptides 1 to 6 appeared only in small amounts. If the isolation of denatured HR would not remove all impurities, peptides 1 -6 could arise from contaminanting proteins not present in native HR preparations. An alternative and preferable explanation for the lack of peptides 1 - 6 is the following: Native HR is isolated in Lubrol PX, a detergent containing oxidizing impurities in a rather high concentration. Methionines localized on the outside of the protein could be oxidized to methionine sulfone and only the methionines located inside the protein structure are protected.

Since two peptides (7 and 8) are obtained from native HR at least 3 (if peptides 7 and 8 are contiguous) or possibly 4 methionines must be protected. This hypothesis is confirmed by the appearance of a methionine sulfone in the sequences shown above (peptide 2). Like some of the methionines, the cysteine and part of the chromophorestructure are protected by the protein structure.
With a single specific functional group it should be possible to couple the protein to a solid support, cleave it, wash away generated peptides and to sequence the remaining cysteine peptide while bound to the solid support. Such an approach offers a distinct advantage over liquid phase sequencing in that hydrophobic peptides are no more washed out. Unfortunately this advantage could not be used in our experiments because heterogeneous material was produced by CNBr cleavage. We explain this with the above mentioned oxidation of parts of the methionines and also with special problems of reactions on solid supports with very porous surfaces. This latter point is documented by the fact that methionines are found even after extended treatment with CNBr (11).
The sequencing of HR, coupled to the iodoacetamidopropylFractosil in a gas phase sequencer allowed us to gather some information about the N-terminal part of the molecule although most of the N-terminus is blocked. This could be explained by several assumptions. Either the modification of the N-terminal amino acid is only complete to an extend of 75% or the processing of the HR precursor leads to slight variations around the N-terminus.

Acknowledgement

We wish to thank I.Dunger for carrying out the amino acid analyses and Dr. P.Hegemann for helpful discussions.

5. References

1. Stoeckenius,W.; Bogomolni,R.A.: Ann. Rev. Biochem. 51, 587-616 (1982)

2. Bamberg,E.; Hegemann,P.; Oesterhelt,D.: Biochem. 23, 6216-6221 (1984)

3. Schobert,B.; Lanyi,J.K.: J.Biol.Chem. 257, 10306-10313 (1982)

4. Bogomolni,R.A.; Taylor,M.E.; Stoeckenius,W.: Proc. Natl.Acad. Scie. 81,5408-5411 (1984)

5. Ariki,M.; Lanyi,J.K.: J.Biol.Chem. 259, 3504-3510 (1984)

6. Steiner,M.; Oesterhelt,D.: EMBO J. 2, 1379-1385 (1983)

7. Janatova,J.; Fuller,J.K.; Hunter,M.J.: J.Biol.Chem. 243, 3612-3622 (1968)

8. Chang,J.Y.; Knecht,R.; Braun,D.G.: Biochem.J. 211, 163-171 (1983)

9. Chrestfield,A.M.; Moore,S.; Stein,W.H.: J.Biol.Chem. 238, 622 -627 (1963)

10. Konigsberg,W.H.; Henderson,L.: Meth.Enzym. 91, 254 -259 (1983)

11. Schegk,E.S.: Diplomarbeit Ludwigs Maximilians Universitaet Muenchen (1984)

12. Stark,G.R.: Meth.Enzym. 47, 129-132 (1977)

13. Chang,J.Y.; Creaser,E.H.; Hughes,G.J.: FEBS Lett. 78,147-150 (1977)

14. Cleveland,D.W.; Fischer,S.G.; Kirschner,M.W.; Laemmli,U.K.: J.Biol.Chem. 252, 1102-1107 (1977)

15. Hegemann,P.: Ph.D. Thesis Ludwigs-Maximilians Universitaet Muenchen (1984)

16. Sugiyama,Y.; Mukohata,Y.: J. Biochem. 96, 413 - 420 (1984)

17. Ovchinnikov,Y.A.; Abdulaev,N.G.; Feigina,M.Y.; Kiselev,A.V.; Lobanov,N.A.: FEBS Lett. 100, 219 - 224 (1979)

STRUCTURAL INVESTIGATION OF GTP-BINDING PROTEIN FROM BOVINE RETINAL ROD OUTER SEGMENTS

V.M. Lipkin, T.M. Shuvaeva, K.A.Ishchenko, A.N. Obukhov, I.N. Telezhinskaya, O.G. Shamborant

Shemyakin Institute of Bioorganic Chemistry, USSR Academy of Sciences, Ul. Miklukho-Maklaya 16/10, Moscow V-437, USSR

Introduction

The GTP-binding protein or transducin is a peripheral protein located in retinal rod outer segments. Transducin is an amplifier and one of the transducers of a visual impulse, which performs the coupling between rhodopsin and cGMP-phosphodiesterase (1). It consists of three subunits: α - (39 kDa), β - (37 kDa) and γ - (8 kDa). This study is a part of the investigation on the primary structures and mechanism of function of the proteins involved in visual signal transduction.

Results

Transducin was isolated according to Kuhn's procedure (2). Two methods underlay separation of its subunits: electrophoresis in the acetate cellulose block under denaturing conditions and HPLC (3). Electophoresis is characterized with high productivity; a few dozens of milligrams of transducin can be separated simultaneously on the 30x40 cm block, 5 mm in width.
However, because of close mobility pure α- and γ-subunits cannot be often obtained. HPLC on the column with reversed phase Silasorb-Phenyl turned out to be mostly productive for pure α,γ subunits. The subunits were eluted in acetonitrile gradient in 0.1% trifluoroacetic acid.
The amino acid composition was determined for every subunit (Table 1). Despite the close molecular weights, the amino acid compositions of the α- and β-subunits have many differences. In particular, the content of glutamic acid, methionine, and arginine resi-

Chemistry of Peptides and Proteins, Vol. 3

Table 1. Amino Acid Composition of Transducin α-, β- and γ-Subunits.

Amino Acid	Number of Residues		
	α-subunit	β-subunit	γ-subunit
Asp	40	43	8
Thr	20	24	2
Ser	22	25	2
Glu	40	25	13
Pro	5	7	4
Gly	18	32	4
Ala	23	29	-
Cys	8	7	2
Val	19	15	6
Met	12	9	2
Ile	28	20	3
Leu	30	30	7
Tyr	13	8	1
Phe	17	10	2
His	9	7	-
Lys	26	11	10
Arg	17	20	3
Trp	2	6	-
Total number of residues	349	328	69

dues differs much. The γ-subunit is characterized by the absence of alanine, histidine, and triptophan residues.

The amino acid sequence of the transducin γ-subunit was established as follows. By the automated sequencing 34 N-terminal amino acid residues were determined. Use was made of the block technique to define the complete amino acid sequence of the γ-subunit. To fragmentize the polypeptide chain of the γ-subunit hydrolyses with *Staphylococcus aureus* protease and chymotrypsin as well as cyanogen bromide cleavage were used. By HPLC on the reversed phase Nucleosil C18, 9 and 12 peptides, respectively, were isolated from enzymatic hydrolyzates; N- and C-terminal cyanogen bromide fragments were also obtained. As a result a complete amino acid sequence of the γ-subunit was determined (4).

The clostripain hydrolyzate of the γ-subunit was analyzed on a ZAB(VG) mass-spectrometer by using ionization with fast atom bombardment. The obtained mass-spectrum mostly consisted of molecular ions of clostripain peptides. Their molecular masses coincided with

1 10
Pro-Val-Ile-Asn-Ile-Glu-Asp-Leu-Thr-Glu-Lys-Asp-Lys-Leu-Lys-Met-
20 30
Glu-Val-Asp-Gln-Leu-Lys-Lys-Glu-Val-Thr-Leu-Glu-Arg-Met-Leu-Val-
40
Ser-Lys-Cys-Cys-Glu-Glu-Phe-Arg-Asp-Tyr-Val-Glu-Glu-Arg-Ser-Gly-
50 60
Glu-Asp-Pro-Leu-Val-Lys-Gly-Ile-Pro-Glu-Asp-Lys-Asn-Pro-Phe-Lys-
69
Glu-Leu-Lys-Gly-Gly

Fig. 1. Amino acid sequence of transducin γ-subunit.

those expected, that confirmed the structure obtained.

The transducin γ-subunit consists of 69 amino acid residues, M_r ~8008 (Fig. 1). The peculiarity of the γ-subunit is the presence of two neighbouring cysteine residues in positions 35 and 36, probably, linked with a disulfide bond. Perhaps such an unusual bond plays an essential part in supporting specific conformation of the protein. The γ-subunit seems to have the β-turn formed by an ionic pair of lysine-34 and glutamic acid-37 along with cysteine residues.

Not long ago Khorana and Hurley (5,6) established the nucleotide sequence of the structural gene of the γ-subunit. The amino acid sequence deduced coincides with our structure. However, the sequence determined by Khorana and Hurley contains an additional residue of methionine in the N-terminus and in the C-terminus - four additional residues of cysteine, valine, isoleucine, and serine, which are evidently cleaved off at the protein processing.

At present we are engaged in investigation of the primary structure of the transducin large α- and β-subunits, simultaneously the work on isolation of the protein structural gene is performed. Freshly isolated protein preparations were found out to be easily hydrolyzed with both proteolytic enzymes and cyanogen bromide. When stored the α- and β-subunits of transducin are quickly denatured and become poorly soluble even in 70% formic acid or in 6M urea.

For cyanogen bromide cleavage the most suitable is the mixture of the α- and β-subunits obtained by dialysis against ammonia water of the fresh transducin preparation solubilized in buffer with 8M guanidine hydrochloride. The α- and β-subunits precipitated, and the γ-subunit remained in solution. The precipitate obtained was centrifuged, dissolved in formic acid and cleaved with cyanogen bromide. The primary fractionation of cyanogen bromide peptides by gel filtration on Sephadex G-50 equilibrated with 20% acetic acid yielded 4 fractions.

Further separation was performed by HPLC on reversed phase Nucleosile C18 in 0.1% trifluoroacidic acid, peptides were eluted in acetonitrile gradient. In total 21 peptides covering ~95% of the amino acid sequence of transducin α- and β-subunits were isolated from cyanogen bromide hydrolyzate. The α- and β-subunits of transducin were also cleaved with trypsin and protease from *Staphylococcus aureus*. Nucleotide probes, used for isolation of transducin α- and β-subunits structural genes from cDNA library prepared from bovine retinal RNA were synthesized on the basis of the peptide structure.

Recently Numa (7) and Lochrie (8) almost simultaneously published the nucleotide sequence of the complementary DNA coding for the transducin α-subunit. The deduced amino acid sequences of the α-subunit differ considerably: of 349 amino acid residues 69 do not coincide. The sequence determined in our laboratory agrees with that found by Numa and covers ~50% of the α-subunit polypeptide chain (Fig. 2).

It seems that cDNA isolated by Lochrie et al. corresponds to the gene for some GTP-binding protein homologous to transducin, but not to the gene for the transducin α-subunit.

Moreover, there is one difference between our results and those by Numa. We isolated a cyanogen bromide peptide having an alanine residue as the N-terminal amino acid residue (its structure corresponds to the α-subunit sequence 2-48). Specificity of the cyanogen bromide cleavage points to methionine as a residue preceeding the alanine-2 residue in the polypeptide chain of the α-subunit. However, the Met-Gly sequence precedes alanine according to Numa's structure. The error took place, probably, because thymidine was missed between guanine-5 and guanine-6 in the Numa nucleotide sequence. Perhaps that resulted from compression; then the missed thymidine was placed between adenine-1 and guanine-3. After thymidine insertion the GTP triplet was formed which is known to be a second initiating codon, coding for the methionine residue at initiation. We established the α-subunit N-terminal amino group was

Met-Gly-Ser Asp Glu-Leu-Ala-Lys-Arg Lys Gln
1-20 Met-Ala-Gly-Ala-Ser-Ala-Glu-Glu-Lys-His- Ser-Arg-Glu-Leu-Glu-Lys-Lys-Leu-Lys-Glu-

Asp Glu Lys
21-45 Asp-Ala-Glu-Lys-Asp-Ala-Arg-Thr-Val-Lys-Leu-Leu-Leu-Leu-Gly-Ala-Gly-Glu-Ser-Gly-Lys-Ser-Thr-Ile-Val-

Pro Tyr-Lys
46-70 Lys-Gln-Met-Lys-Ile-Ile-His-Gln-Asp-Gly-Tyr-Ser-Leu-Glu-Glu-Cys-Leu-Glu-Phe-Ile-Ala-Ile-Ile-Tyr-Gly-

Val Ile Pro Gly Asp Ala-Glu-Val-Ser-Cys
71-95 Asn-Thr-Leu-Gln-Ser-Ile-Leu-Ala-Ile-Val-Arg-Ala-Met-Thr-Thr-Leu-Asn-Ile-Gln-Tyr-Gly-Asp-Ser-Ala-Arg-

Val Asn-Gly Gln Asn-Asn-Leu Ser Pro Leu-Val-Glu
96-120 Gln-Asp-Asp-Ala-Arg-Lys-Leu-Met-His-Met-Ala-Asp-Thr-Ile-Glu-Glu-Gly-Thr-Met-Pro-Lys-Glu-Met-Ser-Asp-

Val Arg-Lys Gly Val Ala
121-145 Ile-Ile-Gln-Arg-Leu-Trp-Lys-Asp-Ser-Gly-Ile-Gln-Ala-Cys-Phe-Asp-Arg-Ala-Ser-Glu-Tyr-Gln-Leu-Asn-Asp-

Ser Asn-Gln Asp Ile-Thr-Ala Asp Leu Asn
146-170 Ser-Ala-Gly-Tyr-Tyr-Leu-Ser-Asp-Leu-Glu-Arg-Leu-Val-Thr-Pro-Gly-Tyr-Val-Pro-Thr-Glu-Gln-Asp-Val-Leu-

Lys Val
171-195 Arg-Ser-Arg-Val-Lys-Thr-Thr-Gly-Ile-Ile-Glu-Thr-Gln-Phe-Ser-Phe-Lys-Asp-Leu-Asn-Phe-Arg-Met-Phe-Asp-

Cys
196-220 Val-Gly-Gly-Gln-Arg-Ser-Glu-Arg-Lys-Lys-Trp-Ile-His-Cys-Phe-Glu-Gly-Val-Thr-Cys-Ile-Ile-Phe-Ile-Ala-

221-245 Ala-Leu-Ser-Ala-Tyr-Asp-Met-Val-Leu-Val-Glu-Asp-Asp-Glu-Val-Asn-Arg-Met-His-Glu-Ser-Leu-His-Leu-Phe-

Lys-Phe Ala Leu Glu
246-270 Asn-Ser-Ile-Cys-Asn-His-Arg-Tyr-Phe-Ala-Thr-Thr-Ser-Ile-Val-Leu-Phe-Leu-Asn-Lys-Lys-Asp-Val-Phe-Ser-

Val Glu Asp Asn Ser
271-295 Glu-Lys-Ile-Lys-Lys-Ala-His-Leu-Ser-Ile-Cys-Phe-Pro-Asp-Tyr-Asn-Gly-Pro-Asn-Thr-Tyr-Glu-Asp-Ala-Gly-

Ser Asp Lys
296-320 Asn-Tyr-Ile-Lys-Val-Gln-Phe-Leu-Glu-Leu-Asn-Met-Arg-Arg-Asp-Val-Lys-Glu-Ile-Tyr-Ser-His-Met-Thr-Cys-

321-345 Ala-Thr-Asp-Thr-Gln-Asn-Val-Lys-Phe-Val-Phe-Asp-Ala-Val-Thr-Asp-Ile-Ile-Ile-Lys-Glu-Asn-Leu-Lys-Asp-

346-349 Cys-Gly-Leu-Phe

Fig. 2. Amino acid sequence of transducin α-subunit. The given structure is deduced from cDNA nucleotide sequence, determined by Numa (1) and confirmed by the peptide structure analysis performed at our laboratory. The peptide amino acid sequences are underlined. Noncoinciding amino acid residues corresponding to the Lochrie structure (2) are italicized. The sequence Met-Ala- corresponding to our data is presented for Met-Gly-Ala-, published by Numa, as the N-terminal sequence of the α-subunit.

blocked probably as a result of acetylation. The blocking group nature is being investigated.

To study the antigen structure of the transducin molecule and to elucidate the protein function mechanism and the role of its individual subunits, monoclonal antibodies were produced (9). The native transducin served as an antigen in all hybridizations, for immunization mice of Balb/c or C57 Bl/6 lines were used, hybridization of immune splenocytes was performed with the cells of myeloma Ag8.653x63. Sharply positive clones were produced as ascyte. All the antibodies were of $\gamma_1 k$ type. For their purufication affinity chromatography on protein A-Sepharose or ion exchange chromatography on DEAE-cellulose was applied.

The specificity of the obtained monoclonal antibodies in relation to the transducin subunits was studied by immunoelectroblotting (Table 2). The results showed, after electrophoresis under nondenaturing conditions two monoclonal antibodies, A3G7 and A3C10, were specifically bound to the α-subunit, A3E4 and 3B3 interacted with the complex βγ, monoclonal antibodies IC3 do not show a clear interaction. According to immunoelectroblotting after SDS electrophoresis in gradient polyacrylamide gel A3G7 and A3C10 were bond to the α-subunit, A3E4 - to the γ-subunit, 3B3 and IC3 do not interact with the denatured protein.

Interaction of monoclonal antibodies with isolated subunits was studied by immunoenzymatic analysis. The α,γ-subunits and the βγ

Table 2. Specificity of the transducin monoclonal antibodies.

Monoclonal antibody	3B3	IC3	A_3G_7	A_3E_4	A_3C_{10}
Immunoelectroblotting under nondenaturating conditions	βγ	βγ + α	α	βγ	α
Immunoelectroblotting in the presence of SDS	-	-	α	γ	α
Immunoenzymatic analysis	γ	native transducin	α	γ	α
Influence in GTPase activity	-	-	100% inhibition	-	80% inhibition

complex were used as antigens. A3G7 and A3G10 were shown to be directed against α, A3E4 and 3B3 - against γ, and IC3 interacts neither with isolated subunits nor with the βγ complex, however, it is bound to the native transducin.

Thus A3G7, A3C10 and A3E4 are directed against linear determinants of the α- and γ-subunits, 3B3 recognizes the determinant of non-linear type present only in the native γ-subunit. As to IC3 it, apparently, recognizes complex topographic determinant, formed with all the subunits of transducin, that provides use of IC3 to determine the efficiency of the transducin reconstruction from the subunits.

We started investigating the influence of obtained monoclonal antibodies on the transducin molecule function. Monoclonal antibody against α was shown to suppress GTPase activity of transducin almost completely. All the five antibodies interact with the native transducin, so they recognize antigen determinants located on the protein surface and, in principle, can catch the molecule regions responsible for interaction of transducin with other components at transduction of the visual impuls, i.e. with rhodopsin and phosphodiesterase. Influence of monoclonal antibodies on transducin interaction with the given proteins is under study. Its interaction with antibodies A3G7 and A3C10 suppressing GTPase activity is of special interest. We are eager to find out whether the antibodies are directly bound to the enzyme catalytic site or it prevents formation of the specific complex of transducin with photolyzed rhodopsin leading to the protein activation.

References

1. Stryer, L., J.B. Hurley, K.-K. Fung. 1981. Trends in Biochem. Sci. 6, 245-247.

2. Kuhn, H. 1980. Nature 283, 587-589.

3. Ovchinnikov, Yu.A., V.M. Lipkin, I.N. Telezhinskaya, T.M. Shuvaeva, A.N. Obukhov, K.A. Ishchenko, V.V. Shemyakin. 1985. Bioorg. Khim. 11 (in press).

4. Ovchinnikov, Yu.A., V.M. Lipkin, T.M. Shuvaeva, A.P. Bogachuk, V.V. Shemyakin. 1985. FEBS Lett. 179, 107-110.

5. Yatsunami, K., B.V. Pandya, D.D. Oprian, H.G. Khorana. 1985. Proc. Natl. Acad. Sci. USA 82, 1936-1940.

6. Hurley, J.B., H.K.W. Fong, D.B. Teplow, W.J. Dreyer, M.I. Simon. 1984. Proc. Natl. Acad. Sci. USA 81, 6948-6952.

7. Tanabe, T., T. Nukada, Y. Nishikawa, K. Sugimoto, H. Suzuki, H. Takahashi, M. Noda, T. Haga, A. Ichiyama, K. Kangawa, N. Minamino, H. Matsuo, Sh. Numa. 1985. Nature 315, 242-245.

8. Lochrie, M.A., J.B. Hurley, M.I. Simon. 1985. Science 228, 96-99.

9. Shamborant, O.G., M.I. Kolosov, R.G. Vasilov, V.M. Lipkin, Yu.A. Ovchinnikov. 1985. Bioorg. Khim. 11 (in press).

EFFECT OF INSULIN MODIFICATION ON ITS CONFORMATION AND ENZYMATIC HYDROLYSIS

G.P.Vlasov ,N.G.Illarionova ,N.L.Izvarina and I.G.Denisov

Institute of Macromolecular Compounds of the USSR Academy of Sciences , Leningrad , U S S R

Introduction

A two-point modified A1,B29-azo-isobutyroyl insulin and its polymer derivatives have been synthesized.The effect of the chemical nature and the molecular weight of the polymer modifier on the biological activity,enzymatic hydrolysis due to action of trypsin and the conformational properties of this hormon were investigated.

Results

We have previously proposed a method for the preparation of "starlike" polymer derivatives of proteins , in particular insulin (1). The biological activity of A1,B29 -azo-izobutyroyl insulin (AIB-insulin) and A1,B29- polymer derivatives of insulin with polyvinylpyrrolydone (PVP-insulin) and polyacrylyc acid (PAA-insulin) has been studied.It was shown that this activity is only slightly dependent on the chemical nature of the polymer modifier but decreases with increasing molecular weight of the polymer (Table 1). In order to study the effect of the chemical nature and molecular weight of the polymer modifier on the rate of enzymatic hydrolysis of polymer derivatives under the influens of trypsin,the values of the maximum rate of enzymatic hydrilysis (V_{max}) and the apparant Michaelis constant (K_M $)_{app}$ were measured for AIB-insulin ond for the polymeric derivatives of insulin (Table 2).Trypsin is known to split two peptide bonds in insulin betweenlysine and alanine B29-B30 and between arginine and glycine B 22 - B 23(2).

Chemistry of Peptides and Proteins, Vol. 3

Table I.
Main Characteristics of the Samples under Investigation.

Insulin derivatives	Mol.weight of the modifier, $\cdot 10^3$ Da	Biol.activity,%	% of α-helicity[b)] at various pH of the medium pH I.5	pH 9.0	pH II.0
Insulin	-	I00	42	40	24
Treated[a)] insulin	-	I00	-	43	26
AIB-insulin	0.2	I00	38	38	24
PVI-insulin	I.5	I50	32	-	27
PVI-insulin	5	56	32	-	33
PVI-insulin	I4	45	32	-	32
PVP-insulin	3	70	32	-	27
PVP-insulin	6	50	32	-	33
PVP-insulin	I4	0	32	-	33
PAA-insulin	4	80	30	-	22
PAA-insulin	6	59	-	-	-

[a)] Insulin that passed all the stages of treatment except covalent modification.
[b)] According to optical rotation dispersion (ORD) in the range of 350 - 590 nm.

Table 2.
Kinetic Parameters of Enzymatic Hydrolysis of Modified Forms of Insulin under the Influence of Trypsin.

Insulin derivatives	Mol.weight of the modifier, $\cdot 10^3$ Da	$(K_M)_{app}$, $M \cdot 10^3$	v_{max}, $\cdot 10^{-10}$ mole/(min·mg)
Insulin	-	4.5	I.9
AIB-insulin	0.2	2.2	3.I
PVI-insulin	6	3.5	3.2
PVI-insulin	I0	2.0	I.4
PVI-insulin	I4	6.3	4.I
PAA-insulin	4	$1.8 \cdot 10^{-3}$	I.8
PAA-insulin	6	$6.6 \cdot 10^{-3}$	2.5
PAA-insulin	8	$5.4 \cdot 10^{-3}$	0.8
PVP-insulin	22	$3.8 \cdot 10^{-3}$	I.0
PVP-insulin	6	$1.3 \cdot 10^{-2}$	2.8

Because of the modification of the ε-amino group of lysin B 29 with a fragment of azo-bis-isobutyric acid or a polymer fragment, trypsin-induced hydrolysis of the B29-B30 bond is ruled out and that greatly simplifies interpretation of the kinetic data enzymatic hydrolysis of insulin derivatives.
The data obtained indicate that the modification mainly affects formation of the enzyme-substrate complex (the Michaelis complex), i.e. the value of $(K_M)_{app}$,and virtually does not effect the hydrolysis itself (V_{max}) .Moreover , the chemical nature of the polymer modifier is the major factor,whereas its molecular weight only slightly effects the hydrolysis.Passing from initial insulin, AIB-insilin and PVI-insulin to PAA-insulin,the value of $(K_M)_{app}$ changes by three orders of magnitude.In connection with thees results it is of interest to consider the effect of the chemical nature and molecular weight of the polymer modifier on the conformation stsbility of insulin.
Table 1 shows that the crosslinking of insulin at the A1,B29 amino groups virtually does not influence the content of amino acids in the α-helical conformation determined according to λ_c (from ORD data) ,whereas polymer modification leads to a certain decrease in α-helicity and the result is in fact independent of molecular weight in the range of up to 24 kDa for PVI and only slightly depends on the chemical nature of the polymer modifier.
The effect of conditions of insulin modification on its conformational properties was investigated.Fig.1 shows the CD (circular dichroism) spectra of initial insulin,crosslinked at the A1,B29 amino groups and medified by PAA.The modification was carried out at the two temperatures : 65° and 75°C.It is clear that the difference in the modification conditions has a certain effect on the CD spectra.An attempt was made to determine the effect of modification conditions on the conformational properties of modified insulin .In order to determine conformational state of insulin at modification temperatures (65° and 75°C) its CD spectra were measured (Fig. 2). At these temperatures insulin's conformation differs considerably from the native one and thermal denaturation is completely reversible .For the A1,B29-crosslinked insulin and insulin in PAA solution (tenfold exeess of PAA by weight) denaturation is also observed at 65° and 75° C (Fig.3).The CD spectra show that for both samples the conformational changes are

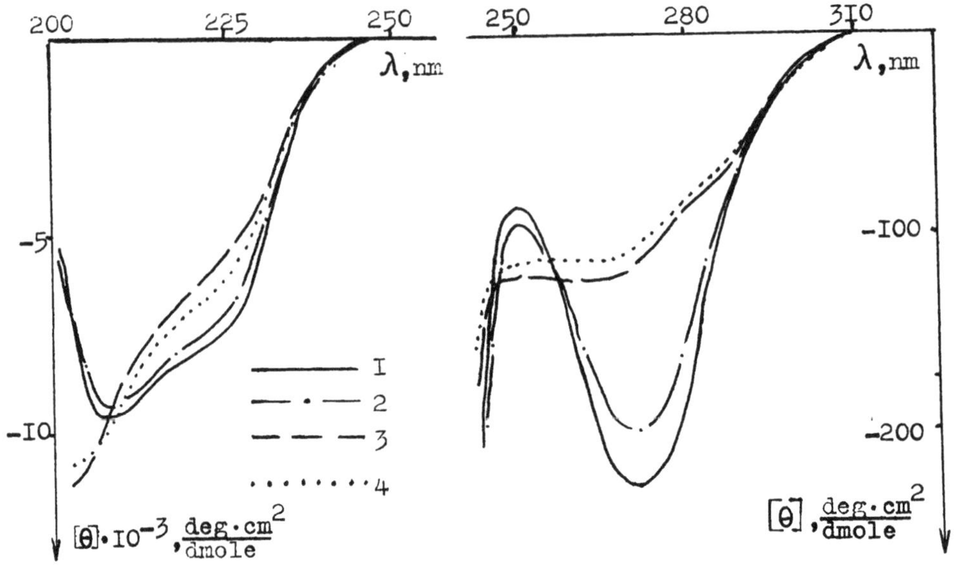

Fig. 1. CD spectra of insulin derivatives: I – insulin, 2 – AI, B29-crosslinked insulin, 3 – PAA-insulin (65°C polymerization), 4 – PAA-insulin (75°C polymerization), 0.2 M NaCl, pH 9.0.

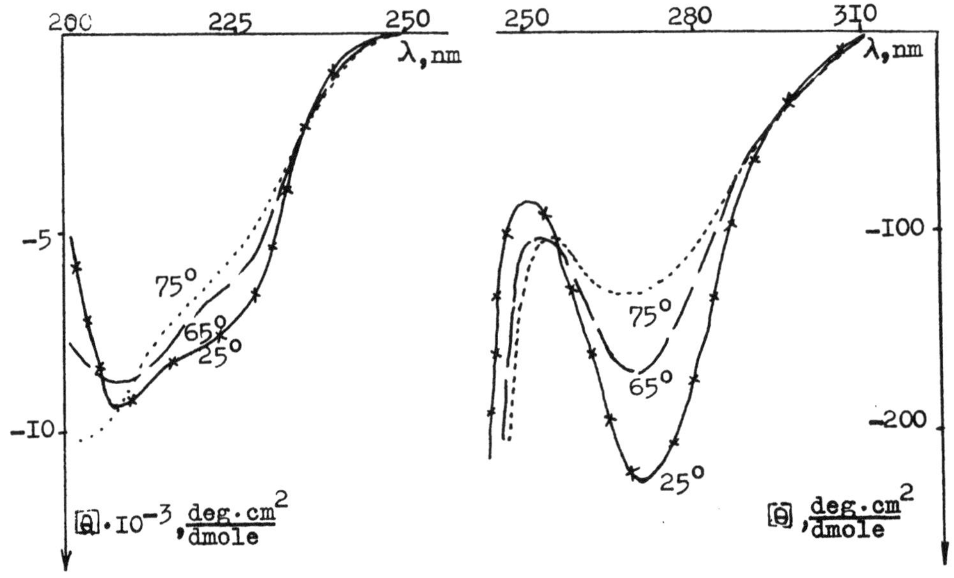

Fig. 2. CD spectra of insulin in solution, 0.2 M NaCl, pH 9.0. Thermal denaturation and renaturation (25°C, 16 hours, -×-).

somewhat more pronounced that for insulin.After equilibrium is attained under renaturation conditions the system does not return to the initial state , i.e. for the insulin modified in this manner denaturation is practically partially irreversible (for A1,B29 -crosslinked insulin it may be due to recombination of two carbon radicals at high temperature). It was found that the degree of thermal denaturation (for the initial protein) increases with the denaturation temperature.The covalent modification of insulin and its noncovalent interaction with polymer in solution probably restrict the possibility of complete restoration of its native conformation during the renaturation from the extended state of the modified sample at high temperature.In the case of polymer modifications of proteins at temperatures in the range of thermal denaturation these factors may profoundly effect the conformational state of the final product.

Fig.4 shows the dependences of specific optical rotation $[\alpha]$ at $\lambda = 450$ nm on the pH of the medium for insulin , as well as insulin crosslinked at the A1,B29 amino groups and modified by PVI various molecular weights. At pH > 9 insulin undergoes a cooperative conformational transition due to a decrease in the content of α-helical conformation from 40 % to 20 % (according to the change in λ_c). In alkaline denaturation of crosslinked insulin the beginning of conformational transition is displaced towards a more alkaline range.For the PVI-insulin a pronounced effect of stabilization of the residual structure of the protein component is observed.With increasing molecular weight of the modfying polymer the degree of cinformational stability of the insulin derivative also increases.Similar data were obtained for the A1,B29 - PVP-insulin.Comparison of dependences of $[\alpha]_{450}$ on the pH of medium for PVI- and PVP-insulin with those of PAA-derivatives (Fig.5) shows that PAA has a destabilizing effect on the residual structure of the modified hormon.Hence the chemical nature and the molecular weight of the polymer modifier virtually do not effect the α - helicity of insulin (in the range of 30 - 33 %) but profoundly effect the conformational stability or destabilization of its residual structure to the action of different denaturants.

Comparision of data on the enzymatic hydrolysis with the conformational properties and conformational stability of modified in-

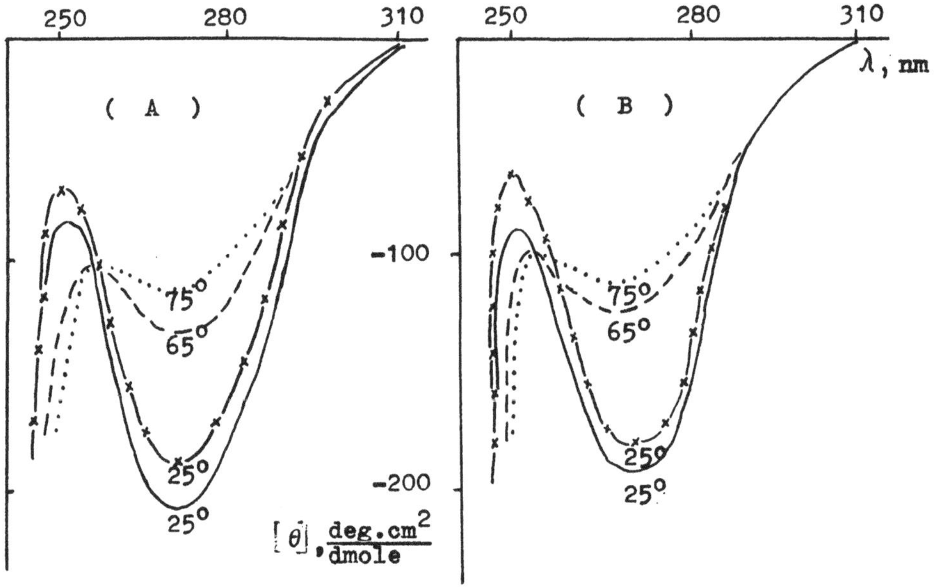

Fig.3. CD spectra of insulin + PAA (1 : 10 by weight) (A) and A1,B29-crosslinked insulin (B) solutions , 0,2 M NaCl , pH 9.0.Thermal denaturation and renaturation (25°C,16hrs -x-).

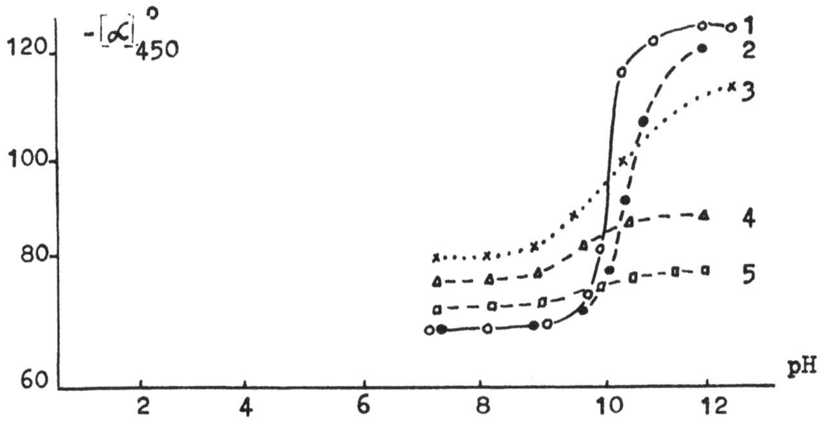

Fig.4. pH stability of insulin derivatives : 1 - insulin ; 2 - A1,B29-crosslinked insulin; 3 -PVI-insulin,M.W.of polymer 8 kDa ; 4 - PVI -insulin,M.W.of polymer 11 kDa ; 5 -PVI-insulin , M.W.of polymer 24 kDa. All solutions contain 0.2 M NaCl.

sulin derivatives to the action of alkaline pH shows that the chemical nature of the polymer effects these parameters in different ways.Herce the change in the conformational properties of insulin during modification cannot be the main cause of the change in the capacity of insulin for undergoing biodegradation.

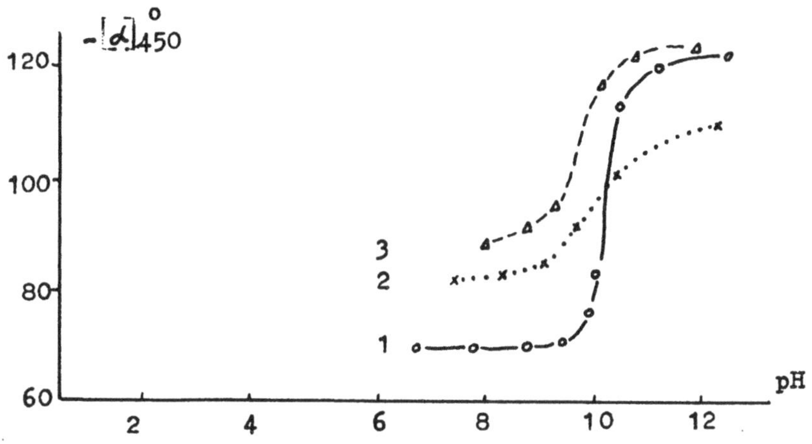

Fig.5. pH stability of polymer derivatives of insulin : 1 -insulin ; 2 - PVI-insulin ; 3 - PAA-insulin . Molecular weight of polymer modifiers 8 kDa.

Evidently in enzymatic hydrolysis the interaction between the polymer component of the polymer derivatives and the enzyme (trypsin) has a greater effect on the change in $(K_M)_{app}$ .The total negative charge of PAA-insulin (as compared to that of insulin and insulin modified by PVI) facilitates the formation of the Michaelis complex,because the isoelectric point of trypsin is 10.8 (3).The increase in the bonding of PVP-insulin to trypsin may be due to high ability of PVP and its copolymers to form complexes with proteins (4). Hence , in our case the chemical nature of the modifying polymer determines the conformational stability and the stability to enzymatic hydrolysis of the modified insulin.

The synthesis of additional polymeric derivatives of insulin and their properties will be reported elsewhere.

References

I. Vlasov,G.P., N.L. Izvarina, N.G.Illarionova. I982. In: Chemistry of Peptides and Proteins (K.Blāha and P.Malon, eds.). Walter de Gruyter&Co., N.Y., v.I, p. 337

2. Carpenter, F.H., W.E. Baum. I962. J.Biol.Chem. 237, 237.

3. Berezin, I.V., V.K. Antonov, K.A. Martinek. I976. Immobilizovannye Fermenty, Ed. Moscow University, Moscow, v. I, p. I35.

4. Kirsh, Yu.E., L.V.Sokolova. I983. Khim.Farm.Zhur., n. 6, 7II.

THE CHEMISTRY OF GLYCOPEPTIDE ANTIBIOTICS OF THE VANCOMYCIN GROUP

G.S. Katrukha, A.B. Silaev

Department of Chemistry, Moscow State University, Moscow, 119899, USSR

The last few years have seen an ever growing interest in structural and functional studies of biologically significant glycopeptides and glycoproteins. Glycopeptide antibiotics have become the object of most extensive investigations, and the glycopeptide antibiotics of the vancomycin group [1,2] - are the subject-matter of this report. Here we survey the literature and sum up our results.

Over 15 antibiotics of the vancomycin group are known to date - named so, after the first antibiotic, vancomycin, discovered in 1956 by McCormick and co-workers [3]. Japanese and American researchers are carrying out a special screening program on these antibiotics, because of considerable interest to experts in the fields of chemistry, medicine and biochemistry. This interest is due to a variety of reasons. First, the vancomycin-type antibiotics possess high bactericidal activity against a broad spectrum of Gram-positive bacteria and are used for treating staphylococcal and streptococcal infections; second, the vancomycin group antibiotics have low toxicity, they are active with respect to bacteria resistant to other antibiotics and do not produce a habituation effect; third, they possess high affinity for the growing peptidoglycane of the bacterial cell wall; fourth, they stimulate the grouth and productivity of animals, and finally, they are remarkable for their unusual composition and chemical structure, which is of indisputable interest to chemists. These properties of the vancomycin-type antibiotics are determined by their structure,

Chemistry of Peptides and Proteins, Vol. 3

Table 1. Composition and Structure of the Vancomycin Group of Antibiotics.

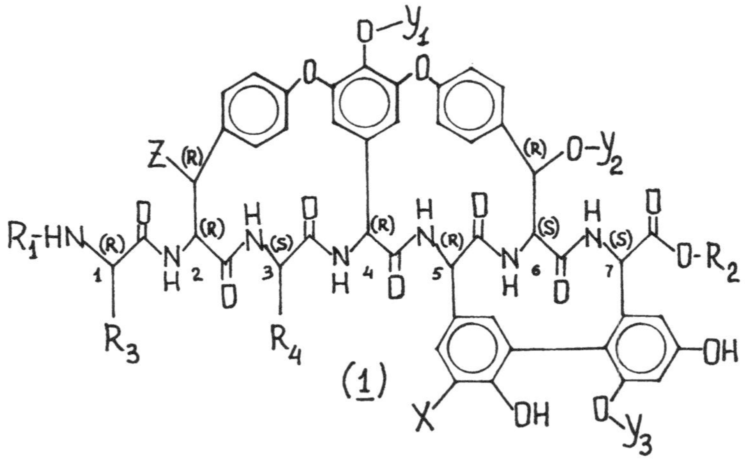
(1)

Abbreviations:

L-Aca = L-actinosamine; L-Aco = L-acosamine; Act = actinoidinic acid; D-Araf = D-arabinose; Galp = galactose; D-Glcp = D-glucose; GlcNAc = N-acetylglucosamine; D-Manp = D-mannose; L-Ria = L-ristosamine; L-Rhap: L-rhamnose; Ris = ristomycinic acid; L-Van = L-vancosamine.

Antibiotics	Ref.	R_1	R_2	R_3	R_4	X_1	X_2	X_3	Sugars	Z
1. Vancomycin	[3-5]	CH_3	H	$CH_2CH(CH_3)_2$	CH_2CONH_2	Cl	Cl	H	Y_1= D-Glcp←L-Van Y_2 = Y_3 = H	OH
2. N-Demethylvancomycin	[6]	H	H	analogous		Cl	Cl	H	analogous	OH

Table 1. (continued).

Antibiotics		R_1	R_2	R_3	R_4	X_1	X_2	X_3	Sugars	Z
3. Actinoidin [7-10]	A	H	H		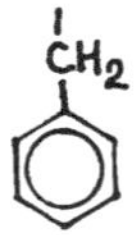	H	Cl	H	Y_1= D-Glcp←L-Aco; Y_2= L-Aca; Y_3 = D-Manp	OH
	B	H	H	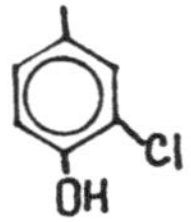		H	Cl	H	analogous	OH
4. Avoparcine [11-13]		CH_3	H	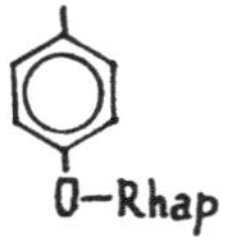		H	Cl	H	Y_1= D-Glcp←D-Ria; Y_2= Ria; Y_3= H	O-Manp
		CH_3	H	analog.	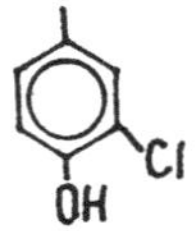	H	Cl	H	analogous	analog.
5. Ristomycin [14-18] (Ristocetin [19-23])	A	H	CH_3	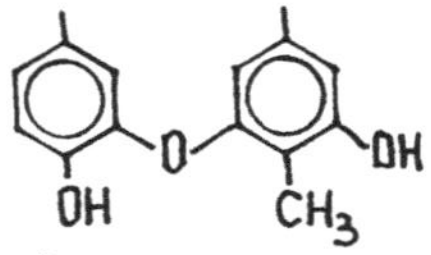		H	H	H	Y_1= D-Glcp←D-Manp (L-Rhap→D-Glcp; D-Araf→D-Manp) Y_2= L-Ria; Y_3=D-Manp	OH
	B	H	CH_3	analogous		H	H	H	Y_1= D-Glcp←L-Rhap; Y_2= L-Ria; Y_3=D-Manp	OH

Table 1. (continued).

Antibiotics		R_1	R_2	R_3	R_4	X_1	X_2	X_3	Sugars	Z
6. A35512B [24–26]		H	CH_3	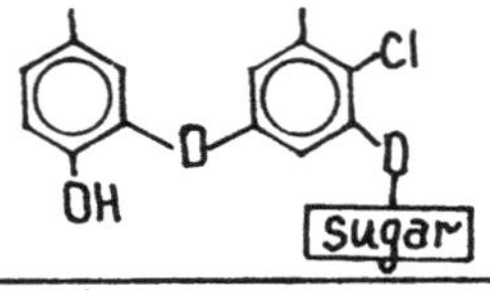		H	H	H	Y_1 = sugars; Y_2 = 3-epi-L-Van; Y_3 = sugars.	OH
7. Actaplanin (A4696) [27,28]	A	H	CH_3	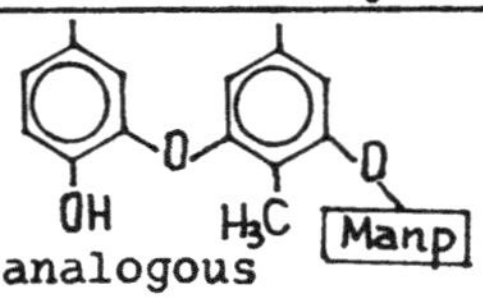		H	Cl	H	Y_1=Glcp←Manp; Y_2=Ria; Y_3=D-Manp	H
	B_2	H	CH_3	analogous		H	Cl	H	Y_1=Glcp; Y_2=Ria; Y_3=D-Manp	H
8. A41030 [29,30]	A	H	H	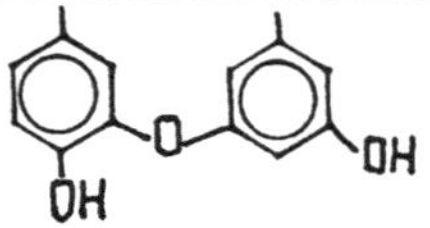		Cl	Cl	Cl	Y_1 = Y_2 = Y_3 = H	H
	B	H	H	analogous		Cl	Cl	H	Y_1 = Y_2 = Y_3 = H	H
	C	H	H	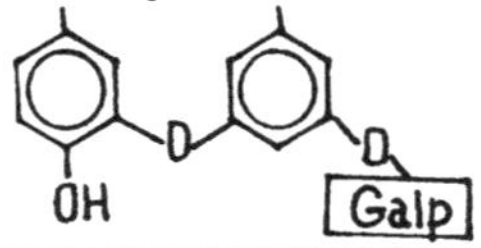		Cl	Cl	Cl	Y_1 = Y_2 = Y_3 = H	H
9. A47934 [30,31]		H	H	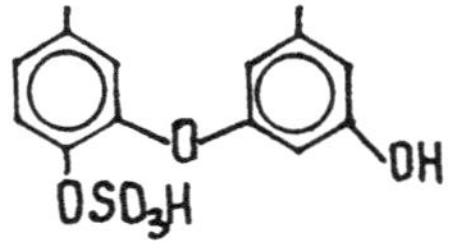		Cl	Cl	Cl	Y_1 = Y_2 = Y_3 = H	H

Table 1. (continued).

Antibiotics	R_1	R_2	R_3	R_4	X_1	X_2	X_3	Sugars	Z
10. Teicoplanin (Teichomycin A_2) [32–35] (5 closely related substances)	H	H	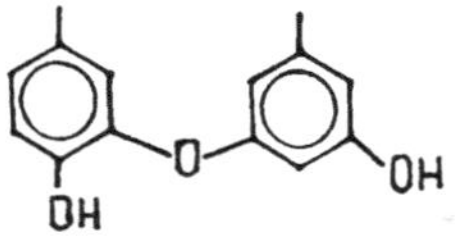		Cl	Cl	H	Y_1= GlcN-isodecanoyl; Y_2= D-GlcNAc; Y_3= D-Manp	H
11. Azureomycins A and B [36]									

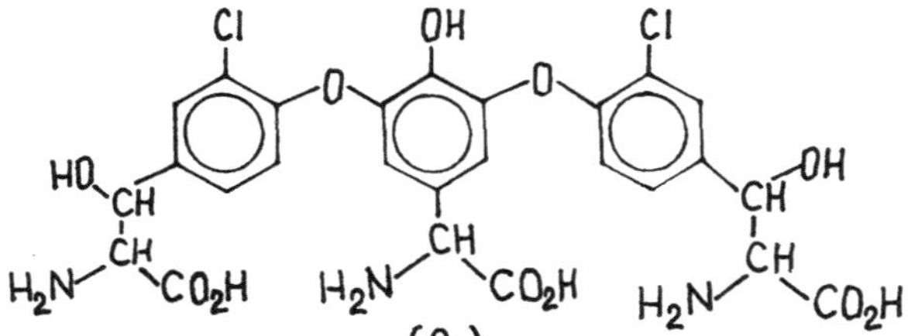

(2)

Vancomycinic acid

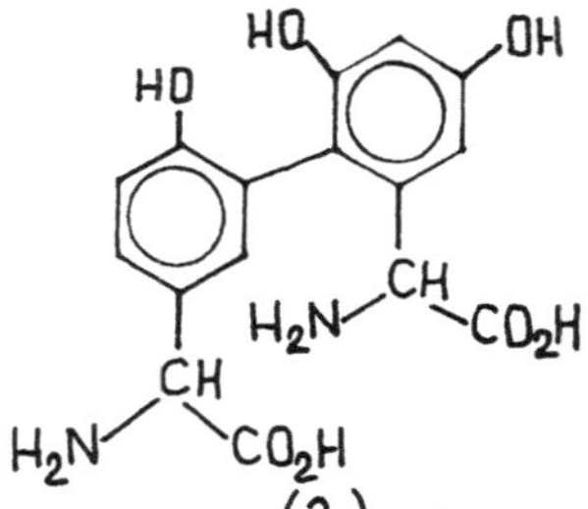

(3)

Actinoidinic acid

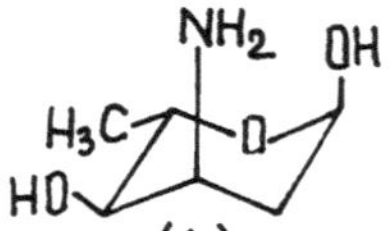

(4)

L-Ristosamine

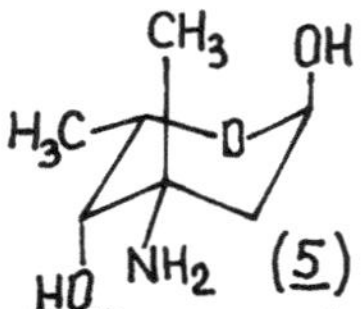

(5)

L-Vancosamine

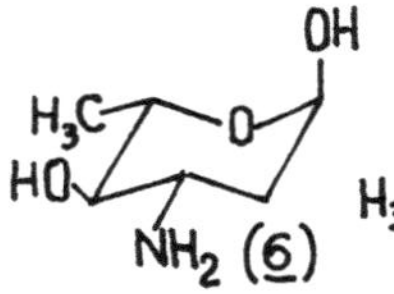

(6)

L-Acosamine

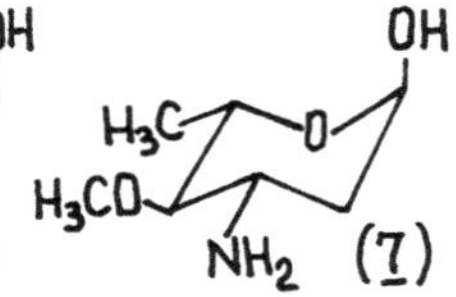

(7)

L-Actinosamine

which is indicated in the rough outline below in (1) (see Table 1). All these antibiotics are glycopeptides, containing either a tri-amino-tricarboxylic acid - vancomycinic acid (2) or mono- or didechlorovancomycinic acid in which two phenylserines are symmetrically linked with the central β-hydroxyphenylglycine moiety through diphenyl ether bonds. They all hold actinoidinic acid (3) or its analogs, as well as different amounts of neutral sugars and unusual amino sugars - 2,3,6-trideoxy-3-aminohexoses (4)-(7), detected only in this group of antibiotics, and are divided into three subgroups, depending on the structure of amino acids with radicals R_3 and R_4 in (1). The first subgroup: R_3 and R_4 - are aliphatic amino acids (in vancomycin and demethylvancomycin). The second subgroup: R_3 and R_4 - are aromatic amino acids (in actinoidins A and B and avoparcines α and β). The third subgroup is the most numerous. It includes R_3 and R_4 - modified phenolglycines, bound at the meta position to the glycine moieties by the ester bond. Diamino-dicarboxylic acid - the ristomycinic acid (8) - is formed. It is found in ristomycins A and B (rystocetins A and B) actaplanins and other antibiotica of the ristomycin's subgroup.
Mention must be made of the fact that the actaplanins differ from the antibiotics described earlier in that they have only one benzyl hydroxygroup in aglycone, i.e. in (1) Z = H.
The actaplanins also differ in the number and distribution of sugars bound to the peptide moiety of the antibiotic.
The unusual aminoacid composition, polycyclicity, polyfunctionality, the presence of one to four carbohydrate branches and the extreme lability of vancomycinic acid (2) - constitute the reasons why the structure of the first antibiotic - vancomycin - was determined only 25 years after its isolation.
Related studies have been undertaken in the Soviet Union (Lomakina et al., Katrukha et al.), in Great Britain (Williams et al.), in the United States (Harris et al., Debono et al., Ellestade et al.), in Hungary (Sztaricskai et al.) and in Italy (Coronelli,Cavalleri).

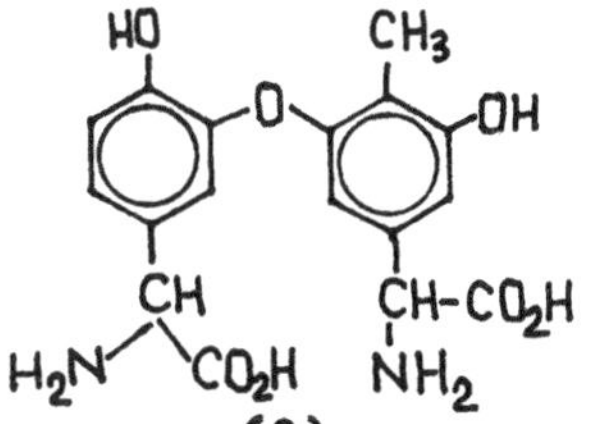

(8)
Ristomycinic acid

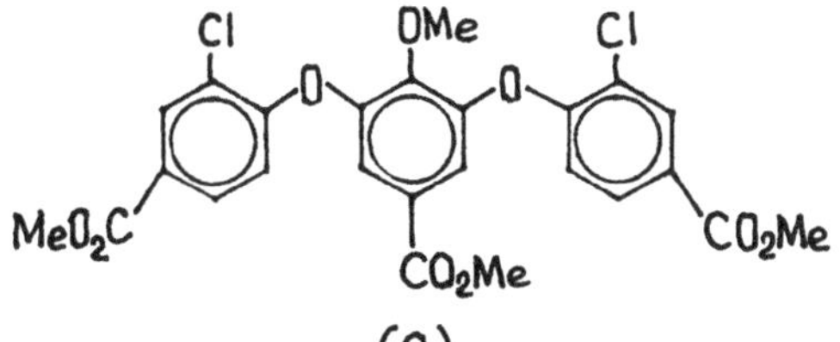

(9)

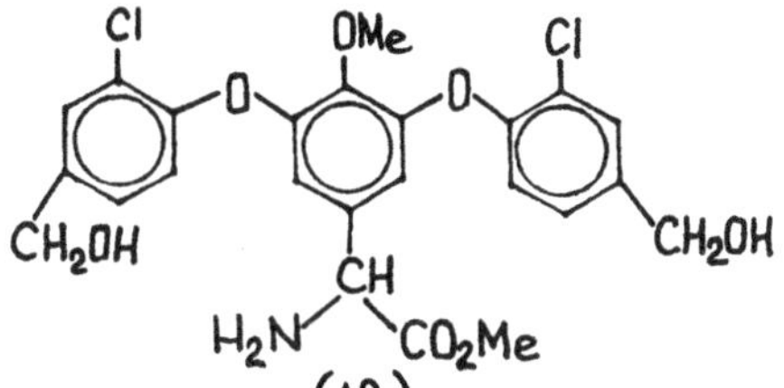

(10)

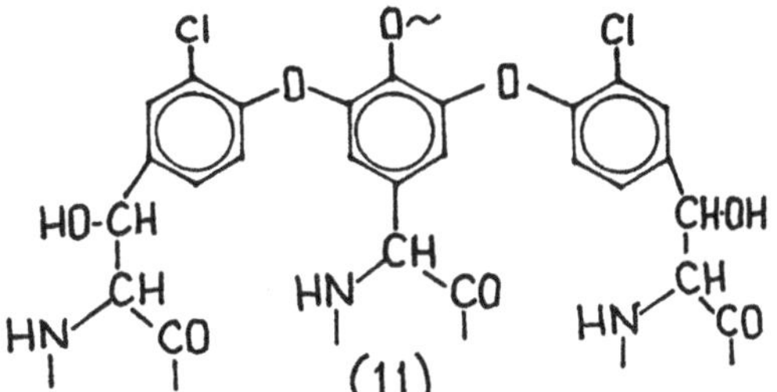

(11)

The first attempts at elucidating the structure of ristomycinic and actinoidinic acids were made in 1968–1972 by Lomakina in Soviet Union [37,38], and of vancomycinic acid (2) – by Smith and Williams in Great Britain only in 1975–1978 [4,39] .
The correct structure of actinoidinic acid as (3) and ristomycinic acid as (8) has been established by work carried out by Williams et al. [39,40], Katrukha et al. [41,42], and Harris et al. [43,44]. While studying the products of oxidative or reductive alkaline hydrolysis of methylated aglucovancomycin, Smith and Williams [39, 45] isolated after methylation acidic products a three-nucleus acid (9) and (10), accordingly. The discovery in the alkaline hydrolyzate of two glycine residues, absent in the vancomycin acidic hydrolyzate, suggested the presence of two serine residues in the precursor (11) [39], that is the residue of vancomycinic acid (2). These conclusions were confirmed in the same laboratory [4] during X-ray analysis of the crystalline substance CDP-1, obtained by heating vancomycin at 90° in water at pH 4.2 for several days.

But the structure of vancomycin as had been proposed previously by Williams et al. [4,46] was revised and consequently corrected in 1982 by Harris et al. [47], who showed that the residue of asparagine, rather than of isoasparagine, is present in the native antibiotic vancomycin: the isoasparagine residue is formed if vancomycin is heated under the above indicated conditions, as a result of intramolecular rearrangement of asparagine residue to isoaspartic acid group, which is well documented in the literature.

A similar type of structure has been detected in actinoidins A and B by Berdnikova et al. [10], with the aid of chemical analysis methods, as well as avoparcines α and β by McGaren [12]. Unlike vancomycin, there is one atom of chlorine in the three-nucleus amino acid (monodechlorovancomycinic acid).

The ristomycin (ristocetin) subgroup of glycopeptide antibiotics has a different type of structure. They consist of didechlorovancomycinic acid (2), actinoidinic acid (3) or its analogs, and ristomycinic acid (8) or its analogs. These antibiotics are: A3512B, A41030 A, B and C, A47934, actaplanins, teicoplanins and finally, ristomycins (ristocetins) A and B. Remarkably, an O-SO_3-group has been detected in the antibiotic A47934, but no carbohydrates.

Let us consider the investigations into the structure of ristomycin A (12), as conducted by the Lomakina and our laboratories. The ristomycin A carbohydrate composition has also been studied by Sztaricskai et al. [2,18,48] and Williamson et al. [49].

We have succeeded in determining the full amino acid composition of the ristomycin peptide core by using reductive acid hydrolysis of the antibiotic in 57% HI containing red P and heated at 107° in vacuo for 20 hr. In the hydrolyzate we detected and characterized triamino-triphenoxy-tricarboxylic acid (13)[50] formed as a result of the reductive substitution of β-hydroxy groups in didechloro-(2) for the hydrogen (Schema 1, A). This method was subsequently applied for studying the structure other antibiotics of this group [47,51,52].

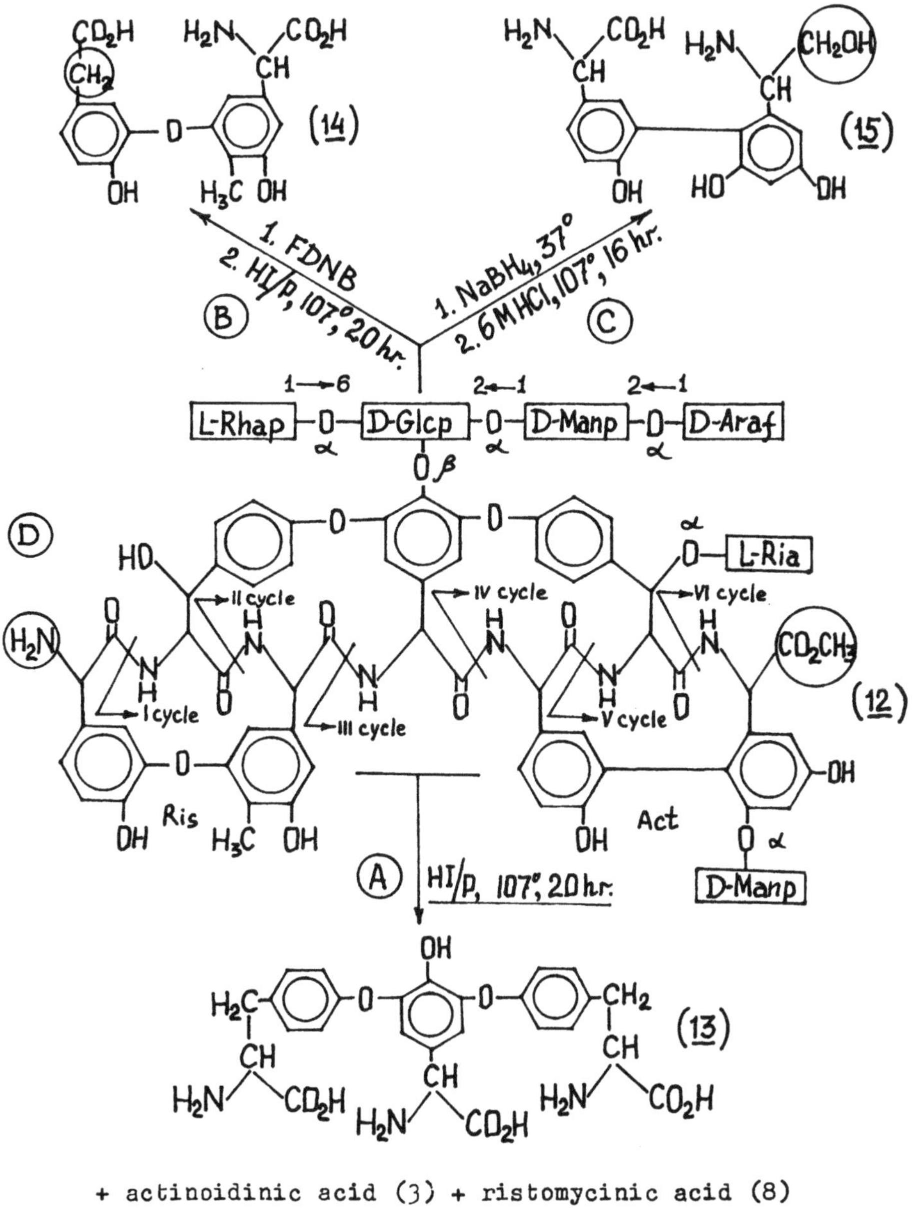

\+ actinoidinic acid (3) + ristomycinic acid (8)

Scheme 1. Investigation of Composition and Structure of Ristomycin A.

The knowledge of the full amino acid composition of ristomycin A enabled us to research the structure of its aglycone and find the binding sites of the three carbohydrate branches.

To begin with, we identified the N- and C-terminal amino acids [16]. To determine the N-terminal amino acid in the aglycone portion we applied 2,4-dinitrophenylation of the antibiotic, as well as complete hydrolysis of the N-2,4-Dnp-derivative and mass spectrometric analysis of monodesamino-ristomycinic acid (14) (Scheme 1, B). It was shown that the free NH_2-group is contained in the di-3,4-hydroxyphenylglycine residue of (8).

We determined the location of the C-terminal methyl ester group upon its reduction with $NaBH_4$, complete hydrolysis and analysis of monocarboxy-actinoidinic acid (15) (Schema 1, C). The methyl ester group was found to belong to the di-m-hydroxyphenylglycine residue of (3). The other amino and carboxylic groups of three complex amino acids (2), (3) and (8) are involved in peptide bonds. We determined their order with the aid of the subtractive Mth-method of Edman degradation by performing, after each cycle, HI/P-hydrolysis of the Mth-derived antibiotic and analyzing changes in its amino acid composition. In this way, we succeeded in sequencing the seven amino acid residues of ristomycin aglycone [16]. A chemical and NMR analysis of the same antibiotic ristocetin A, performed by Harris et al. [53] and by Williams et al. [40], showed the structural identity of the aglycones.

Lomakina et al. [15] deduced the sequence of sugars in ristomycin A and B by identifying the oligosaccharides formed in series of acid hydrolyses.

Location of carbohydrate chains in the antibiotic aglycone moiety posed a problem. The main method used (Tokareva et al. [54]) was analysis of the composition of different glycosides, obtained during a partial acid detachment of carbohydrates. The binding site of tetrasaccharide to the aglycone of ristocetin A (ristomycin A) was also determined by Sztaricskai et al. [18], who isolated the fragment (16) from the mild alkaline hydrolysate of O-methylated ristocetin A (ristomycin A) in the presence of $NaBH_4$.

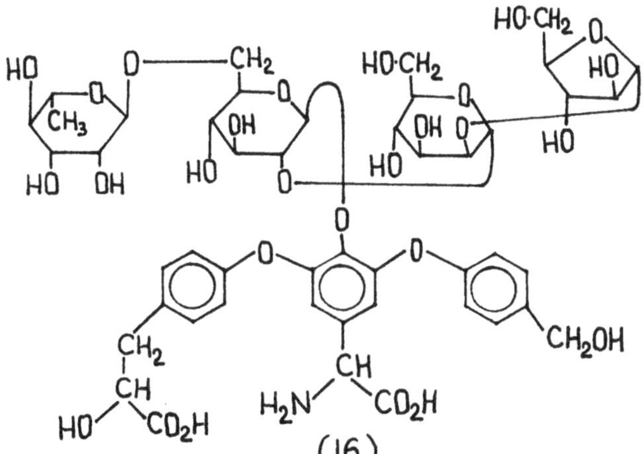

(16)

The site connection of the ristosamine (4) in ristomycin A was determined in our laboratory [17]. After replacing the aminosugar by a methyl amino group, two cycles of Edman degradation followed by acid hydrolysis Mth-aglycone we found out mono-Mth-actinoidinic acid. This experiment establishes that an aminosugar is glycosidically attached to the β-oxygen atom of β-hydroxytyrosine in C-terminal part of didechloro-(2). Thus, the full structure of the ristomycin A (12) was found: it is a linear-cyclic heptapeptide with two free amino groups and three carbohydrate branches [17]. Ristomycin A possesses high bactericidal activity with respect to Gram-positive microorganisms. We performed selective chemical modification to elucidate the role of functional groups in its antibacterial and thrombogenic activity. With this aim in view, we obtained a number of N- and O-derivatives of ristomycin A and its aglycone, and determined their antimicrobic properties and capacity to induce aggregation of platelets in human blood plasma [55,56]. The biological testing results are given in the Table 2, from which

Table 2. Bioassay of Ristomycin A and its Derivatives.

Antibiotics	MIC, µg/ml (Bac. subtilis)	% Platelet aggregation
Ristomycin A	0,8	72
(O-Gly)$_2$-Ristomycin A	1,0	49
N-L-Ile-Ristomycin A	1,4	-
N,N'-(Gly)$_2$-Ristomycin A	3,3	10
N-Ac-Ristomycin A	3,3	-
N,N'-(Ac)$_2$-Ristomycin A	22,7	8
N,N'-(Suc)$_2$-Ristomycin A	100	10
The aglycone of Ristomycin A	0,8	0
(O-SO_3H)$_1$-Aglycone	1,0	0
(O-Gly)$_2$-Aglycone	3,0	0
(N,O-Gly)$_2$-Aglycone	6,6	-
Vancomycin	0,3	0
The Aglycone of Vancomycin	0,8	-
Actinoidin A	0,8	0
The Aglycone of Actinoidin A	1,6	-

the following conclusions are deduced: a) The blockage of one amino group of ristomycin A by a neutral acetyl residue or of both amino groups by the α-aminoacyl residue decreases the antimicrobic activity from 2 to 5-fold, yet the minimum inhibitory concentration (MIC) values of these derivatives persist at a high level (1,0-6,6 µg/ml); b) Acetylation of both amino groups decreases their activity 20-fold, and succinylation inactivates the antibiotic; c) The aglycones of the antibiotics investigated exhibit high antimicrobic activity; d) Selective modification of both ristomycin A and its aglycone likewise produces active derivatives. It is remarkable that the N-derivatives of ristomycin A, the aglycone itself and its derivatives do not induce platelet aggregation in human blood plasma. Ristomycin A as well as ristocetin A are capable of inducing plasma blood factor-VIII-dependent aggregation of platelets and this is employed as useful tool in diagnosing the von Willebrand's disease [57].

We found also that in the course of alkaline inactivation of ristomycin A, two reactions take place: 1) β-elimination of ristosamine (4) and 2) retro-aldol cleavage of C_α- and C_β-bonds in the N-terminal β-hydroxytyrosine fragment of didechlorovancomycinic acid [58]. These two reactions lead to a significant alteration of the aglycone conformation and its consequent inactivation. The results of the chemical modifications of ristomycin A as well as its aglycone, point at the significant role that the α-amino group and the specific conformation of the aglycone play in the mechanism of action of the antibiotic. Obviously, the free amino groups of ristomycin A are responsible for the recognition of the target and primary binding to it; subsequently, the aglycone forms a tight complex with the receptor on the bacterial cell, and this leads to its death by lysis.

As clearly indicated in the works of Nieto,Perkins [59-61] and Williams et al. [62,63], vancomycin group of antibiotics form a complex with the fragment -D-Ala-D-Ala-OH of the growing peptidoglycan of the bacterial cell wall, and this halts the completion of its construction. Thus, it was shown that the model peptide Ac-D-Ala--D-Ala-OH is linked to the vancomycin aglycone pocket via three and in the case of ristocetin aglycone via four intermolecular hydrogen bonds to the -NH-CO-groups of aglycone, with the high association binding constant $2{,}0-7{,}2 \times 10^4\ l \cdot mol^{-1}$. Also, the hydrophobic and steric interactions of the peptide CH_3-groups with the aromatic nuclei of aglycone amino acids play a significant role in stabilizing the complex [64].

This mechanism of action of these antibiotics sheds light on their two properties mentioned above, i.e., 1) low toxicity, and the

2) failure of various microorganisms to develop resistance to the antibiotic. Ristomycin A is not toxic to man and animals, because the structure of animal cells are substantially different, both morphologically and in chemical composition, from that of bacterial cells. Human cells, thus, do not serve as a target for the aglycone of this antibiotic. To alter bacterial stability to vancomycin-type antibiotics would require an alteration of the structure of the cell wall biopolymer and consequently of the entire system of the genetically conditioned enzymatic reactions, which is extremely difficult, if not impossible.

Proceeding from an analysis of the literature and our data, we may conclude, that further studies of the chemistry of the vancomycin group antibiotics are a promising area of research. No doubt, it would enable us to gain an insight into peptide-peptide interactions in biological systems and obtain new and valuable medicinal preparations.

References

1. Williams, D.H., V.Rajananda, M.P.Williamson, G.Bojesen. 1980. In: Topics in Antibiotic Chemistry. Vol. 5, Part B. (P.Sammes, ed.), Ellis Horwood, Chichester, U.K., pp. 119-158.

2. Sztaricskai, F., R.Bognar. 1984. In: Recent Developments in the Chemistry of Natural Carbon Compounds. Vol. 10 (R.Bognar, Cs. Szantay, eds.). Akademiai Kiado, Budapest, pp. 91-209.

3. McCormick, M.H., W.M.Stark, G.E.Pittenger, R.C.Pittenger, J.M. McGuire. 1956. Antibiotics Annual. 1955-1956, pp. 606-611.

4. Sheldrick, G.M., P.G.Jones, O.Kennard, D.H.Williams, G.A.Smith. 1978. Nature 271, 223-225.

5. Harris, C.M., T.M.Harris. 1982. J.Am.Chem.Soc., 104, 4293-4295.

6. Boeck, L.D., F.P.Mertz, R.K.Wolter, C.E.Higgens. 1984. J.Antibiotics 37, 446-453.

7. Shorin, V.A., S.D.Judintsev, I.A.Kunrat, L.E.Goldberg, N.S.Pevzner, M.G.Brazhnikova, N.N.Lomakina, E.F.Oparisheva. 1957. Antibiotiki 2, 44–49.

8. Lomakina, N.N., I.A.Spiridonova, Yu.N.Scheinker, T.F.Vlasova. 1973. Khim.Prir.Soedin. 1973, 101–108.

9. Berdnikova, T.F., N.N.Lomakina. 1977. Antibiotiki 22, 1077–1081.

10. Berdnikova,T.F., N.N.Lomakina, N.P.Potapova. 1982. Antibiotiki. 27, 252–258.

11. Kunstmann, M.P., L.A.Mitscher, J.N.Porter, A.J.Shay, M.A.Darken. 1968. Antimicrob. Agents Chemother. 1968, 242–245.

12. McGahren, W.J., R.A.Leese, F.Barbatschi, G.O.Morton, N.A.Kuck, G.A.Ellestad. 1983. J.Antibiotics 36, 1671–1682.

13. Ellestad, G.A., W.Swenson, W.J.McGahren. 1983. J.Antibiotics 36, 1683–1690.

14. Gause, G.F., E.S.Kudrina, R.S.Ukholina, G.V.Gavrilina. 1963. Antibiotiki 8, 387–392.

15. Lomakina, N.N., L.I.Muravyeva, M.Puskas. 1968. Antibiotiki 13, 867–871.

16. Silaev, A.B., B.Diarra, Zh.P.Trifonova, G.S.Katrukha. 1979. Khim.Prir.Soedin., 1979, 742–744.

17. Lomakina, N.N., G.S.Katrukha, M.G.Brazhnikova, A.B.Silaev, L.I.Muravyova, Zh.P.Trifonova, N.L.Tokareva, B.Diarra. 1982. Antibiotiki 27, 248–252.

18. Sztaricskai, F., C.M.Harris, A.Neszmelyi, T.M.Harris. 1980. J.Am.Chem.Soc., 102, 7093–7099.

19. Grundy, W.E., A.C.Sinclair, R.J.Theriault, A.W.Goldstein, C.J. Rickher, H.B.Warren, Jr., T.J.Oliver, J.C.Sylvester. 1957. Antibiot.Annu., 1956–1957, 687–692.

20. Harris, T.M., J.R.Fehlner, A.B.Raabe, D.S.Tarbel. 1975. Tetrahedron Lett., 1975, 2655–2658.

21. Kalman, J.R., D.H.Williams. 1980. J.Am.Chem.Soc.,102,897–905.

22. Williamson, M.P., D.H.Williams. 1981. J.Chem.Soc., Perkin I 1981, 1483–1491.

23. Harris, C.M., T.M.Harris. 1982. J.Am.Chem.Soc., 104, 363-365.

24. Michel, K.H., R.M.Shah, R.L.Hamill. 1980. J.Antibiotics 33, 1397-1406.

25. Harris, C.M., T.M.Harris. 1983. Tetrahedron 39, 1661-1666.

26. Hunt, A.H. 1983. J.Am.Chem.Soc., 105, 4463-4468.

27. Debono, M., K.E.Merkel, R.M.Molloy, M.Barnkart, E.Presti, A.H. Hunt, R.L.Hamil. 1984. J.Antibiotics 37, 85-95.

28. Hunt, A.H., T.K.Elrey, K.E.Merkel, M.Debono. 1984. J.Org.Chem., 49, 641-645.

29. Eggert, J.H., K.H.Michel, L.D.Boeck, W.M.Nakatsukasa, R.E.Kastner. 1983. Program and Abstracts, 23rd Intersci. Conf. Antimicrob. Agents Chemother., Las Vegas, NV, Oct 1983, Abstr. 440.

30. Hunt, A.H., J.L.Occolowitz, M.Debono, R.M.Molloy, G.M.Maciak. 1983. Program and Abstracts, 23 rd Intersci. Conf.Antimicrob. Agents Chemother., Las Vegas, NV, Oct. 1983, Abstr. 441.

31. Hamill, R.L., L.D.Boeck, R.E.Kastner, R.M.Gale. 1983. Program and Abstracts, 23rd Intersci.Conf.Antimicrob. Agents Chemother., Las Vegas, NV, Oct. 1983, Abstr. 443.

32. Parenti, F., G.Baretta, M.Berti, V.Arioly. 1978. J.Antibiotics 31, 276-283.

33. Hunt, A.H., R.M.Molloy, J.L.Occolowitz, G.G.Marconi, M.Debono. 1984. J.Am.Chem.Soc., 106, 4891-4895.

34. Barna, J.C.J., D.H.Williams, D.J.M.Stone, T.-W.Ch.Leung, D.M. Doddrell. 1984. J.Am.Chem.Soc., 106, 4895-4902.

35. Malabarba, A., P.Strazzolini, A.Depaoli, M.Landi, M.Berti, B.Cavalleri. 1984. J.Antibiotics, 37, 988-999.

36. Omura, S., H.Tanaka, Y.Tanaka, P.Spiri-Nakagawa, R.Oiwa, Y.Takahashi, K.Matsuyama, Y.Iwai. 1979. J.Antibiotics 32, 985-994.

37. Lomakina, N.N., V.A.Zenkova, R.Bognar, F.Sztaricskai, Yu.N. Sheinker, K.F.Turchin. 1968. Antibiotics 13, 675-682.

38. Lomakina, N.N., M.S.Yurina, Yu.N.Sheinker, K.F.Turchin. 1972. Antibiotiki 17, 488-492.

39. Smith, G.A., K.A.Smith, D.H.Williams. 1975. J.Chem.Soc., Perkin I, 1975, 2108-2115.

40. Williams, D.H., V.Rajananda, J.R.Kalman. 1979. J.Chem.Soc., Perkin I, 1979, 787-792.

41. Katrukha, G.S., P.B.Terentiev, B.Diarra, E.S.Gershtein. 1978. Khim.Prir. Soedin., 1978, 141-144.

42. Katrukha, G.S., A.B.Silaev, Zh.P.Trifonova, B.Diarra, N.N.Lomakina, N.L.Tokareva, L.A.Baratova, L.P.Belyanova. 1978. Antibiotiki, 23, 38-43.

43. Harris, T.M., J.R.Fehlner, A.B.Raabe, D.S.Tarbell. 1975. Tetrahedron Lett. 1975, 2655-2658.

44. Harris, C.M., J.J.Kibby, T.M.Harris. 1978. Tetrahedron Lett., 1978, 705-708.

45. Smith, K.A., D.H.Williams, G.A.Smith. 1974. J.Chem.Soc., Perkin I, 1974, 2369-2376.

46. Williamson, M.P., D.H.Williams. 1981. J.Am.Chem.Soc., 103, 6580-6585.

47. Harris, C.M., T.M.Harris. 1982. J.Am.Chem.Soc., 104, 4293-4295.

48. Sztaricskai, F., A.Neszmelyi, R.Bognar. 1980. Tetrahedron Lett. 1980, 2983-2986.

49. Williamson, M.P., D.H.Williams. 1980. Tetrahedron Lett. 1980, 4187-4188.

50. Katrukha, G.S., B.Diarra, A.B.Silaev, Zh.P.Trifonova, B.V.Rozynov, O.S.Reshetova. 1979. Antibiotiki 24, 179-181.

51. Trifonova, Zh.P., G.S.Katrukha, A.B.Silaev, B.Diarra, B.V.Rozynov, O.S.Reshetova. 1979. Khim.Prir.Soedin., 1979, 875-877.

52. Berdnikova, T.F. 1981. Antibiotiki 26, 816-820.

53. Harris, C.M., T.M.Harris. 1979. Tetrahedron Lett. 1979, 3905-3908.

54. Tokareva, N.L., T.F.Berdnikova, N.N.Lomakina. 1981. Antibiotiki 26, 821-824.

55. Trifonova, Zh.P., I.G.Smirnova, G.B.Fedorova, G.S.Katrukha. 1983. 7th Soviet-Indian Symposium on Chemistry of Natural products, Tbilisi, Abstracts, pp. 118-119.

56. Trifonova, Zh.P., M.B.Kobrin, G.B.Fedorova, A.B.Silaev, G.S. Katrukha. 1983. 6th All-Union Symposium on Chemistry of Proteins and Peptides. Riga, Abstracts, p. 308.

57. Boda, Z., N.O.Solum, F.Sztaricskai, K.Rak. 1979. Thromb. and Haemost., 42, 1164-1180.

58. Trifonova, Zh.P., E.R.Eganian, A.B.Silaev, G.S.Katrukha. 1984. Antibiotiki 28, 170-175.

59. Nieto, M., H.R.Perkins. 1971. Biochem. J., 123, 773-787.

60. Nieto, M., H.R.Perkins. 1971. Biochem.J., 124, 845-852.

61. Perkins, H.R., M.Nieto. 1972. In: Proceeding of the Symposium on Molecular Mechanisms of Antibiotic Action on Protein Biosynthesis and Membranes, Granada, Elsevier, Amsterdam, pp.363-387.

62. Kalman, J.R., D.H.Williams. 1980. J.Am.Chem.Soc., 102, 906-912.

63. Williamson, M.P., D.H.Williams, S.J.Hammond. 1984. Tetrahedron 40, 569-577.

64. Williamson, M.P., D.H.Williams. 1984. Eur.J.Biochem., 138, 345-348.

STRUCTURE ELUCIDATION OF THE ANTIMYCETIC GLYCOLIPODEPSIPEPTIDE HERBICOLIN A

W.A. König*, M. Aydin, N. Lucht
Institut für Organische Chemie der Universität Hamburg,
D-2000 Hamburg 13

G. Winkelmann
Institut für Biologie I der Universität Tübingen,
D-7400 Tübingen

R. Lupp, G. Jung
Institut für Organische Chemie der Universität Tübingen,
D-7400 Tübingen

Introduction

In a screening program for antifungal agents herbicolin A and B were isolated from a bacterial strain identical with *Erwinia herbicola*. Both antibiotics are highly active against yeasts and filamentous fungi but not active against bacteria (1,2). Inhibition was observed of *Trichophyton rubrum*, *Epidermophyton floccosum*, *Microsporum canis*, *Ureaplasma* and *Spiroplasma* species. Herbicolin A may act by inhibition of the growth of sterol-requiring organisms. Also herbicolin A activates to a very small extent membrane bound Ca^{2+}/calmodulin-dependent guanylate cyclase of *Paramecium* (3). Fermentation, isolation and purification of herbicolin by counter-current distribution in a neutral system has been described (1,4). We report here on the structure elucidation of herbicolin A.

Results

The presence of unusual constituents, D-enantiomers, N-methylamino acids and cyclic structures requires the application of highly specific techniques for the structure investigation of a peptide

Chemistry of Peptides and Proteins, Vol. 3

antibiotic. Sequential degradation according to Edman, end group determination and, particularly, enzymatic partial cleavage can usually not be applied because of numerous structural peculiarities. As an alternative we have utilized successfully gas phase analytical methods like gas chromatography and mass spectrometry together with micro-chemical techniques in the past (5-8).

There are roughly three tasks to be encountered in a systematic approach to structure analysis of a peptide antibiotic:

a) The identification of all constituents,
b) the solution of stereochemical problems and
c) the determination of the amino acid sequence and of intramolecular ring structures.

a) Constitutents of Herbicolin A

Amino acid analysis showed the presence of 2 Gly, 2 Thr, 1 Glx, 1 Leu and 1 Arg. The more specific gas chromatographic analysis of volatile trimethylsilyl (TMS) and isopropyl ester/N,O-trifluoroacetyl derivatives of herbicolin, hydrolyzed for 18 h at 110°C in 6 N HCl, revealed that Thr *and* aThr was present. In addition in the gas chromatographic/mass spectrometric investigation 3-hydroxy-n-tetradecanoic acid and another amino acid constituent was detected, which was tentatively characterized as "homothreonine" by its mass spectrum. Comparison with synthetic reference compounds showed identical mass spectra of the unknown amino acid with N-methyl-threonine (Fig. 1). However, coinjection of the herbicolin hydrolyzate and the synthetic N-methyl-threonine gave clearly different retention times. Only after comparison with synthetic N-methyl-*allo*threonine identical mass spectra *and* retention times were observed.

A further constituent of herbicolin A, D-glucose, could be detected by gc/ms after partial hydrolysis of herbicolin A (10 min, 6 N HCl, 100°C). Complete degradation of the glucose residue is observed, probably by Maillard reaction with the amino acids, after prolonged hydrolysis.

In the partial hydrolyzate 3-hydroxy-tetradecanoic acid is present only in traces, while a large peak of a compound appears in the gas chromatogram showing a mass spectrum of the TMS derivative with a

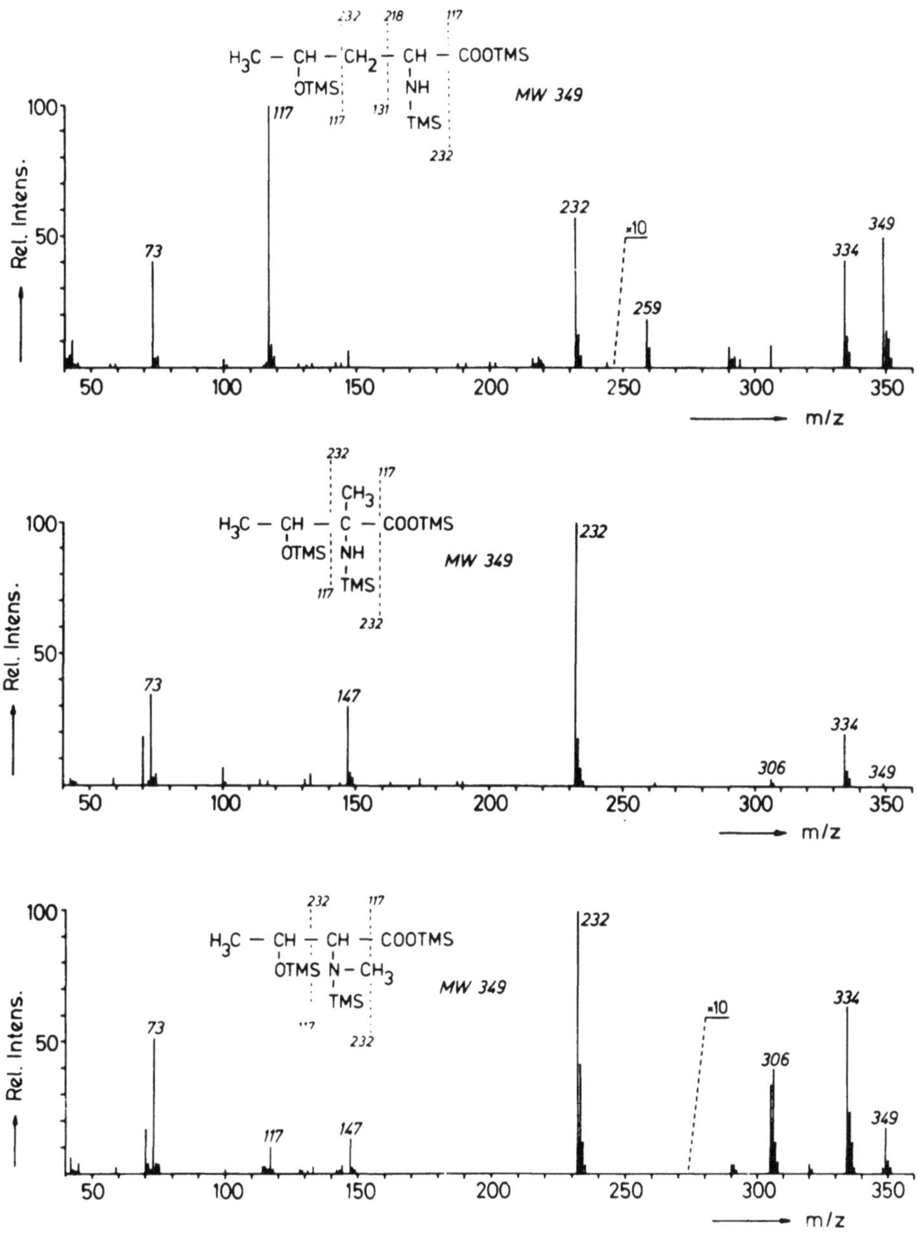

Figure 1. 70eV-Mass spectra (gc/ms) of the trimethylsilyl derivatives of a) homothreonine, b) α-methyl-threonine and c) N-methyl-threonine. The spectrum of N-methyl-threonine is identical with the unknown constituent of herbicolin.

molecular weight of 387, which is by one mass unit smaller than the molecular mass of 3-hydroxy-tetradecanoic acid and which was found to be identical with 3-hydroxy-tetradecanoic acid amide. This amide is formed according to scheme 1 by hydrolysis of a peptide bond next to a dehydro-aminobutyric acid moiety. This amino acid was detected as TMS derivative after mild hydrolysis as N-(3-hydroxy-tetradecanoyl)2,3-dehydro-2-amino-n-butanoic acid by gc/ms (see Fig. 4) but undergoes complete degradation during total hydrolysis (9).

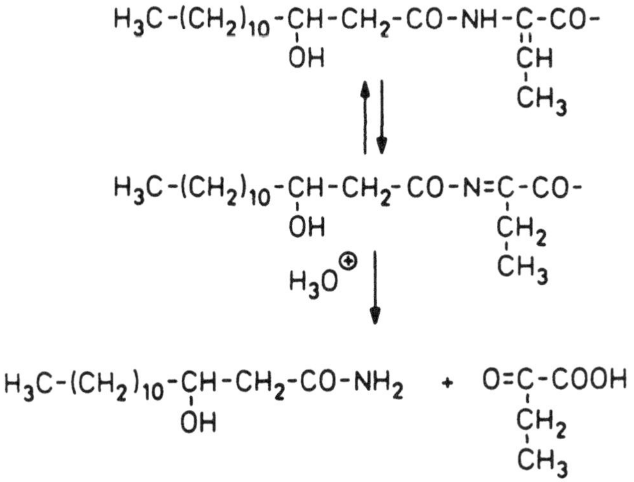

Scheme 1. Hydrolytic cleavage of N-(3-hydroxy-tetradecanoyl)-2-amino-2,3-dehydro-n-butanoic acid residue.

b) Stereochemical Investigations

The occurrence of D-amino acids in peptide antibiotics is very common. The most elegant way to determine the configuration of amino acids is their direct enantiomeric resolution by capillary gas chromatography on chiral stationary phases. This technique was introduced by E. Gil-Av et al. (10) and considerably improved by Frank, Nicholson and Bayer (11) in using chiral polysiloxanes as thermostable and highly enantioselective stationary phases (Chirasil-val). With this technique the D-configuration of Leu, aThr, Glx and the L-configuration of Thr and Arg was confirmed.

The configuration of hydroxy acids can be determined after reaction with isopropyl isocyanate yielding the isopropyl urethane/ isopropyl amide derivatives (12,13). Enantiomeric resolution of these derivatives is possible on the chiral polysiloxane XE-60-L-valine-(S)-α-phenylethylamide (14). In this way the (R)-configuration of 3-hydroxy-tetradecanoic acid of herbicolin was established. The D-configuration of glucose was confirmed by analysis of the trifluoroacetylated methyl glycoside and by coinjection with a racemic mixture of glucose on XE-60-L-valine-(S)-α-phenylethylamide (Fig. 2) (14,15).

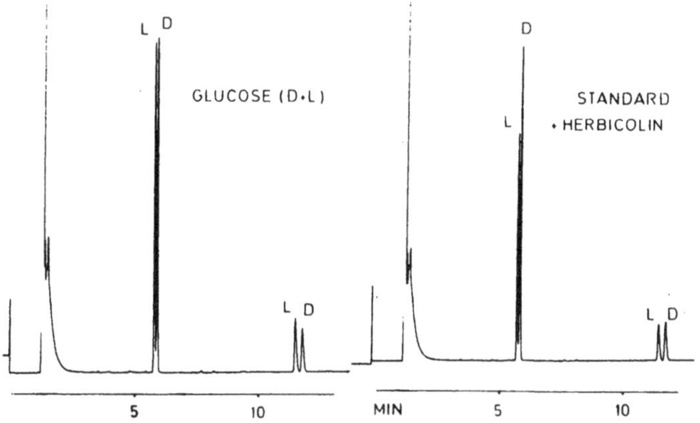

Figure 2. Gas chromatography of a racemic mixture of glucose after formation of trifluoroacetylated methyl glycoside (two pairs of enantiomers = α- and β-pyranosides) and coinjection of herbicolin A sample. 35 m fused silica capillary column with XE-60-L-valine-(S)-α-phenylethylamide. Column temperature: 120°C.

Finally the configuration of the N-methyl-aThr residue was determined by enantioselective gas chromatography after esterification with isopropanol and subsequent reaction with phosgene. The N-methyl-oxazolidin-2-ones of all four stereoisomers can be separated on columns with XE-60-L-valine-(R)-α-phenylethylamide (Fig. 3) (16,17). In this experiment the L-configuration of the natural product was proved.

c) Amino Acid Sequence and Intramolecular Ring Formation

The concept of amino acid sequence determination by gas chromato-

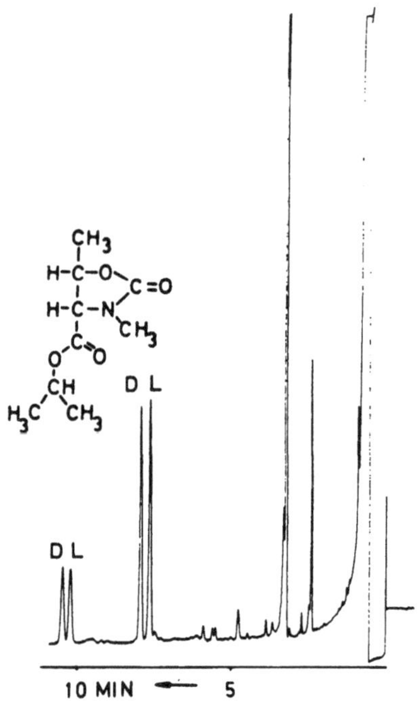

Figure 3. Gas chromatography of the stereoisomers of N-methyl-Thr (first pair) and N-methyl-aThr (second pair) after formation of isopropyl ester and reaction with phosgene. 25 m fused silica capillary column with XE-60-L-valine-(R)-α-phenylethylamide. Column temperature: 140°C.

graphy and mass spectrometry is based on partial cleavage of the peptide chain and formation of derivatives with an optimum of volatility and structural information of their mass spectra. Since enzymatic cleavage by proteases is inhibited by the presence of D-enantiomers in the antibiotic peptide bond hydrolysis is achieved by hydrochloric acid at moderate temperatures (37°C, 12 h, conc. HCl) (5,6). Peptide fragments in the range of dipeptides up to heptapeptides are formed under these conditions.

Three types of volatile derivatives were investigated by gc/ms, namely the trifluoroacetylated peptide methyl esters (18), trimethylsilylated peptides (19) and trimethylsilylated polyamino alcohols (20). The latter derivatives were obtained by N-acetylation, reduction with diborane and O-trimethylsilylation according to Scheme 2.

As shown in Table 1 only the combination of all three types of derivatization procedures was sufficient to derive the complete amino acid sequence. Particularly the dehydroamino butyric acid residue could be verified as such by identification of the trimethylsilylated derivative (Fig. 4).

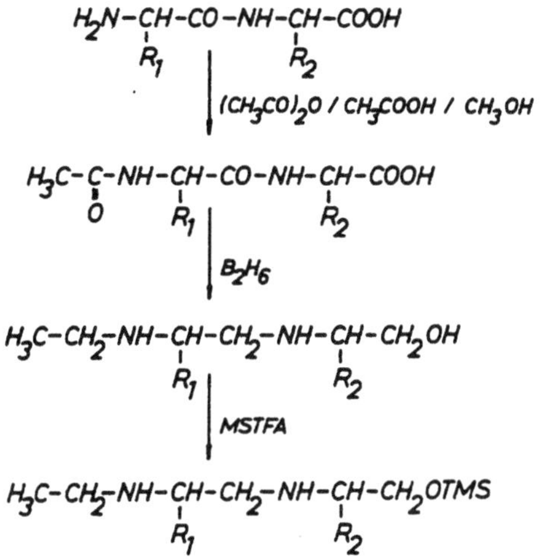

Scheme 2. Conversion of peptide fragments to O-trimethylsilyl polyamino alcohols.

(R)-3-OH-TETRADEC. - DH-ABU	(TMS)
(R)-3-OH-TETRADEC. - DL-ABU	(B_2H_6)
(R)-3-OH-TETRADEC. - DL-ABU - L-THR	(B_2H_6)
DL-ABU - L-THR	(B_2H_6)
L-THR - D-aTHR	(TMS; B_2H_6)
L-THR - D-aTHR - D-LEU	(TMS; TFA; B_2H_6)
L-THR - D-aTHR - D-LEU - GLY	(B_2H_6)
D-aTHR - D-LEU	(TFA; TMS; B_2H_6)
D-aTHR - D-LEU - GLY	(TMS; B_2H_6)
D-aTHR - D-LEU - GLY - D-GLX	(TMS)
GLY - D-GLX	(TFA; TMS)
D-GLX - GLY	(TMS; TFA; B_2H_6)
D-GLX - GLY - L-N-ME-aTHR	(B_2H_6)
GLY - L-N-ME-aTHR	(TMS)
L-N-ME-aTHR - L-ARG	(TMS, B_2H_6)
(R)-3-OH-TETRADEC. - DH-ABU - L-THR - D-aTHR - D-LEU - GLY - D-GLX - GLY - L-N-ME-aTHR - L-ARG	

Table 1. Overlapping peptide fragments identified by gc/ms of volatile derivatives (in parentheses) and reconstructed total sequence of herbicolin A.

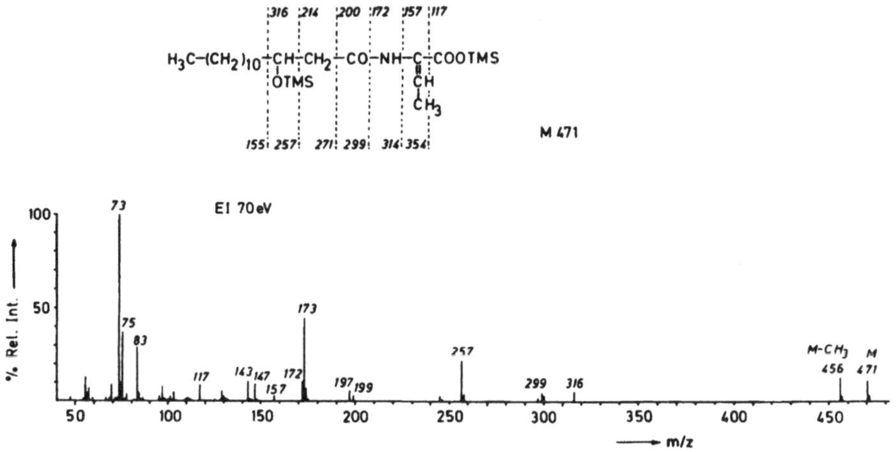

Figure 4. 70eV-Mass spectrum (gc/ms) of 3-hydroxy-tetradecanoyl-2-amino-2,3-dehydrobutyric acid from the hydrolyzate of herbicolin A after trimethylsilylation.

Excellent properties for sequence determination are displayed by the polyaminoalcohol derivatives. As an example Fig. 5 shows the mass spectrum of the tetrapeptide Thr-aThr-Leu-Gly. Prominent N-terminal sequence ions (m/z = 174, 333, 432) and C-terminal sequence ions (m/z = 146, 245, 404) facilitate the identification of the structure.

Usually special problems of volatility are caused by the presence of arginine in a peptide. In the case of herbicolin A the diborane method yielded a volatile dipeptide derivative of N-methyl-aThr-Arg. The arginine residue is converted to an N-methyl-ornithine derivative by this derivatization procedure (Fig. 6). In this example the benefit of chemical ionization for the identification of the molecular ion is demonstrated (21).

From the gc/ms results an unambiguous sequence could be derived for herbicolin A except for the possibility to discriminate between glutamine and glutamic acid and except for the position of the L-Thr and the D-aThr residues. The latter problem could be settled by direct comparison of the herbicolin hydrolyzate with reference dipeptides L-Thr-D-Leu and D-aThr-D-Leu. In the capillary gas chromatogram the D-aThr-D-Leu derivative co-eluted with

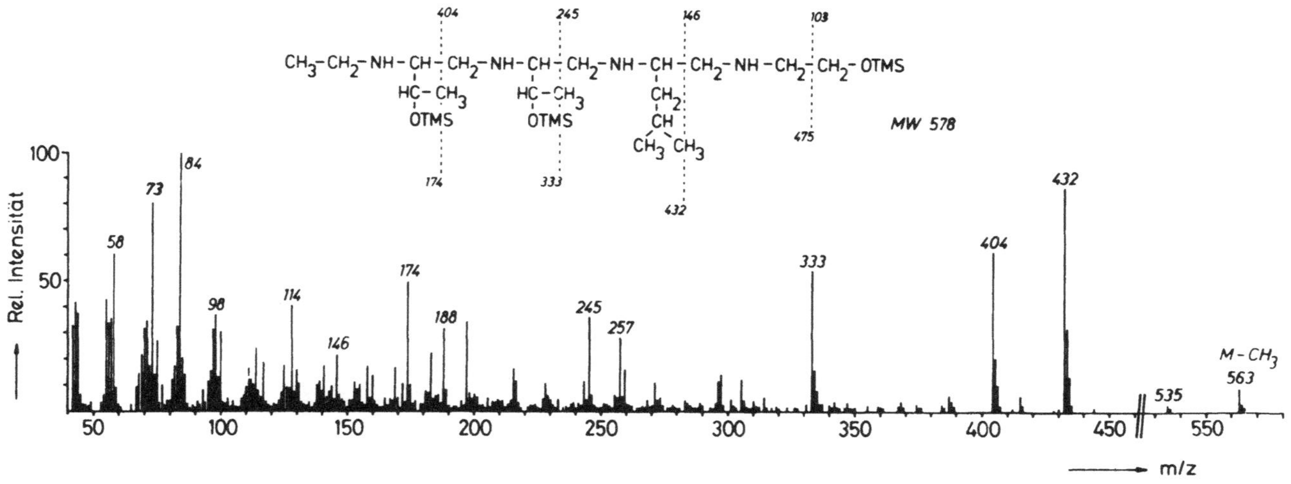

Figure 5. 70eV-Mass spectrum (gc/ms) of tetrapeptide Thr-aThr-Leu-Gly from herbicolin A hydrolyzate after N-acetylation, reduction with diborane and trimethylsilylation.

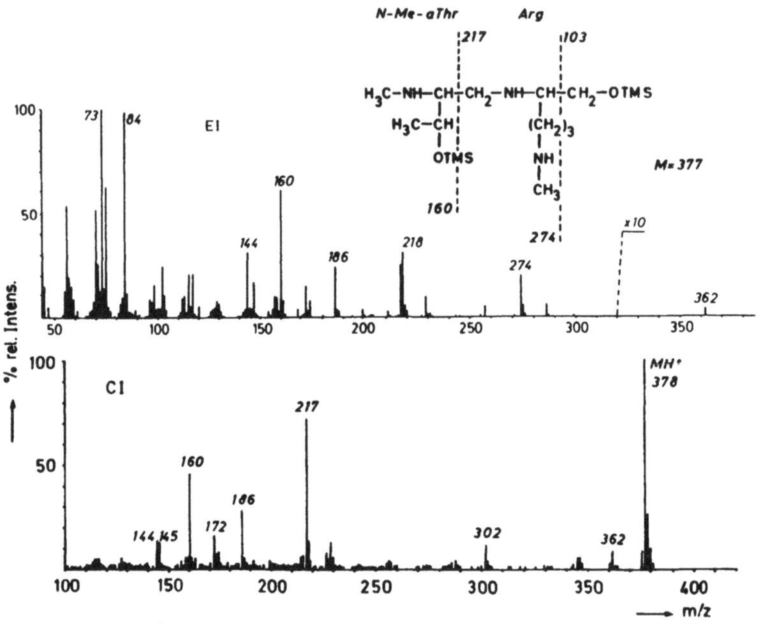

Figure 6. 70eV-Mass spectrum and isobutane chemical ionization mass spectrum (gc/ms) of the derivative of N-methyl-aThr-Arg from herbicolin A hydrolyzate after N-acetylation, diborane reduction and trimethylsilylation.

the corresponding peak of the hydrolyzate (24).

Fast atom bombardment (FAB), a new technique in mass spectrometry, has gained major importance in recent years, especially for the investigation of polar and high molecular weight natural compounds (22,23). In Fig. 7 the negative ion FAB spectrum of herbicolin is shown. Except for the $[M - H]^-$ ion at m/z = 1298 an $[M + Cl]^-$ ion at m/z = 1334 and cluster ions $[(M - H) + \text{thioglycerol}]^-$ at m/z = 1406 and $[(M - H) + \text{thioglycerol} + Cl]^-$ at m/z = 1442 confirm the molecular weight of 1299 for herbicolin A. The signal at m/z = 1118 arises from the loss of a glucose molecule from $[M - H]^-$ (isotope signals are not shown).

The molecular weight of 1299 indicates that herbicolin A must be a cyclic peptide because of the difference of 18 mass units from the molecular weight calculated from the constituents identified by gc/ms and assuming a glutamine and not a glutamic acid residue

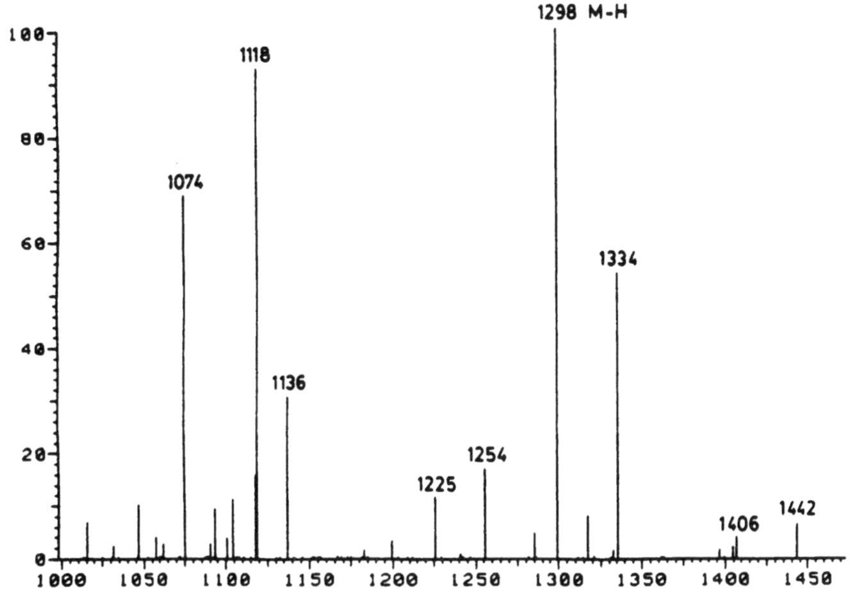

Figure 7. Negative ion FAB mass spectrum of herbicolin A, high mass range. 1-Thioglycerol was used as matrix substance and Xenon neutral atoms for secondary ion formation.

for Glx. The lactone structure can indeed be proved by FAB mass spectrometry after mild alkali treatment of herbicolin by a mass shift of the $[M - H]^-$ ion to 1316, due to the hydrolysis of a lactone structure.

It was now necessary to identify the constituents involved in the formation of the lactone structure. It is known that $NaBH_4$ selectively reduces the lactone carbonyl function (5). After treatment of herbicolin A with $NaBH_4$ in methanol/water solution at room temperature in the amino acid analysis of a total hydrolyzate except for arginine all the other amino acids were present. This is a strong indication that the carboxylic group of the arginine residue participates in the lactone ring formation.

To detect the hydroxy function involved in the lactone structure a sample of herbicolin A was oxidized with chromic acid in pyridine and acetic acid (25,26). In this reaction all free hydroxy groups are destroyed, while the lactone structure remains unchanged. After total hydrolysis of this sample the gc/ms investigation proved that Thr was the only hydroxy amino acid still present, whereas aThr and N-methyl-aThr had disappeared, strongly indicating that the L-Thr residue is esterified with the C-terminal L-Arg residue.

The last open question is now to locate the binding site of the glucose. From the oxidation experiment a glycosidic bond to one of the amino acids except L-Thr can be excluded. The only hydroxy group left is located in the 3-hydroxytetradecanoic acid residue. Herbicolin B, a congener of herbicolin A, differs from herbicolin A only by the absence of glucose. Chromic acid oxidation of herbicolin B reveals that L-Thr also is preserved. This makes it very probable that the binding site for the glucose residue in herbicolin A is indeed the hydroxy group of the fatty acid. Permethylation of herbicolin A with CD_3I after treatment with NaH/dimethylsulfoxide, subsequent hydrolysis and trimethylsilylation gave a molecular weight of 320 (gc/ms). The same compound was obtained from saccharose as a reference compound. This unequivocally proves the glycosidic linkage of glucose in herbicolin A. The α-glycosidic bond was established by comparative ^{13}C-NMR studies.

In conclusion the structure of herbicolin A can be described by the formula shown in Fig. 8.

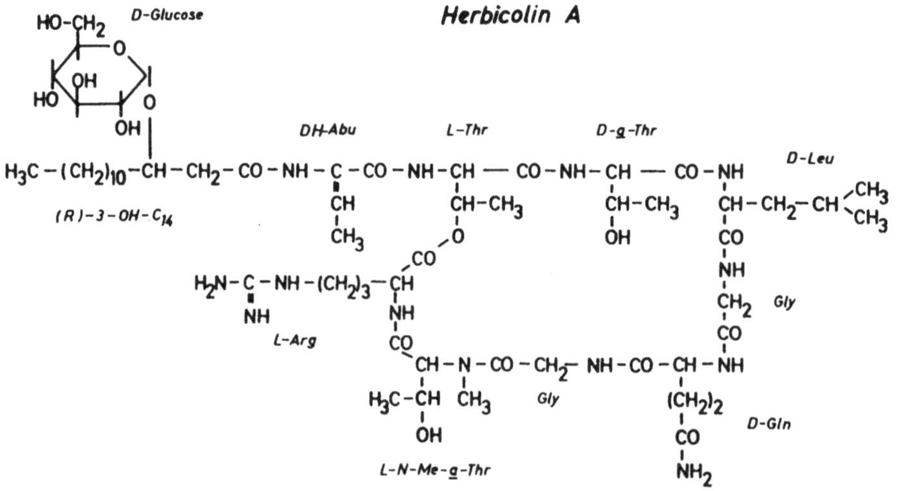

Figure 8. Structure of herbicolin A.

References

1. Winkelmann, G., Lupp, R., Jung, G.: J. Antibiotics 33, 353 (1980).

2. Freundt, E.A., Winkelmann, G.: Antimicrobial Agents and Chemotherapy 26, 112 (1984).

3. Klumpp, S., Jung, G., Schultz, J.E.: Biochim. Biophys. Acta 800, 145 (1984).

4. Winkelmann G., Glück, S., Lupp, R., Jung, G. in: Structure and Activity of Natural Peptides", Selected Topics (W. Voelter, G. Weitzel, Eds.), Walter de Gruyter, Berlin (1981), p. 237.

5. Rinken, M., Lehmann, W.D., König, W.A.: Liebigs Ann. Chem. 1984, 1672.

6. Brückner, H., König, W.A., Aydin, M., Jung, G.: Biochim. Biophys. Acta 827, 51 (1985).

7. Katz, E., Schmitt, H., Aydin, M., König, W.A., Jung, G.: Liebigs Ann. Chem. 1985, 365.

8. Katz, E., Aydin, M., Lucht, N., König, W.A., Ooka, T., Jung, G.: Liebigs Ann. Chem. 1985, 1041.

9. Liesch, J.M., Millington, S., Pandey, R.C., Rinehart jr., K.L.: J. Am. Chem. Soc. 98, 8237 (1976).

10. Gil-Av, E., Feibush, B., Charles-Sigler, R. in: A.B. Littlewood (Ed.), "Gas Chromatography 1966", Institute of Petroleum, London (1966), p. 227.

11. Frank, H., Nicholson, G.J., Bayer, E.: J. Chromatogr. Sci. 15, 174 (1977).

12. Benecke, I., König, W.A.: Angew. Chem. 94, 709 (1982); Angew. Chem. Int. Ed. Engl. 21, 709 (1982).

13. König, W.A., Benecke, I., Lucht, N., Schmidt, E., Schulze, J., Sievers, S.: J. Chromatogr. 279, 555 (1983).

14. König, W.A., Benecke, I., Sievers, S.: J. Chromatogr. 217, 71 (1981).

15. König, W.A., Benecke, I., Bretting, H.: Angew. Chem. 93, 688 (1981); Angew. Chem. Int. Ed. Engl. 20, 693 (1981).

16. König, W.A., Steinbach, E., Ernst, K.: Angew. Chem. 96, 616 (1984); Angew. Chem. Int. Ed. Engl. 23, 527 (1984).

17. König, W.A., Steinbach, E., Ernst, K.: J. Chromatogr. 301, 129 (1984).

18. Weygand, F., Prox, A., Fessel, H.H., KunSun, K.: Z. Naturforsch. 20 b, 1169 (1960).

19. Baker, K.M., Shaw, M.A., Williams, D.H.: J. Chem. Soc. Chem. Commun. 1969, 1108.

20. Kelley, J.A., Nau, H., Förster, H.-J., Biemann, K.: Biomed. Mass Spectrom. 2, 39 (1982).

21. König, W.A., Krohn, H., Greiner, M., Brückner, H., Jung, G. in: "Advances in Mass Spectrometry", Vol. 8 B, A. Quayle, Ed., Heyden & Son Ltd. (1980), p. 1109.

22. Barber, M., Bordoli, R.S., Sedgwick, R.D., Tyler, A.N.: J. Chem. Soc. Chem. Commun. 1981, 325.

23. König, W.A., Aydin, M., Schulze, U., Rapp, U., Höhn, M., Pesch, R., Kalikhevitch, V.N.: Int. Journ. Mass Spectr. Ion Phys. 46, 403 (1983).

24. Aydin, M., Lucht, N., König, W.A., Lupp, R., Jung, G., Winkelmann, G.: Liebigs Ann. Chem. 1985 (in press).

25. Sheehan, J.C., Nakamura, S., Stock, J.A., Maeda, K.: J. Am. Chem. Soc. 90, 462 (1968).

26. Laatsch, H.: Liebigs Ann. Chem. 1982, 28.

Miscellaneous and Biological Activity of Peptides and Proteins

SUGGESTION TO THE MODE OF PHALLOIDIN'S MOLECULAR TOXICITY

Theodor Wieland
Max-Planck-Institute for Medical Research, Jahnstr. 29
D-69 Heidelberg, F.R.G.

Introduction

Since more than 30 years the author and his colleagues have been engaged in revealing the mode of toxic action of phalloidin from the toadstool Amanita phalloides. At first the chemical structure of the fungal toxin has been elucidated, then several similarly acting components, phallotoxins, could be isolated from the green mushroom.

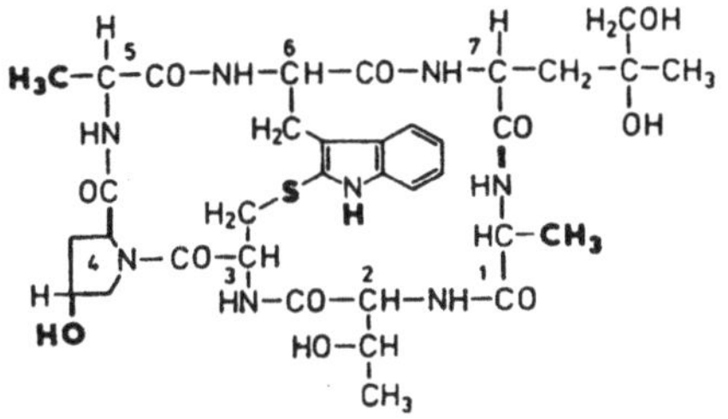

Fig. 1. Formula of phalloidin.

Pathological observations by the electron microscope pointed to a weakening effect of liver cell membrane structures by the toxin that was related with the interaction of phallotoxins with actin. Recent reviews about phallotoxin research have appeared in 1978 (1) and 1983 (2).
Phalloidin, the first toxic component isolated from the poisonous toadstool is a bicyclic heptapeptide (Fig. 1). Intraperitonially administered to experimental animals (white mouse albino rat) it causes within one to two hours severe swelling

Chemistry of Peptides and Proteins, Vol. 3

of the liver due to accumulation of blood in the liver parenchymal cells. The blood enters the liver cells by invaginations of the lipid membrane which are separated to form vacuoles inside the cell.

The molecular mechanism of phalloidin's pathological effect is its very specific and tight binding to F-actin, the polymeric form of G-actin, a protein of the molecular mass of 43000 d, from which it is assembled, establishing an equilibrium

$$\text{(i)} \qquad n(\text{G-actin}) \rightleftharpoons \text{F-actin}$$

The interaction of phalloidin, exclusively with F-, not with G-actin, results in a dramatic increase in the stability of actin filaments towards depolymerizing conditions. Phalloidin bound to F-actin maximally in a molar ratio of one to one protomer (but also in a lesser proportion) shifts the equilibrium (i) practically to the polymer side. Along with this stabilizing property, phalloidin accelerates the polymerization of G-actin thus transforming cell actin entirely into the F-state within the relatively short time of several minutes. In order to understand the mechanism of these phenomena and, herewith, the mechanism of its toxicological effect on liver cells, we attempted to find out the binding site of actin (3). This was achieved by the use of affinity labelling.

Affinity labelling of F-actin (A. Deboben, M. Nassal)

In the phallotoxins the side chain no. 7 (γ,δ-dihydroxy-L-leucine) can be chemically modified without altering the affinity of the drug to F-actin. Thus, an amino group was introduced by two different ways:

a) condensation of the keto side chain (obtained by periodate oxidation) with 1-amino-2,3-dimercapto propane) (4)

b) tosylation of the δ-hydoxyl group and replacement of OTos by NH_2 (5)

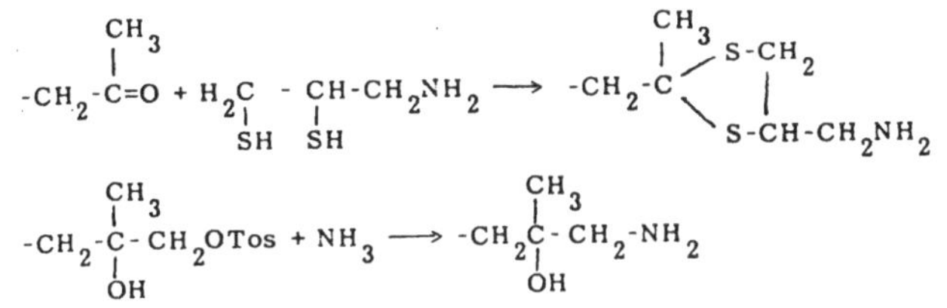

a)

or

b)

The amino derivative of a) was equipped with a radioactive alkylating group, the [^3H]-containing iodoacetyl residue, to

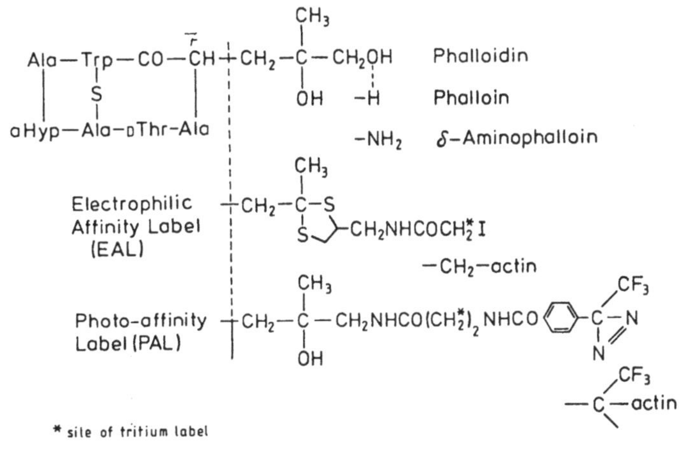

Fig. 2. Electrophilic affinity label (EAL) and photo-affinity label (PAL).

the amino group of b) was attached a radioactive photolabile diazirine (6). The iodoacetyl compound (EAL) as well as the photolabile compound (PAL) of Fig. 2 were reacted with F-actin, (the former after protection of thiol groups by NEM) by keeping the mixture for several hours, the latter by irradiating for several minutes.

The Binding Sites (J. Vandekerckhove)

The covalently labelled actin samples were digested with trypsin and thermolysin, and the radioactive fragments isolated by electrophoresis and chromatography on paper.
From the reaction with EAL two preferentially labelled peptides P1 and P2 were obtained. PAL-actin yielded the informative peptide P3. Edman-sequencing yielded for P1: PheGlnGln<u>Met</u>Trp (residues 352-356) P2: <u>Met</u>ThrGlu (119-121)
(PAL)P3: AlaAsnArg<u>Glu</u>Lys (114-118)

10 20 30
Ac-Asp-Glu-Asp-Glu-Thr-Thr-Ala-Leu-Val-Cys-Asp-Asn-Gly-Ser-Gly-Leu-Val-Lys-Ala-Gly-Phe-Ala-Gly-Asp-Asp-Ala-Pro-Arg-Ala-Val-
40 50 60
Phe-Pro-Ser-Ile-Val-Gly-Arg-Pro-Arg-His-Gln-Gly-Val-Met-Val-Gly-Met-Gly-Gln-Lys-Asp-Ser-Tyr-Val-Gly-Asp-Glu-Ala-Gln Ser-
70 80 90
Lys-Arg-Gly-Ile-Leu-Thr-Leu-Lys-Tyr-Pro-Ile-Glu-Ile-Gly-Ile-Ile-Thr-Asn-Trp-Asp-Asp-Met-Glu-Lys-Ile-Trp-His-His-Thr-Phe-
100 110 120
Tyr-Asn-Glu-Leu-Arg-Val-Ala-Pro-Glu-Glu-His-Pro-Thr-Leu-Leu-Thr-Glu-Ala-Pro-Leu-Asn-Pro-Lys-Ala-Asn-Arg-[Glu]-Lys-[Met]-Thr-
130 140 150
Gln-Ile-Met-Phe-Glu-Thr-Phe-Asn-Val-Pro-Ala-Met-Tyr-Val-Ala-Ile-Gln-Ala-Val-Leu-Ser-Leu-Tyr-Ala-Ser-Gly-Arg-Thr-Thr-Gly-
160 170 180
Ile-Val-Leu-Asp-Ser-Gly-Asp-Gly-Val-Thr-His-Asn-Val-Pro-Ile-Tyr-Glu-Gly-Tyr-Ala-Leu-Pro-His-Ala-Ile-Met-Arg-Leu-Asp-Leu-
190 200 210
Ala-Gly-Arg-Asp-Leu-Thr-Asp-Tyr-Leu-Met-Lys-Ile-Leu-Thr-Glu-Arg-Gly-Tyr-Ser-Phe-Val-Thr-Thr-Ala-Glu-Arg-Glu-Ile-Val-Arg-
220 230 240
Asp-Ile-Lys-Gln-Lys-Leu-Cys-Tyr-Val-Ala-Leu-Asp-Phe-Glu-Asn-Glu-Met-Ala-Thr-Ala-Ala-Ser-Ser-Ser-Ser-Leu-Glu-Lys-Ser-Tyr-
250 260 270
Glu-Leu-Pro-Asp-Gly-Glu-Val-Ile-Thr-Ile-Gly-Asn-Glu-Arg-Phe-Arg-Cys-Pro-Glu-Thr-Leu-Phe-Gln-Pro-Ser-Phe-Ile-Gly-Met-Glu-
280 290 300
Ser-Ala-Gly-Ile-His-Glu-Thr-Thr-Tyr-Asn-Ser-Ile-Met-Lys-Cys-Asp-Ile-Asp-Ile-Arg-Lys-Asp-Leu-Tyr-Ala-Asn-Asn-Val-Met-Ser-
310 320 330
Gly-Gly-Thr-Thr-Met-Tyr-Pro-Gly-Thr-Ala-Asp-Arg-Met-Gln-Lys-Glu-Ile-Thr-Ala-Leu-Ala-Pro-Ser-Thr-Met-Lys-Ile-Lys-Ile-Ile-
340 350 360
Ala-Pro-Pro-Glu-Arg-(Lys)-Tyr-Ser-Val-Trp-Ile-Gly-Gly-Ser-Ile-Leu-Ala-Ser-Leu-Ser-Thr-Phe-Gln-Gln-[Met]-(Trp)-Ile-Thr-Lys-Gln-
370
Glu-Tyr-Asp-Glu-Ala-Gly-Pro-Ser-Ile-Val-His-Arg-Lys-Cys-Phe

Fig. 3. Amino acid sequence of rabbit muscle actin with windowed amino acids 117, 119, and 355.

The amino acids underlined are those having reacted covalently with the respective affinity phallotoxins. EAL has alkylated the sulfur atoms of the methionines 355 and 119 out of 16 Met-residues of actin. PAL has reacted (as carbene, most probably by CH-insertion) with glutamic acid-117 (Fig. 3).

Vandekerckhove's results indicate that phalloidin binds to F-actin around residues 117-119 and 355. The sequences in these regions exhibit a remarkable degree of conservation as revealed by comparison of more than 30 actins from such different sources as plants, yeast, amoebae, invertebrates, and vertebrates (3). Although these results suggest that the relevant amino acids are surface-oriented, it is not clear whether they belong to one F-actin monomer or to two monomers adjacent in the double-helical F-actin chain. Supposing the former situation a suggestion can be offered on the site of phalloidin binding to F-actin.

In a recent paper (7) a structure has been proposed for the G-actin-DNase I-complex derived from 6 Å x-ray resolution. These studies reveal the existence of two domains in the actin monomer, a smaller one, approx. 1/3 of the molecule comprising the amino terminal part, reasonably from amino acid 1 to 120 ± 10 and a larger one, possibly with the carboxyl terminal part, the ATP being located in the cleft between the two domains. A conformational change is likely to occur during the transition of G-actin to F-actin, as it, for instance, can be concluded from differences of proteolytic digestibility of monomeric actin in the absence and presence of K^{+}-ions (8). One can assume that the change of conformation mainly involves the cleft region making accessible Met 355 at the cleft side of the large domain and at the same time setting free the region around Met 117 and Glu 119 at the opposite side. Phalloidin, then would bind to between the domains.

Stabilization of F-actin and acceleration of polymerization of G-actin

The tight binding of phallotoxins seems to stabilize a conformation of the monomers in the F-actin filament, which naturally provides the contact regions between the units. Without the drug F-actin chains will break and heal permanently at single points and lose monomers from the ends by (reversible) dissociation. If the proper conformation of the monomer, however, is stabilized by its bound phallotoxin, then the contact between the monomers is augmented to the maximum. The critical concentration of G-actin i.e. its concentration in

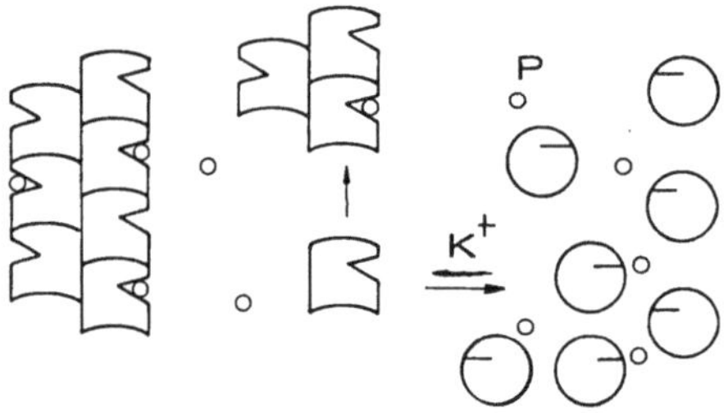

Fig. 4 , Suggested mechanism of phallotoxin action on G-actin .

the F-G-equilibrium will decrease to practically zero (9). The accelerating effect of phalloidin on the polymerization of G-actin could be understood, according to Fig. 4. G-actin monomers do not assemble, neither in the absence nor in the presence of phallotoxins unless K^+ ions are present. Potassium ions are required to provide a sufficiently high concentration of a polymerizable G-actin species denoted as F-ATP monomer actin by Rich and Estes (8,10) here called fG-actin

(ii) G-actin $\overset{K^+}{\rightleftharpoons}$ fG-actin

(iii) n·fG-actin $\rightleftharpoons$ $\underset{\text{nuclei}}{(\text{fG-actin})_n}$ (n = 3,4)

fG-Actin monomers are assumed to be the true polymerizing entities starting with the formation of oligomers (nuclei) which then grow by stepwise adding fG-monomers, to F-actin. It is now tempting to postulate that fG-actin are being stabilized by binding of phalloidin. By this way the number of assembling fG-actin-PHD molecules would increase giving rise to a rapid formation of F-actin, possibly without nucleation - but also without the possibility of simply proving this suggestion.

The pathological event

The outer lipid membrane of liver cells are reinforced by a layer of closely adhering felt-like F-actin. Phallotoxins that enter the membrane via a carrier system (11), locally will shift the actin equilibrium to the F-actin side, thus giving rise to inhomogeneity of the shoring layer, even to actin-free spots. The lipid membrane, unprotected at those sites will be bulged inwards thus forming the vacuoles that cause the swelling of the organ. It is very probable, although not proven unambigously, that excessive bleeding into the liver eventually leads to death. The destruction of the protecting F-actin layer underneath hepatocyte membranes has been visualized by Jahn (12) using a fluorescent phallotoxin (Fig. 5).

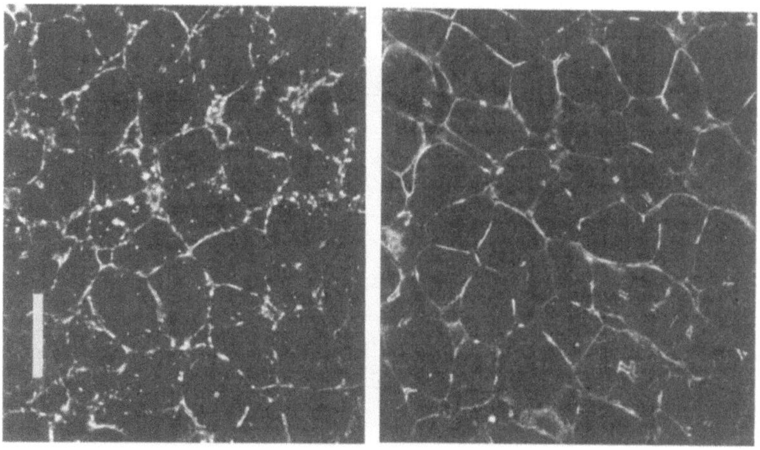

Fig. 5. Fluorescence microscopic picture of actin in normal (left) and phalloidin poisoned (right) rat liver.

Sections from a control rat liver and from a phalloidin-treated one were stained with a fluorescent phallotoxin.
As shown in Fig. 5 the section of the untreated liver shows the picture of a polygonal pattern of the cells each of which

with bright fluorescent F-actin along the border. In the poisoned liver the F-actin layers are disrupted, a part of the F-actin is concentrated in clusters, the space between is free. The bright spots between the cells mark F-actin surrounding bile canaliculi. This region is disturbed by phalloidin, too. A stop of bile flow is an additional symptom of the intoxication by phalloidin.

References

1. Wieland, T., Faulstich, H. 1978. Crit. Rev. Biochem. 5, 185-260.

2. Wieland, T., Faulstich, H. 1983. In: Handbook of Natural Toxins (R.F. Keller and A.T. Tu eds.) Marcel Dekker, Inc., New York and Basel, pp. 585-635.

3. Vandekerckhove, J., Deboben, A., Nassal, M., Wieland, T. 1985. Ms. in preparation.

4. Wieland, T., Deboben, A., Faulstich, H. 1980. Liebigs Ann. Chem., 416-424.

5. Wieland, T., Hollosi, M., Nassal, M. 1983. Liebigs Ann. Chem., 1533-1540.

6. Nassal, M. 1983. Liebigs Ann. Chem. 1510-1523.

7. Suck, D., Kabsch, W., Mannherz, H.G. 1981. Proc. Natl. Acad. Sci. USA., 4319-4323.

8. Faulstich, H., Schäfer, A.J., Weckauf, M. 1977. Hoppe-Seyler's Z. Physiol. Chem. 358, 181-186.

9. Rich, S.A., Estes, J.E. 1976. J. Mol. Biol. 104, 777-792.

10. Estes, J., Selden, L., Gershman, L. 1981. Biochemistry 20, 708-712.

11. Wieland, T., Nassal, M., Kramer, W., Fricker, G., Bickel, U., Kurz, G. 1984. Proc. Natl. Acad. Sci. USA 81, 5232-36.

12. Jahn, W., Faulstich, H., Deboben, A., Wieland, T. 1980. Z. Naturforsch. 35c, 467-469.

RAISE OF A MONOCLONAL ANTIBODY AGAINST α-AMANITIN AND STUDIES ON ITS MOLECULAR INTERACTION WITH AMATOXIN PEPTIDES

H. Faulstich, K. Kirchner
Max-Planck-Institut für Mediznische Forschung, Abteilung Physiologie, D-6900 Heidelberg, F.R.G.

Introduction

Amatoxins account for more than 90% of all fatal cases of mushroom poisoning in Europe and hence represent the most dangerous mushroom toxins. They are a family of 9 bicyclic peptides, the structures of which have been elucidated by Th. Wieland and collaborators [1,2]. α-Amanitin is the main toxin and is widely established as a tool in molecular biology [3].

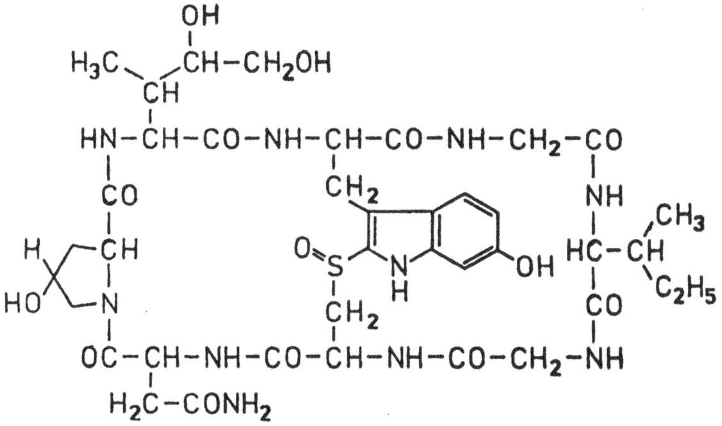

Figure 1. Structure of α-amanitin.

Antibodies against amatoxins have been raised in rabbits and in rats, predominantly for establishing radioimmunoassays for the detection of amatoxins in urine samples of patients [4,5]. A radioimmunoassay on a nylon support has recently been established in most emergency clinics of Europe [6,7].

Chemistry of Peptides and Proteins, Vol. 3

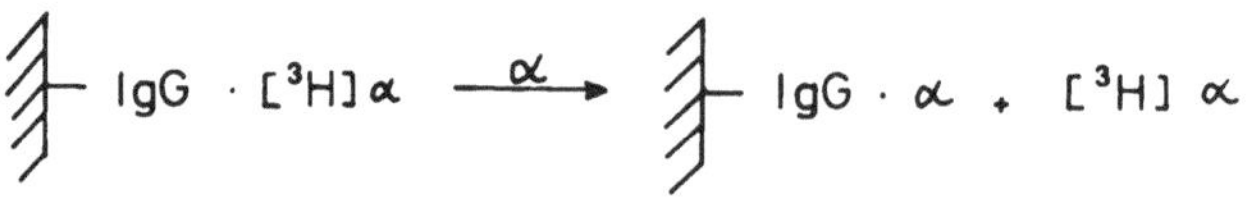

Figure 2. Radioimmunoassay for amatoxins .

As Figure 2 shows, the IgG fraction of an immunized rabbit, 5-10% of which is amatoxin-specific, is covalently attached to a nylon surface and complexed with tritiated amatoxin, $[^3H]\alpha$. In contact with native amatoxin, α, the labelled toxin is released into solution and counted. For the preparation of such assays we wanted to have an amatoxin-specific immunoglobulin of standard quality, such as a monoclonal antibody. Unfortunately, it turned out that the large scale preparation of a monoclonal antibody is cumbersome. Therefore, such radioimmunoassays are still prepared using the IgG fraction of rabbit sera.

A second aim for the preparation of amatoxin-specific antibodies was to try the therapy of amatoxin poisoning with amatoxin-binding immunoglobulins. Attempts to immunize animals, e.g. a horse, were made as early as 1930, and the antibodies tentatively produced were supposed to neutralize the toxins in a way similar to snake venoms. The result, however, was ambiguous and the question remained open until monoclonal antibodies became available [8]. When mice, poisoned with α-amanitin, were treated intravenously with the equimolar amount of an amatoxin-specific monoclonal IgG, an increase of toxicity was found instead of a decrease. The effect was even greater when Fab-fragments of the immunoglobulin were applied. This is, to our knowledge, the only example that antibodies do not detoxify, but enhance the toxicity of a poison. Accordingly, prospects are rather remote to establish an immunotherapy of human mushroom poisoning. However, interesting enough, the effect is under study at present.

First of all, the monoclonal antibody against amatoxins was wanted for studying the physical interaction between the cyclic amatoxin peptides and an amatoxin-binding protein. Such studies have been done with the target enzyme of the amatoxins, RNA polymerase II [9,10]. For a great number of amatoxin peptides the inhibiting constants, K_i, and the apparent equilibrium dissociation constants, K_D, were measured, and correlated to single structural features of the peptides. In these studies several hydroxy groups were identified, which contribute to RNA polymerase inhibition, above all of them the hydroxy group attached to proline, which is essential for inhibitory activity. However, RNA polymerase II is a complex enzyme, and a better insight into the interaction of the cyclic peptides with a protein was expected from studies with an amatoxin-specific monoclonal antibody. Final aim of the work is the preparation of a crystalline amatoxin/Fab complex suitable for X-ray analysis. In the present study we measured the affinities of various amatoxin derivatives to the monoclonal immunoglobulin and correlated these values to single structural features of the peptide.

Preparation and characterization

Amatoxins are too small to be immunogenic by themselves. Two sites of the peptide proved suitable for coupling the peptides to proteins, the phenolic group of α-amanitin (Fig. 3II) and the carboxyl group of β-amanitin (Fig. 3I).

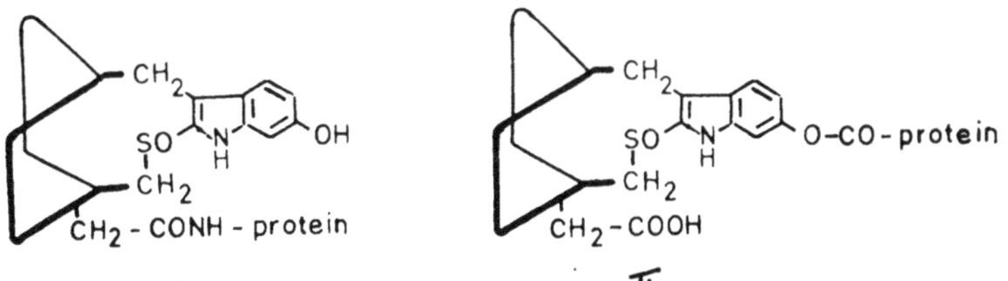

I

II

Figure 3. Protein conjugates of β-amanitin (I) and α-amanitin (II).

After immunization of rats with a fetuin conjugate of β-amanitin, spleen cells of these animals were fused with a mouse myeloma cell line. After cloning and subcloning hybridoma cells producing amatoxin-specific antibodies were detected by an enzyme-linked immunosorbent assay (ELISA). Limit of detection was the immunoglobulin production of 4-8 cells in a volume of 0.2 ml. The cells were expanded in culture flasks, or, with a 100fold higher yield, in the peritoneal cavity of a nude mouse. Immunoincompetent mice had to be used, because the cloned cells were hybrids of mouse and rat cells. Mouse albumin was the predominant protein in the ascites fluid of such mice, while the amatoxin-specific IgG accounted

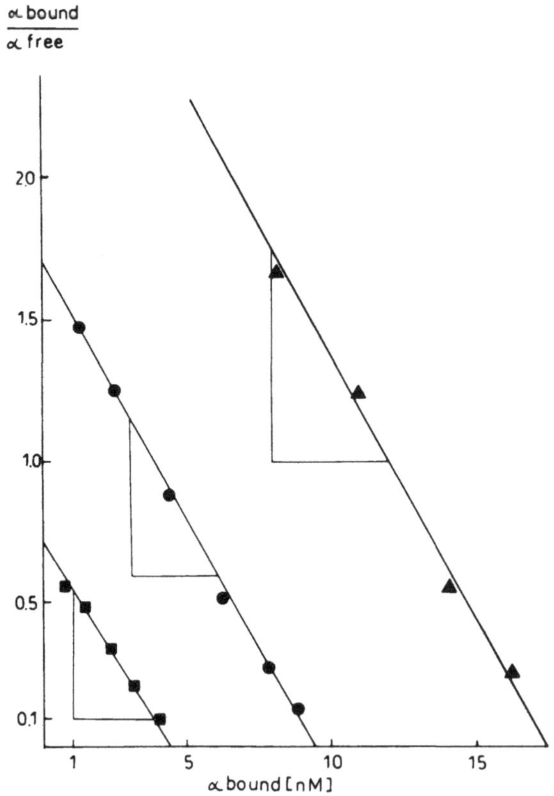

Figure 4. Scatchard plots of 3 equilibrium dialysis experiments using different concentrations of immunoglobulin.

for ca. 5% of the total protein. On the other hand, mouse IgG was absent, and the monoclonal IgG could be separated from the other ascites components by affinity chromatography on Protein A Sepharose. Details of the procedure will be published elsewhere [8].

The affinity of the monoclonal IgG for amatoxins was determined by equilibrium dialysis. Since α-amanitin is not available with a radioactive label, we used an amatoxin with a modified and tritiated side chain, 6'-O-methyl-dehydroxymethyl-α-amanitin, abbreviated $[^3H]\alpha$. The results were plotted according to Scatchard [11], as shown in Fig. 4.

The equilibrium dissociation constant, K_D, is directly correlated to the slopes of the plots, and was calculated to be ca. 5 nM. Obviously, the affinities of $[^3H]\alpha$ for the monoclonal IgG and for RNA polymerase II are comparable.

Structure-affinity relationship

The affinity of the monoclonal IgG for 4 natural and 5 semisynthetic amatoxin derivatives was determined, using polyethylene glycol precipitation of the complex for separating $[^3H]\alpha$ (bound). It was made certain that the separation procedure did not disturb complex formation. IgG and $[^3H]\alpha$ were incubated with increasing amounts of the amatoxin derivative to be assayed, precipitated with PEG 6000, and the amount of amatoxin required for 50% substitution of $[^3H]\alpha$ in the complex was defined as measure for affinity.

Structural modifications of the amatoxin derivatives were located in the side chain of the dihydroxylated isoleucine, which in native toxins may have 2 OH (α-amanitin), 1 OH (γ-amanitin), or none (amanullin). We further assayed two amatoxins methylated in 6'-position: O-methyl-α-, and O-methyl-γ-amanitin. Finally, β-amanitin was studied, a native toxin with a carboxyl moiety instead of the carboxamide in α-amanitin.

As expected, α-amanitin had the highest affinity for the immunoglobulin. Since all biologically active amatoxin derivatives have conformations similar to α-amanitin [12], as proven by their circular dichroism spectra, any decrease in the affinity of a derivative can be reasonably correlated to its deviation from α-amanitin structure. For example, the lower affinity of β-amanitin (0.65), related to α-amanitin (1.0), must be due to the presence of carboxylate ($-COO^-$, in β) instead of carboxamide ($-CONH_2$ in α). Likewise, the continuing decrease of affinity in the series α-amanitin, γ-amanitin, and amanullin, which were determined as 1.0, 0.79, 0.26, respectively, can be attributed to the presence of 2 OH, 1 OH, or none, respectively, in the isoleucine side chain. According to the decrease in affinity values we defined relative amounts of binding energy. For example, 21% of the total binding energy is contributed by the δ-OH and 53% by the γ-OH of the isoleucine side chain. Methylation of the 6'-OH group decreases the affinity by 46% in α-amanitin, and 48% in γ-amanitin, indicating that the hydroxylated indole moiety is an important antigenic site and contributes about one half of the binding energy. The fact that the decrease in affinity is nearly identical for both compounds proves the validity of the method.

In general, we found that binding of the amatoxins to the amatoxin-specific immunoglobulin is based upon hydrogen bonding. For example, the carboxamide in α-amanitin probably acts as a hydrogen donor, a part that cannot be taken by the carboxyl in β-amanitin. Presence of carboxyl lowers the affinity by about 37%. The two hydroxy groups of isoleucine may act as hydrogen bond donors or acceptors as well, and the same role may be attributed to the 6'-hydroxy group of the indole moiety. In the latter case, however, it may also occur that the disc-shaped indole nucleus of the toxin is fitted in between rigid lipophilic structures of the immunoglobulin binding site, an interaction possibly sterically disturbed by methylation of the hydroxy group in 6'-position.

Of great interest, but not elucidated so far, is the question, whether the hydroxy group attached to the proline ring contributes to immunoglobulin binding. This OH group is essential for RNA polymerase binding as shown for proamanullin, a native precursor molecule isolated from the mushroom. Proamanullin lacks this OH group and does not bind to the enzyme [10]. Unfortunately, proamanullin is very rare and was not available when these experiments were done. Likewise unknown is the part contributed by the lipophilic side chain of isoleucine. From an X-ray analysis [13] of β-amanitin we know that this side chain is exposed and fixed in its position by two adjacent glycine residues. It is thus expected to contribute to binding of both the immunoglobulin and the RNA polymerase II.

Acknowledgement

This research was supported by a grant of Deutsche Forschungsgemeinschaft.

References

1. Wieland, Th., Faulstich, H., 1978. Crit. Rev. Biochem. 5, 88
2. Wieland, Th., Faulstich, H., 1983. In: Handbook of Natural Toxins, Vol. 1: Plant and Fungal Toxins (R.F. Keeler and A.T. Tu eds.) Marcel Dekker, New York, p. 585
3. Faulstich, H., 1980. In: Progress in Molecular and Subcellular Biology, Vol. 7 (F.E. Hahn ed.) Springer, New York, 88
4. Faulstich, H., Trischmann, H., Zobeley, S., 1975. FEBS-Lett. 56, 312
5. Fiume, L., Busi, C., Campadelli-Fiume, G., Franceschi, C., 1975. Experientia, 31, 1233
6. Faulstich, H., Zobeley, S., Trischmann, H., 1982. Toxicon, 20, 913
7. Faulstich, H., 1984. In: Diagnose, Verlaufskontrolle und Therapie schwerer exogener Vergiftungen (E. Deutsch, G. Kleinberger, R. Ritz and H.P. Schuster, eds.) F.K. Schattauer, Stuttgart, p. 181
8. Kirchner, K., Faulstich, H., submitted
9. Cochet-Meilhac, M., Chambon, P., 1974. Biochim. Biophys. Acta. 353, 160
10. Wieland, Th., Götzendörfer, C., Zanotti, G., Vaisius, C., 1981. Europ. J. Biochem. 117, 161
11. Scatchard, G., 1949. Ann. N.Y. Acad. Sci. 51, 660
12. Faulstich, H., Bloching, M., Zobeley, S., Wieland, Th., 1973. Experientia 29, 1230
13. Kostansek, E.C., Lipscomb, W.N., Yocum, R.R., Tiessen, W.E., 1977. J. Am. Chem. Soc. 99, 1273

SYNTHESIS AND REACTIVITY OF PEPTIDES - SUBSTRATES AND INHIBITORS OF THROMBIN

V.K.Kibirev, A.A.Hershkovich, S.A.Poyarkova, V.P.Romanova, S.B.Serebryany
Bio-Organic Department of the Institute of Organic Chemistry, Academy of Sciences of the Ukrainian SSR, Kiev, USSR

Introduction

Thrombin is a trypsin-like serine protease of blood clotting. Coagulation of blood is known to be a very complex cascade of enzymic reactions. Thrombin plays the leading part in hemostasis. The principal function of the enzyme is to cleave fibrinopeptides A and B from fibrinogen and to convert it into fibrin-monomer,which is spontaneously assembled to a clot. Thrombin also activates factor V, factor VIII and factor XIII; it causes platelets aggregation and granular contents release. All these processes speed up thrombogenesis. As far as we know, thrombin may also participate in the biochemical processes causing hemostasis retardation. Thus,thrombin should be advisably regarded as the enzyme with bioregulatory functions (1-3). The present work aims to study of the physico-chemical basis of thrombin regulatory action.

Results

Thrombin reactivity analysis has been carried out using synthetic substrates and inhibitors. All these compounds are derivatives of arginine or arginine-containing oligopeptides.The guideline for obtaining these peptides was the amino acid sequence at the thrombin-susceptible bond in A α-chain of fibrinogen - the physiological substrate of thrombin. The sim-

Chemistry of Peptides and Proteins, Vol. 3

plest substrates or inhibitors of serine proteases are known to be the derivatives of amino acids, corresponding to subsite P_1, the primary site of the natural substrates. Arylsulfonylargininemethyl esters are such simple substrates of thrombin. That's why we began the study of the enzyme using these compounds. The esters, represented by the general formula 1, have been prepared using the well known procedures (4).

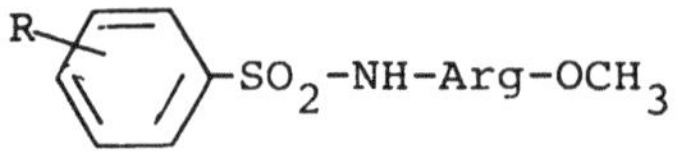

R: CH_3, C_2H_5, i-C_3H_7, s-C_4H_9, NH_2, OCH_3, F, Cl, Br, $COOCH_3$

1

The use of amino acid derivatives as substrates or inhibitors has, however, been mainly limited due to their low specificity. One would suggest that an amino acid sequence rather than a single amino acid residue could imitate the secondary site of the natural substrate. In order to investigate the role of these secondary binding interactions the thrombin-catalyzed hydrolysis of some oligopeptides, ranging in length from P_4 to P'_2-subsites, has been studied. The following substrates and inhibitors have been assayed: (i) peptides of increasing length; (ii) diastereoisomeric pairs of peptides in which the particular L-amino acid residue has been replaced by its D-analogue; (iii) pairs of peptides in which hydrophobic side chains have been changed for the polar ones.

Figure 1 shows the synthesis of di- and tripeptides of general formulae 2 and 3. These compounds have been obtained by the interaction of N-hydroxysuccinimide esters of different tosyl amino acids with unprotected arginine. Glycine-containing peptides have been synthesized using the activated pentafluorophenyl ester method.

Tos-P_2-Arg-OMe

P_2: Gly, L-Ala, L-Asn, L-Phe, L(D)-Val, L-MeVal

2

TOS-P_3-P_2-Arg-OMe

P_2: Gly, Ala, Val; P_3: Gly, Pro, Sar.

3

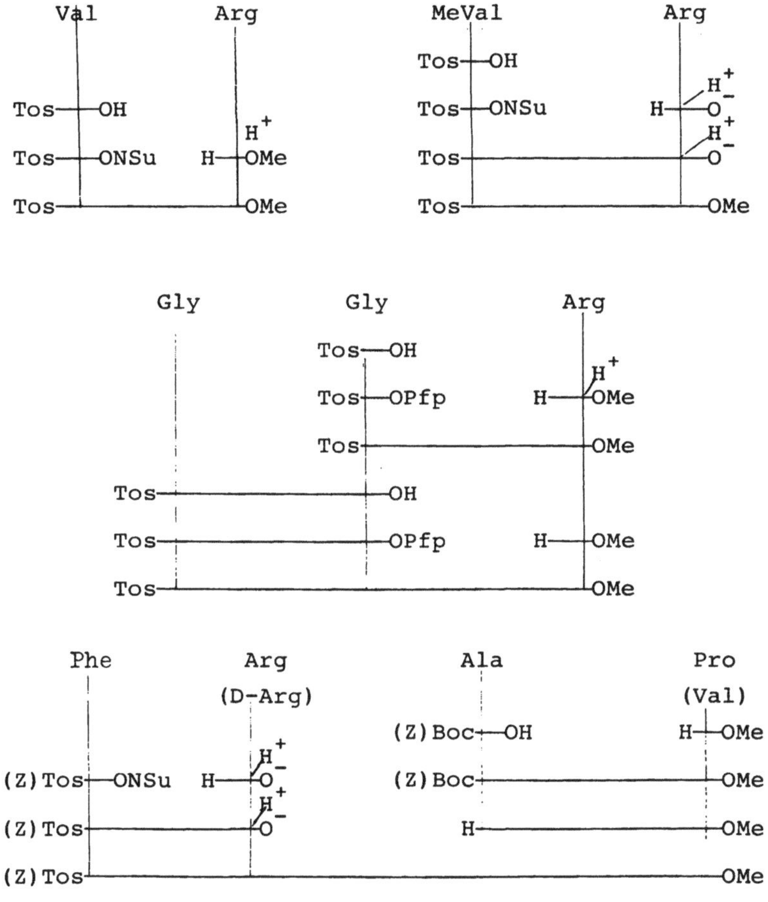

Fig.1. Synthesis of arginine-containing tri- and tetrapeptides.

Method for the preparation of tri- and tetrapeptides, containing stereoisomeric amino acids, incorporates, both, stepwise and block condensation. The polypeptide chain has been elongated in the presence of N-hydroxysuccinimide as a nucleophile additive, used to avoid racemisation when coupling peptide fragments made by DCC-method.
It should be noted that even the most extended synthetic peptides are split by thrombin dozens times less readily than the physiological substrate fibrinogen (5).
Blombäck et al.(6) and Scheraga and coworkers (5) showed that the amino terminal part of A α-chain of fibrinogen, incorporating the first 51 amino acids (fig.2), does actually contain all the residues required for the specific interaction with thrombin. The rate of thrombotic hydrolysis falls only threefold after shortening of A α (1-51) fragment by seven carboxy-terminal amino acids, but further removal of sequence 34-44 causes dramatic reduction of the reaction rate (6).
We suggest that the sequence 34-44 is capable of interacting with the so-called exosite of thrombin, which is located outside the active centre of the enzyme. It would be very interesting to synthesize the fragment 30-44 of A α-chain and to check its ability to bind with thrombin.
Two compounds, hexapeptide <u>4</u> and pentadecapeptide <u>5</u>, relevant to the amino acid sequence 30-44 (fig.2), have been synthesized by solid phase procedure. These peptides contain substitutions of some amino acids. Thus, Cys^{36} in both analogues was replaced by Ala, $Trp^{33,41}$ were replaced in pentadecapeptide <u>5</u> by Phe; lysine-44 in this analogue was omitted.
All the synthesized compounds <u>1</u> - <u>5</u> were used for the kinetic and inhibitory analysis of the relationships between the chemical structure and its reactivity towards thrombin.
The effectivity of thrombin-catalyzed hydrolysis of esters <u>1</u> is found to depend on the nature of substituents R (Table 1). The reaction rate decreases when the size of R increases and its polarity drops. The following equation was derived on the basis of Hammett type correlation (7):

$$\log (k_{cat}/K_m) = (1.70 \pm 0.59)\delta + (0.66 \pm 0.25)E_s + (4.71 \pm 0.14)$$

```
 1             5                   10                  15
H-Ala-Asp-Ser-Gly-Glu-Gly-Asp-Phe-Leu-Ala-Glu-Gly-Gly-Gly-Val
              20                  25                  30
-Arg-Gly-Pro-Arg-Val-Val-Glu-Arg-His-Gln-Ser-Ala-Cys-Lys-Asp-
              35                  40                  45
-Ser-Asp-Trp-Pro-Phe-Cys-Ser-Asp-Glu-Asp-Trp-Asn-Tyr-Lys-Cys-
              50
-Pro-Ser-Gly-Cys-Arg-Met-OH
```

A α (1-51)

H-Asp-Ser-Asp-Trp-Pro-Phe-Ala-Ser-Asp-Glu-Asp-Trp-Asn-Tyr-Lys-

-Gly-OH

4

H-Asp-Ser-Asp-Phe-Pro-Phe-Ala-Ser-Asp-Glu-Asp-Phe-Asn-Tyr-Gly-

-OH

5

Fig.2. Amino acid sequence of A α (1-51) region of fibrinogen and the structure of two synthetic fragments: hexadecapeptide 4 and pentadecapeptide 5.

Table 1. Kinetic Parameters of Thrombin-catalyzed Hydrolysis of $R-C_6H_4-SO_2$-Arg-OMe at pH = 8.5 and a Temperature 25°C (8).[a]

R	$10^5 K_m$ (M)	k_{cat} (s^{-1})	$10^{-5} k_{cat}/K_m$ ($M^{-1}s^{-1}$)
4-CH_3	3.3±0.47	2.5±0.45	0.76
4-C_2H_5	2.7±0.48	1.3±0.10	0.48
4-i-C_3H_7	0.9±0.22	0.1±0.03	0.11
4-s-C_4H_9	3.2±0.29	0.1±0.03	0.04
4-CH_3O	3.6±0.57	0.3±0.05	0.08
4-NH_2	5.9±0.55	1.4±0.46	0.24
4-$NHCOCH_3$	6.0±0.55	1.5±0.12	0.25
4-F	7.3±0.61	3.9±0.49	0.54
4-Cl	3.4±0.15	4.3±0.17	1.30

Table 1. (continued).

R	$10^5 K_m$ (M)	k_{cat} (s^{-1})	$10^{-5} k_{cat}/K_m$ ($M^{-1}s^{-1}$)
4-Br	3.7±0.20	4.3±0.06	1.16
3-$COOCH_3$	3.6±0.22	6.5±0.74	1.80

[a/] The reaction has been carried out in 0.15 M KCl with pH-stat; $[E_o] \sim 2 \cdot 10^{-8}$M; $[S_o] \sim 5 \cdot 10^{-5}$M. The average values from 4-7 runs and standard deviations are given.

According to this equation the thrombin-catalyzed reaction is believed to be sensitive to steric and polar effects of the substituent R. In contrast with the enzymatic process, the rate of saponification of the esters 1 does not depend on the chemical structure of R.

Sensitivity of the enzymatic reaction towards the nature of R is suggested to account for the alteration of stability of the hydrogen bond formed between arylsulfonylamide group of the substrate and the CO-group of serine-214 of the thrombin active site.

The arylsulfonyl-argininemethyl esters 1 were found to possess antithrombin activity as well (9).

To compare the relative inhibitory effect of these compounds the I_{50} value has been assayed for each ester 1. It was found that hydrogen substitution for the bulky group in the aryl part of the molecule causes an increase in inhibitory power of the ester. The following equation showing that antithrombin activity of the esters 1 is determined by the polarity and the size of R substituent (7) has been found:

$$\log I_{50} = 1.45\,\sigma + 0.56\,E_s - 3.50$$

Table 2. Antithrombin Activity of Arylsulfonylargininemethyl Esters (9).

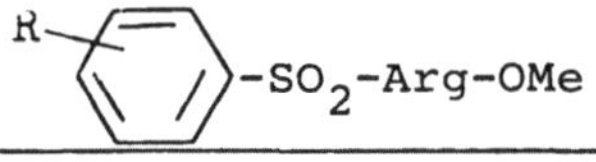

R	I_{50},[a] μM
4-CH_3	350
4-C_2H_5	140
4-i-C_3H_7	50
4-s-C_4H_9	15
4-CH_3O	300
4-Cl	1400
4-F	1000
4-Br	800
2.4-di-NO_2	200
3-$COOCH_3$	650

a/The I_{50} value has been defined as the concentration of ester at which the clotting time of fibrinogen is enhanced twofold comparing to the control. The assay is carried out at pH = 7.0 and temperature 20°C.

The proposed relation might be used in order to find more potent thrombin inhibitors.

Our studies in the series of peptides 2 and 3 show the importance of the configuration and lipophylicity of amino acids located at P_2 position. Table 3 demonstrates that D-isomer or polar side chain (for example, Asn) causes sharp decrease of the catalytic constants. Thrombin appears not to split these substrates which are normally digested by trypsin. Thus, thrombin has a striking secondary stereospecificity. Analysis of the obtained kinetic parameters (Table 3) shows that the perfect substrates would be those compounds which have at their P_2 position the hydrophobic amino acid with relatively small side chain; this residue might be N-methylated. Among natural amino acids proline corresponds to these structural and lipophylic requirements. Indeed, proline or its analogues: pipecolic acid or azetidinecarboxylic acid are the most advantageous P_2 residues in tripeptide p-nitroanalides, (10), used as chromogenic substrates or in tripeptide aldehydes (11) and peptidyl chloromethylketones (12), employed for the inhibition of thrombin.

The structural requirements of the nature of amino acids located at the subsites P_3 and P_4 are investigated less profoundly. In any case one might say that hydrophobic amino acids should be located at these positions. Besides, residue at P_3 may be a D-isomer (10-12).

As to the influence of the residues at more distant subsites than P_4, it might only be noted that a ß-turn is present in the P_7-P_4 region of natural substrates (13). Thus Gly-Gly and Gln-Gly sequences are located within the fragment P_7-P_4 of such physiological substrates as fibrinogen and factor XIII. These sequences are known to have very high potential for ß-turn.

Scheraga and coworkers (14) have studied some peptide fragments of fibrinogen by NMR and shown the presence of ß-bend in its peptide chain.

The studies of thrombin with low molecular weight substrates reveal two features of thrombin action: 1) thrombin is several orders of magnitude less active than such serine proteases as trypsin, chymotrypsin; 2) specificity of thrombin is, however, higher than that of trypsin.

We have suggested that decrease of thrombin activity is dependent on the reduction of hydrophobicity and nucleophilicity of the enzyme active site. That is why only those peptides will be perfect substrates, which are able to create the apolar surrounding of thrombin active site.

Table 3. Kinetic Parameters of Thrombin-catalyzed Hydrolysis of Tosyl-peptidylmethyl Esters at pH 8.5 and 25°C (13)[a].

Peptide	$10^5 K_m$ (M)	k_{cat} (S^{-1})	$10^{-5} K_{cat/Km}$ ($M^{-1}S^{-1}$)
Tos-Gly-Arg-OCH_3	1.12±0.05	4.78±0.29	4.27
Tos-Ala-Arg-OCH_3	1.43±0.30	11.oo±1.3	7.7
Tos-Asn-Arg-OCH_3	no hydrolysis		
Tos-Phe-Arg-OCH_3	0.24±0.11	0.86±0.11	3.6
Tos-Val-Arg-OCH_3	0.45±0.04	2.56±0.20	5.8
Tos-D-Val-Arg-OCH_3	no hydrolysis		
Tos-Me-Val-Arg-OCH_3	1.56±0.09	2.90±0.23	1.86
Tos-Gly-Gly-Arg-OCH_3	1.41±0.4	12.25±2.5	8.69
Tos-Sar-Gly-Arg-OCH_3	1.54±0.7	11.32±4.2	7.35
Tos-Gly-Val-Arg-OCH_3	1.44±0.1	14.08±0.9	9.78
Tos-Sar-Val-Arg-OCH_3	1.78±0.27	12.23±0.56	6.87

a/ The average values from 10 -12 assays and the standard deviation are given. Reaction has been carried out in 0.15 M KCl with a pH-stat. $[E_o] \sim 2.10^{-8}$M, $[S_o] \sim 5.10^{-5}$ M.

It should be emphasized that according to Scheraga and coworkers (5) synthetic peptides are, however, split by thrombin much less than the natural substate fibrinogen. This might be explained by the absence in the low molecular weight compounds of some structural elements which have to interact with exosite of thrombin. When studying two synthetic peptides relevant to the 30-44 sequence of the A α-chain of fibrinogen we have found that, both, the hexadecapeptide 4 and the pentadecapeptide 5 are able to inhibit the clotting activity of thrombin non-competitively (K_i = 2.6 mM). As the pentadecapeptide 5 has neither arginine nor lysine capable of interacting with the primary binding cleft of thrombin, these compounds are suggested to be bound not to the active centre but with the exosite of

thrombin. These findings allow one to think that the 30-44 sequence of the fibrinogen A α-chain is a very important structure of fibrinogen molecule, which is recognized by exosite of thrombin. It can be deduced that the structure of thrombin binding site is rather complex (1,3). This site consists of three loci: 1) the primary binding pocket, accomodating the positively charged side chain of arginine, located at P_1 subsite; 2) the secondary apolar binding site, interacting with the P_9-P_2 region of physiological substrates; 3) the exosite, needed for the interaction with the protein substrates. Thus, our experimental and literature data, concerning the esterase, amidase, protease and clotting activity of thrombin allow to conclude that the substrate specificity of thrombin results from at least two factors: (1) the relatively low nucleophilicity of the catalytic site of thrombin; (ii) a rather complex structure of the binding site of the enzyme.
According to our hypothesis the low reactivity of thrombin towards many peptides results from the decrease of hydrophobicity of its active site region due to substitutions of some hydrophobic amino acids by polar ones. The side chains of substrates might be suggested to participate in the organisation of fully reactive catalytic thrombin site only when peptides contain amino acids with apolar side chains and when these residues screen the catalytic site of thrombin from the solvent.

Acknowledgement

We thank Mrs H.V.Sirotina for her assistance in the preparation of this manuscript.

Rererences

1. Fenton, J.W. 1981. Ann.N.Y. Acad.Sci. 370, 468-495.

2. Strukov, A.I., S.M.Strukova. 1980. Archiv Patholog. 42, 3-16.

3. Serejskaya, A.A., V.K.Kibirev, S.B.Serebryany. 1981. Biokhimiya Zhivotnich i Cheloveka 5, 3-28.

4. Serebryany, S.B., D.M.Fedoryak, V.K.Kibirev. 1975. Khim. Prir.Soedin., 69-73.

5. Hageman, T.C., H.A.Scheraga, 1977.Arch.Biochem.Biophys.179, 506-517.

6. Hogg, D.H., B.Blombäck. 1978. Thromb.Res. 12, 953-964.

7. Kibirev,V.K., A.A.Gershkovich, S.B., Serebryany. 1983. Ukr.Biokhim. Zh. 55, 271-276.

8. Fedoryak. D.M., V.K.Kibirev, A.A.Serejskaya, S.B.Serebryany. 1977. Biokhimiya 42, 1595-1602.

9. Serejskaya, A.A., V.K.Kibirev, D.M.Fedoryak, S.A.Poyarkova, S.B. Serebryany. 1977. Dokladi Akad.Nauk Ukr. SSR, ser. B, N 4, 346-349.

10. Lottenberg, R., J.A.Hall, M.Blinder, E.P.Binder, C.M.Jackson. 1983. Biochem.biophys.acta 742, 539-557.

11. Bajusz,S., E.Barabas, P.Tolney, E.Szell, D. Bagdy. 1978. Int.J. Pept.Prot.Res. 12, 217-221.

12. Kettner, C., E.Shaw. 1977. In: Chemistry and Biology of Thrombin (R.L.Lundblad, J.W.Fenton, K.G. Mann, eds). Ann Arbor Science, p.p. 120-143.

13. Kibirev, V.K., V.P.Romanova, S.B. Serebryany. 1983. Biokhimiya 48, 937-943.

14. Rae, J.D., H.A.Scheraga. 1979. Int.J.Pept.Prot.Res. 13, 304-314.

POTENT B-LYMPHOCYTE MITOGENS AS COVALENTLY BOUND CARRIERS FOR THE PRESENTATION OF ANTIGENS AND ENHANCEMENT OF IMMUNE RESPONSE

G.Jung, G.Becker, J.Metzger, K.-H.Wiesmüller

Institut für Organische Chemie, Universität Tübingen, D-74oo Tübingen, FRG

W.G.Bessler, L.Biesert

Institut für Biologie, Mikrobiologie II, Universität Tübingen, D-74oo Tübingen, FRG

H.-J.Bühring, C.Muller

Medizinische Klinik, FACS-Labor, Universität Tübingen, D-74oo Tübingen, FRG

Introduction

Surface components of bacteria such as lipoprotein (1,2), lipopolysaccharide (LPS) (3), protein I and protein II (4) are potent B-lymphocyte activators. The molecular structure of lipoprotein from E.coli responsible for mitogenicity resides in the N-terminal three fatty acids containing lipopeptide region (5). In order to determine the minimal structural requirements for biological activity, we prepared oligopeptide analogues to the N-terminal lipid part of lipoprotein by chemical synthesis (6-8).
We demonstrated that the synthetic lipopeptide segment S-(2,3-bis(palmitoyloxy)-(2-RS)-propyl)-N-palmitoyl-(R)-cysteinyl-serine (Pam_3Cys-Ser) is fully active in mouse B-lymphocytes to stimulate proliferation and immunoglobulin synthesis (9). The fragment of this molecule lacking the terminal serine (Pam_3Cys-OH or Pam_3Cys-OMe (7)) is by far less active. Segments carrying additionally parts of the N-terminal amino acid sequence of native lipoprotein, i.e. Pam_3Cys-Ser-Ser, Pam_3Cys-Ser-Ser-Asn, Pam_3Cys-Ser-Ser-Asn-Ala (6) exhibit stimulatory activities also comparable to the native lipoprotein.
Analogues carrying less than three fatty acids respectively lacking the S-glyceryl moiety such as Pam-Ser-Ser-Asn-Ala, showed drastically reduced effects in stimulating B-cell proliferation. We

Chemistry of Peptides and Proteins, Vol. 3

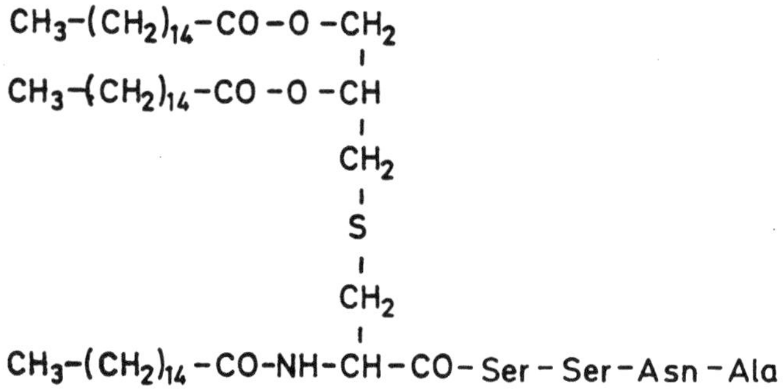

Structure of the mitogen tripalmitoyl-pentapeptide (Pam_3Cys-Ser-Ser-Asn-Ala,TPP)

reported elsewhere (8) on *in vitro* and *in vivo* experiments using synthetic tripalmitoyl-pentapeptide (TPP) (6) as a potent B-lymphocyte activator as the parent lipoprotein, which enhances also nonspecifically antigenic response and resistance to infections.

Most promising results were recently obtained using tripalmitoyl-S-glyceryl-cysteinyl-serine (Pam_3Cys-Ser) as mitogenic carrier for antigens (9-11). We demonstrate here the covalent incorporation of this molecule as the last residue in the solid-phase peptide synthesis of antigens (7). Using appropiate spacers such as hexamethylene diamine or lysine antigenic determinants may be attached also by their carboxy termini. The resulting lipopeptides were found to be highly active mitogens and to induce immunoglobuline production *specifically* against the antigen (1o). Immunization experiments on mice required markedly reduced immunization periods and no addition of any further adjuvants. A more complex approach with the additional use of lipophilic helices for antigen presentation on lymphocytes has been discussed by one of us in context with the design of the "Pam_3Cys-Antigen-Helix System" (11) as a potential synthetic vaccine.

Non-specific enhancement of immune response by tripalmitoyl-pentapeptide

All Pam_3Cys conjugates were tested for mitogenic activity (8,12,13) towards spleen cells from Balb/c, LPS-nonresponder C3H/HeJ, and congenitally athymic C3H/Tif/Bom/nu/nu mice, measuring the incorporation of ^{3}H-thymidine into DNA. Lymphocyte activation was confirmed by determination of the incorporation of ^{3}H-uridine into RNA and ^{3}H-leucine into protein with a maximum at 48-72 hours after mitogen addition. The synthetic lipopeptides were also investigated for their ability to stimulate B-lymphocytes into immunoglobulin secretion, as shown by a hemolytic plaque assay using trinitrophenylated sheep red blood cells (TNP/SRBC). Throughout our studies, the lipopeptides carrying at least 2 amino acids exhibited strong stimulatory activity towards B-lymphocytes comparable to native lipoprotein starting at concentrations around 0.1 µg/ml with an optimum at 30 µg/ml, e.g. in spleenocyte cultures at a cell density of 6.6 x 10^5/ml. All compounds exhibited only a marginal effect on thymocytes. Thus, a series of defined synthetic fragments of a bacterial outer membrane component exhibits a pronounced mitogenic and polyclonally stimulating activity towards B-lymphocytes.

Recently synthetic tripalmitoyl-pentapeptide S-(2,3-bis(palmitoyloxy)-(2-RS)-propyl)-N-palmitoyl-(R)-cysteinyl-(S)-seryl-(S)-seryl-(S)-asparaginyl-(S)-alanine has also been shown to induce the production of interleukin 1 in human mononuclear leukocytes and the release of prostaglandin (PGE_2) from mouse peritoneal macrophages (manuscript in preparation and 14). The three fatty acids in the hydrophobic part of the lipopeptide probably serve to anchor the mitogen in the lymphocyte plasma membrane presenting the lipopeptides to binding proteins there. Binding proteins for lipoprotein, which might be involved in cell activation, were shown to include MHC gene products (15,16). The mitogenic activity seems to be B-cell specific, as pure T-lymphocytes obtained from thymus largely failed to respond to the compounds (M.Cox, W.G. Bessler, unpublished).

Bacterial lipoprotein has also been shown to activate the B-cell tumor line BCL1, accordingly the low molecular weight tripalmitoyl-pentapeptide was also active towards the in vitro and in vivo BCL1 B-cell lines (17) and towards the B-cell lymphoma WEHI 279.1 (18).

One should emphasize, that the diastereomeric TPP molecules differing in their configuration at the S-glycerol moiety differ in their ability to enhance immune response (6,8).

Pam_3Cys-Ser-Ser-Asn-Ala (6) has been found to stimulate immunoglobuline synthesis also in human peripheral blood lymphocytes

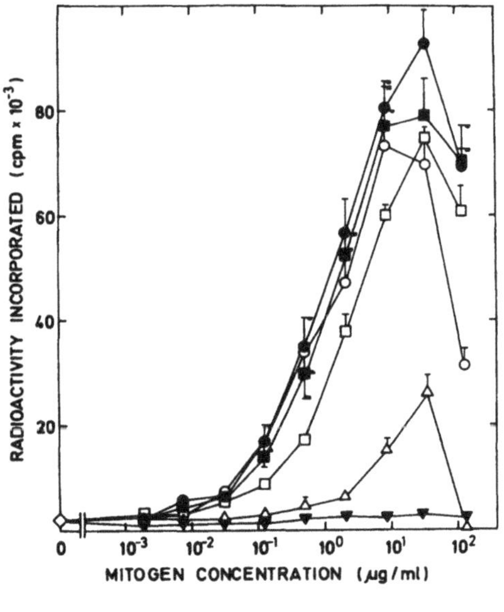

Fig.1. Dose-response curves for the incorporation of ^{3}H-thymidine into DNA in spleen cells from C_3H/HeJ mice induced by Pam_3-Cys-Ser-Ser-Asn-Ala(6)(O), Pam_3Cys-Ser-Ser-Asn(●), Pam_3 Cys-Ser-Ser(□), Pam_3Cys-Ser(■), Pam_3Cys(△) and LPS(▼). Cell cultures were carried out in 3040 microtiter plates (72 h, 37°C, moist CO_2 incubator, 6.6 x 10^5 cells/ml) in 200 µl aliquots. Culture medium: RPMI 1640 supplemented with 10% fetal calf serum (FCS, Seromed, Munich), glutamine (2mM), penicillin (100 units/ml), streptomycin (100 µg/ml) and 2-mercaptoethanol (5x 10^{-5}M). The mitogens were added to the cultures after suspension in Minimal Essential Medium containing 10 % FCS by sonification (6 x 10 sec., 100 W, Labsonic 1510 sonifier, Braun, Melsungen). 24 h before harvesting 1 µCi ^{3}H-thymidine (Amersham, Braunschweig, specific activity 5 Ci/mmol (185 GBq/mmol) was added to each well. The cultures were harvested (Mash-II-harvester), collected on glass fibre filters and measured by liquid scintillation counting. Means of triplicate determinations.

(W.G.Bessler and S.Sager, unpublished). In very recent experiments we additionally could show that the synthetic lipopeptides are able to act directly on single isolated spleenic murine B-lymphocytes to induce clonal proliferation (G.D.Wetzel and W.G.Bessler, unpublished) as has been described before for LPS.

Summarizing our results, the synthetic lipopeptides containing 2-5 amino acids (MW 998 to 1270) are able to polyclonally activate murine B-lymphocytes from both LPS-responder and LPS-nonresponder mouse strains and are, in all aspects tested, mitogenically as active as native lipoprotein (MW 72oo daltons). In all experiments the lipopeptides were more active than the basic structure Pam_3Cys, indicating that in order to obtain full biological activity the presence of a lipophilic part followed by a hydrophilic oligopeptide consisting of a least 2 amino acids is necessary. Interestingly, the active lipopeptides are all diacylglycerol analogues, and indeed there are hints that they might be able to stimulate leukocytes via the activation of protein kinase C.

Our findings suggest that these bacterial surface components may influence the host-parasite interaction, presumably prior to the establishment of specific humoral and cellular responses, Thus, the new synthetic mitogens will be valuable tools for continuing the investigation of the molecular mechanism of B-lymphocyte activation, they appear to be useful as in vivo stimulants of the immune response, and they may non-specifically enhance host-resistance to infections.

Fluorescent mitogenically active lipopeptide

In order to obtain a more water-soluble mitogenic lipopeptide we decided to synthesize a Pam_3Cys-Ser derivative coupled to tetralysine. This lipohexapeptide was prepared by Merrifield synthesis and characterized by chromatographic and instrumental methods (FD-MS, ^{13}C- and ^{1}H-NMR). Due to its amphiphilic properties Pam_3Cys-Ser-$(Lys)_4$-OH was a highly efficient membrane anchor for a variety of agents and labels; it was soluble in water and in organic solvent mixtures, and moreover it showed biological activities as potent as the parent tripalmitoyl-pentapeptide (Fig.2).

The incorporation of FITC to one ε-amino group of Pam_3Cys-Ser-$(Lys)_4$-OH gave a biologically active fluorescent cell marker, which showed the typical capping and patching effects on the membranes of lymphocytes under the fluorescence microscope. The same photographs are observed upon interaction of FITC-labeled antibodies with cells. Furthermore this derivative proved to be usefull for studies in the fluorescence activated cell sorter(FACS).

Most of the lipopeptides prepared by us (6-8) proved to be highly interesting substances for monolayer experiments and liposome preparation. Using a microprocessor-controlled film balance pre-

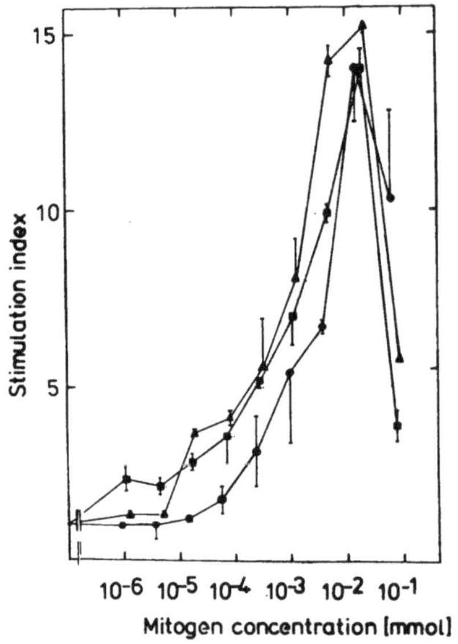

Fig.2. Mitogenic activition of Balb/c spleen cells by Pam_3Cys-Ser-$(Lys)_4$-OH x 3 HCl (triangles), Pam_3Cys-Ser-$(Lys)_4$-OH x 2 CF_3COOH (squares) and the fluorescent cell marker Pam_3Cys-Ser$(Lys)_4$[FITC]-OH (circles). (FITC = fluoresceine isothiocyanate). Ordinate: stimulation index: incorporation of ^{3}H-thymidine into DNA [cpm] divided by control values [cpm].

cision measurements of isotherms and isobars were carried out in monolayer experiments (H.Ringsdorf, W.Praß, K.-H.Wiesmüller and G.Jung, to be published). Phase transitions and areas/molecule were determined for a series of the mitogenic lipopeptides carrying one to three fatty acid residues. Electron microscopy was used to demonstrate the formation of liposomes incorporating tripalmitoyl-pentapeptide.

Enhancement of production of specific antibodies by presentation of lipopeptide-antigen conjugates

We described above that tripalmitoyl-pentapeptide is an active mitogen and polyclonal activator in vitro and in vivo. It constitutes also a potent immune adjuvant increasing the humoral response 1o-1oo fold, if given in combination with antigen (e.g. TNP-SRBC). And it enhanced the protective capacity of bacterial vaccines (Salmonella thyphimurium S-form)(19), thus reduction of the vaccine dose by 1/10 and supplementation by lipopeptide adjuvant led to the same effectiveness as the full vaccine dose. The physico-chemical properties of tripalmitoyl-S-glyceryl-cysteine (Pam_3Cys) derivatives on one side and conformationally stabilized amphiphilic α-helices (alamethicin type, see this volume) on the other side prompted one of us to design the "Pam_3Cys-Antigen-Helix" system (11) for presentation of antigens on the outer membrane of lymphocytes and macrophages. This approach is illustrated in Fig.3, details are described in ref.(11).

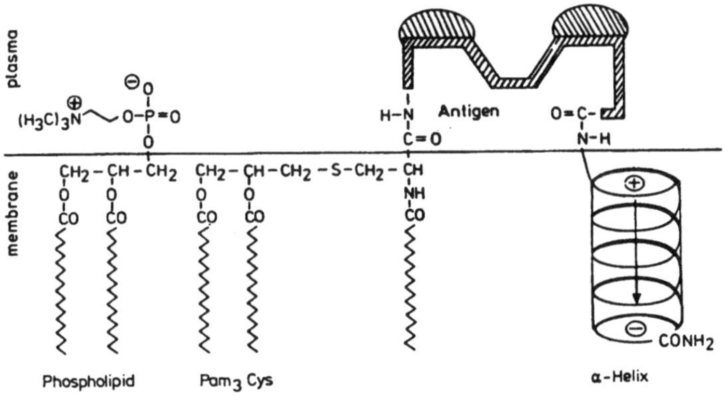

Fig.3. Assembly of the "Pam_3Cys-Antigen-Helix" system designed for a synthetic vaccine with built-in properties of both carrier and adjuvant (11).

Among other properties the following advantages of this kind of antigen presentation seem to be evident. Hydrophilic antigenic sites may be directly presented on macrophages without "processing"step. The assembly is chemically defined and uniform, and it can be prepared in any amount with the same purity in contrast to common immunogens prepared via crosslinking antigens to BSA. Only the formation of desired antibodies is elicited and the amount of heterogenous Ig will be diminished. Due to the N- and C-terminal linkage and the high distribution coefficient versus the lipid layer the bioavailability is enhanced (less degradation in serum). The conjugate contains no toxic constituents, since only fatty acids, ester bonds,amino acids, and peptide bonds, are involved and S-glyceryl-cysteine is a natural metabolite. Furthermore the "Pam_3Cys-Antigen-Helix" system (11) allows the antigen to be presented in a mobile state, which allows a presentation of e.g. ß-turns in a protein-like environment. Finally the assembly provides many built-in modification sites for attachment of fluorescence-,luminescence-, spin-labels and radioactive labels. It is obvious, that also one or two Pam_3Cys residues or α-helices may serve for the same purpose (11).

In order to realize this idea we coupled covalently Pam_3Cys-Ser to the non-immunogenic tetradecapeptide segment 516-529 of the epidermal growth factor receptor (EGF-R 516-529). After administration without further adjuvants and carriers this conjugate (o.2 μmol) elicited within few days specific antibodies against the EGF-R 516-529 peptide as shown by ELISA assay (9,1o). It is noteworthy that a single i.p. administration in mouse was sufficient to obtain antibodies (IgM and IgG) neutralizing this nonimmunogenic peptide of the extracytoplasmic region of the EGF-R. As demonstrated in Fig.4 antigen-specific antibodies were not formed neither with the adjuvant nor polypeptide alone, nor with mixtures of these two compounds. Reproducible results were obtained in two different laboratories both by in vivo and in vitro experiments (19,2o). For the in vitro immunization we used Balb/c murine spleen cells and observed after 5-7 days a high antibody titer, which as commonly observed, was lower than that of the in vivo titers.

Recently we could show that a series of various Pam_3Cys conjugates with fluoresceine and dinitrophenyl derivatives rised antibodies specific for these haptens.

Conclusion

Using the mitogenic synthetic lipopeptides from the N-terminus of the E.coli lipoprotein (6,7) as adjuvant covalently linked to antigen we could show, that a single injection of the Pam_3Cys-Ser-peptide conjugate elicited a high level of specific antibodies (9,1o,2o).

The method of covalently coupling adjuvants to antigen has been described before for muramyldipeptide-antigen conjugates, which were used to amplify the immune response against viral peptide sequences (21). Coutinho et al. (22) coupled bacterial LPS to hapten and thus obtained a specific immune response at a very low hapten dose. Synthetic vaccines have since been described by several groups (23-25). However, these approaches differ in many

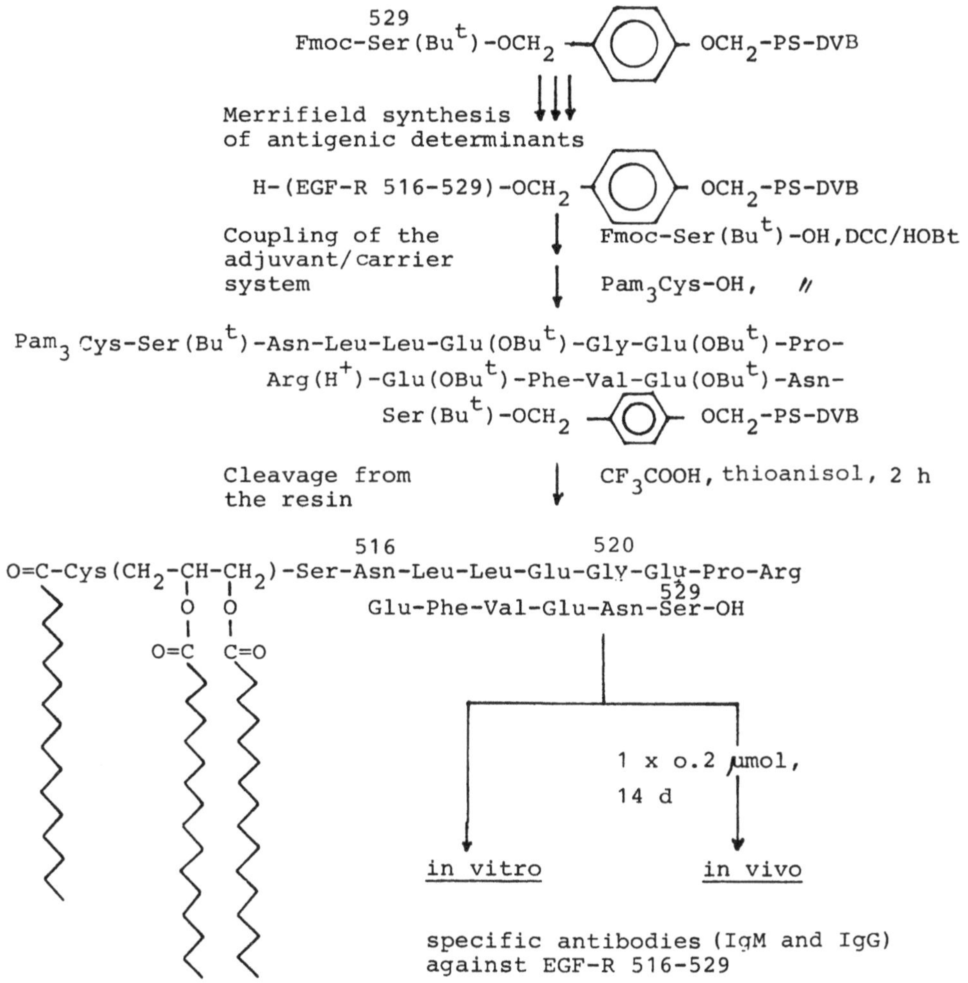

Fig.4. Routine incorporation of the lipopeptide carrier and adjuvant Pam_3Cys-Ser during Merrifield synthesis of the antigenic determinant EGF-R 516-529.

theoretical and practical aspects from our Pam_3Cys-antigen conjugates.

The experimental procedure of covalently coupling weakly immunogenic antigens to mitogenic lipopeptides (6,26) in order to circumvent the processing step may have a wide range of applications. Advantages of our low molecular weight, chemically well-defined carrier-antigen-adjuvant conjugates are: relatively easy prepara-

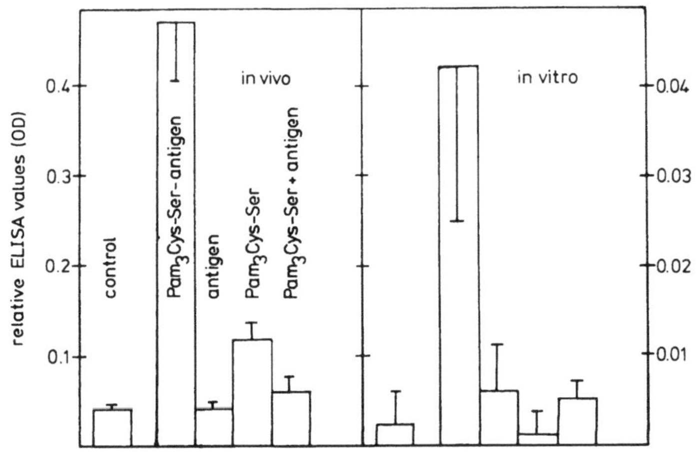

Fig.5. Elicidation of specific antibodies against EGF-R 516-529 peptide after in vivo (left side) and in vitro (right side) immunization (9,1o) with control (a), Pam_3Cys-(EGF-R 516-529) conjugate (b), tetradecapeptide EGF-R 516-529 alone (c) mitogen Pam_3Cys-Ser alone (d), mixture of mitogen and tetradecapeptide (e). For details see refs. (9,1o).

tion, routine incorporation in Merrifield syntheses (9), production of specific antibodies (IgM and IgG) within 1-2 weeks after single administration, no severe skin reactions since no Freund's adjuvant is required, high efficiency also in vitro providing elegant and fast screening possibilities in gene technology. It is also obvious that our approach is not restricted to peptides as antigens.

References

1. Melchers, F., Braun, V. and Galanos, C., J.Exp.Med.142, 473-482 (1975).

2. Bessler, W.G. and Ottenbreit, B.P., Biochem.Biophys.Res.Commun. 76, 239-246 (1977).

3. Andersson, J., Sjöberg, D. and Möller, G., Eur.J.Immunol.2,349-353 (1972).

4. Bessler, W.G. and Henning, U., Immunbiol.155, 387-398 (1979).

5. Bessler, W., Resch, K., Hancock,E., and Hantke, K., Z.Immun. Forsch.153, 11-22 (1977).

6. Wiesmüller, K.-H., Bessler, W., and Jung, G., Hoppe-Seyler's Z.Physiol.Chem. 364, 593-6o6 (1983).

7. Jung, G., Carrera, C., Brückner, H. and Bessler, W.G., Liebigs Ann.Chem.1983, 1608-1622.

8. Bessler, W.G., Wiesmüller, K.-H., Scheuer , W., Johnson, R.B., Fouad, H.H. and Jung, G., in Chemistry of Peptides and Proteins, Vol.2 (Voelter, W., E.Bayer, Y.A.Ovchinnikov, and E.Wünsch), de Gruyter, Berlin, 1984, p. 87-96.

9. Jung, G., Wiesmüller, K.-H., Becker, G., Bühring, H.-J., Bessler, W.G., Angew.Chem. 97, 883-885 (1985).

1o. Bessler, W.G., Suhr, B., Bühring, H.-J., Muller, C.P., Wiesmüller, K.-H., Becker, G. , and Jung, G., Immunbiol. 17o, 239-244 (1985).

11. Jung, G., in Synthetic Antigens, Proc.First Int.Conf.Siena, 29.-3o. Okt.1984, Ann.Sclavo 1984, 191-2o8.

12. Bessler, W.G., Johnson, R.B., Wiesmüller, K.-H. and Jung,G., Hoppe-Seyler's Z.Physiol.Chem. 363, 767-77o (1982).

13. Johnson, R.B., Köhl, S., Wiesmüller, K., Jung, G.,and Bessler, W.G., Immunbiol.165, 27 (1983).

14. Bessler,W.G., Flad, H.-D., Schade, U. and Rietschel, E., Lymphokine research 3,234 (1984).

15. Streck, H., Biesert, L., Wiesmüller, K.-H., Jung, G., and Bessler, W.G., Immunbiol. 168, 1o (1984).

16. Scheuer, W.V., Müller, H.,and Bessler, W.G., Immunbiol.165, 349 (1983).

17. Tittle, T.V. and Bessler, W.G., Immunbiol.165, 24o (1983).

18. Kleine, B., Martinez, C., Bessler, W.G., and Coutinho, A., Immunbiol.167, 4 (1984).

19. Bessler, W.G., Lex, A., Suhr, B., Ortmann, A., Schlecht, S., Bühring, H.-J., Muller, C., Metzger, J., Wiesmüller, K.-H. and Jung, G., Proc.17th Int.Leukocyte Culture Conf., Ithaka, New York, 1985, in press.

2o. Jung, G., Wiesmüller, K.-H., Metzger, J., Bühring, H.-J., Muller, C.P., Biesert, L., and Bessler, W.G., in Peptides, Proc.9th Amer.Pept.Symp., Toronto, 1985, in press.

21. Arnon, R., Sela, M.,Parant, L., and L.Chedid, Proc.Natl.Acad. Sci. 77, 6772 (198o).

22. Coutinho, A., Gronowicz, E., Bullock, W.W.,and Möller, G.,J. Exp.Med.139, 74 (1974).

23.Audibert, F., Jolivet, M., Chedid, L., Arnon, R., and Sela, M., Proc.Natl.Acad.Sci. USA 79, 5o42-5o46 (1982).

24. Delbende, C., in: Peptides, Structure and Function (Hruby,V.J. and Rich, D.H., eds.) p. 877, Pierce, Rockford (USA), 1983.

25. Sela, M., Biopolymers 22, 415-424 (1983).

26. Bessler, W.G., Cox, M., Lex, A., Suhr, B., Wiesmüller, K.-H., and Jung, G., J.Immunol.135, 1353-1358 (1985).

ELECTRON MICROSCOPY STUDIES OF FAST SODIUM CHANNEL STRUCTURE

V.V. Demin, E.V. Grishin, V.A. Kovalenko, S.N. Spadar
Shemyakin Institute of Bioorganic Chemistry, USSR Academy of Sciences, Moscow, USSR

Introduction

Voltage-sensitive sodium channels regulate selective sodium permeability of electroexcitable membranes at generation of the action potential. Many important characteristics of these channels were obtained by electrophysiological techniques. However, their molecule function mechanism is still oscure. Lately important progress was achieved in elucidation of the channel molecular composition (1-5). Use was made of molecular-genetic approaches to establish the protein structure of the large polypeptide which form the sodium channel in the membrane of electric eel electroplax (6). The present paper deals with the elucidation of the spatial arrangement of the sodium channel from rat brain.

Results

The sodium channel was isolated by the procedure described earlier (5). Four main stages underlay the method: ion exchange chromatography of the membranes solubilized by LubrolPX on DEAE-Toyopearl, two-phase distribution in the system dextran T500/polyethylenglycol 6000, affinity chromatography on Sepharose with immobilized wheat germ lectin and chromatography on hydroxylapatite. The isolation was carried out in the presence of the main classes of proteinase inhibitors. To prevent inactivation of receptor activity of the channel molecules the Lubrol-lecithin mixture were used. Table 1 illustrates the yield of the protein and receptor activity in typical preparative isolation of the sodium channel from rat brain (500 g). Specific activity of $|^{3}H|$TTX binding to different preparations is usually within region 200-800 pmol/mg of the protein.

Chemistry of Peptides and Proteins, Vol. 3

However, the experiments on the reconstruction of the channel into phospholipid vesicles showed that rather low TTX binding activity is stipulated by inactivation of the receptor protein during isolation. The inactivation is of reversible character. We observed 4-12 fold increase in binding of $|^{3}H|$TTX to the reconstituted channel preparations. The binding of $|^{3}H|$TTX to the receptor protein embedded into phospholipid vesicles is highly thermostable. Besides, the reconstituted channel can be cleaved by a series of proteolytic enzymes without the considerable decreasing in the number of binding sites for $|^{3}H|$TTX.

Table 1. Purification of Tetrodotoxin Receptor from Rat Brain.

Purification step	Protein		TTX receptor		Specific activity*
	mg	%	nmol	%	pmol/mg
Lubrol extraction	12.760	100	25.5	100	2
DEAE-Toyopearl 650M	2016	16	21.8	88	11
Two-phase partition	-	-	11.0	37.9	-
WGA-Sepharose 4B	41.5	0.3	3.7	14.7	90
Hydrohylapatite	3.8	0.03	1.58	6.2	480

*Calculation of specific activity was carried out wighout regard to the receptor inactivation.

According to electrophoresis and gel-chromatography results in the isolated preparations of the sodium channel three main protein components with molecular weights of ∿260, 39 and 37 kDa are usually present which designated by W. Catterall as α, β_1 and β_2 respectively (2).

Analysis of individual subunits of the channels isolated by HPLC methods demonstrated all the channel components to be sialoglycoproteins (Table 2). The carbohydrate portion is from ∿20 to 32% of the subunit mass. Despite the integral membrane function of the sodium channel all the subunits are of comparatively low hydrophobicity. The less hydrophobic is the β_2 subunit, which is mostly glycosylated. Interestingly, N-terminal amino acid residues of all subunits are blocked.

Table 2. Amino Acid and Carbohydrate Composition of the TTX Receptor Subunits (mole, %).

	α	β_1	β_2
Asx	10.65	9.47	10.82
Thr	5.1	6.24	6.18
Ser	7.82	7.09	8.31
Glx	11.73	14.04	13.66
Pro	4.45	3.06	4.11
Gly	6.50	7.34	9.55
Ala	6.49	7.92	5.37
Cys	0.91	0.92	0.98
Val	5.9	7.66	6.99
Met	2.87	2.30	1.89
Ile	5.12	4.67	3.52
Leu	9.61	8.25	8.34
Tyr	2.57	3.59	3.00
Phe	5.51	4.61	3.91
His	2.34	1.79	1.87
Lys	6.13	5.79	6.43
Arg	4.97	4.45	4.7
Trp	1.32	0.80	0.38
Fuc	4.33	3.29	5.32
Man	13.49	22.29	19.73
Gal	13.75	13.78	19.58
GlcNAc	17.6	19.41	22.17
GalNAc	4.7	4.01	4.51
Sialic acid	46.14	37.22	28.68
Protein %	79.5	71.1	68.12
Carbohydrate %	20.5	28.9	31.88

For electron microscopy the tetrodotoxin receptor fraction (number 6-8) obtained by sedimentation in sucrose concentration gradient was used (Fig. 1). As it shows receptor fractions with sedimentation coefficient ∿9.5S have maximal binding activity and contain only α- and both β-subunits of sodium channel.

For electron microscopy the solubilized protein or reconstituted membrane fragments were absorbed on a thin carbon film and subsequently stained by aqueous solytion either 2% uranyl acetate or 2% $Th(NO_3)_4$. Grids were examined in a IEM-100EX 11 electron microscope at accelerating voltage 80 kV and magnification ∿60.000.

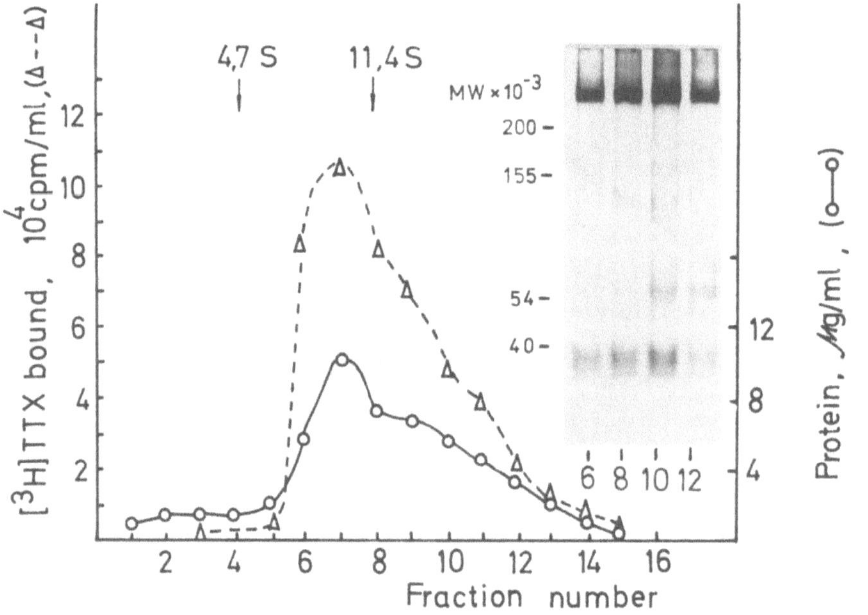

Fig. 1. Sedimentation of purified TTX receptor in sucrose concentration gradient (5-20%). Rotor SW-41, 37000 rpm, 14 h, 4°C. (Insert) Analysis of SDS-gel electrophoresis of obtained receptor fractions.

Electron micrographs of small areas containing single particles were scanned by optical densitometer at resolution ∿35 Å. Each particle was displayed on a computer-link graphic system. The images were aligned by using the cross-correlation function, with one particle selected as a reference (7,8). All images were low-pass filtered to eliminate high-resolution portions of the noise by multiplication of the Fourier transform by the Gaussian filter function.

Electron microscopy of the negative by stained purified protein reveals two structural types (Fig. 2a,b): "rod" shaped and "square" particles structures. To clarify structural details revealed by

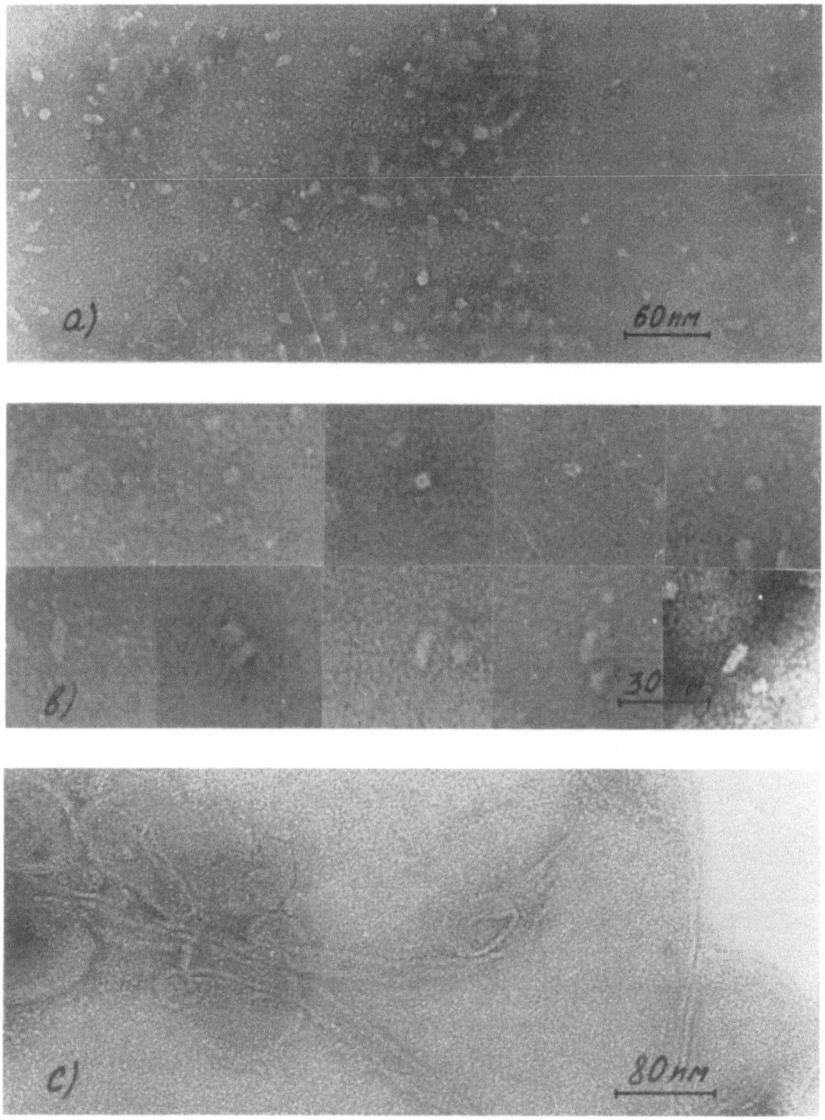

Fig. 2. Negatively stained sodium channels in solubilized (a,b) and membrane reconstituted (c) forms.

electron micrographs the digital image processing was carried out using software originally written by Frank and Vershoor (7) and Orlova and Sherman (8). Computer filtration of the images presented in Fig. 2a,b shows that the "rod" shape structure is narrowed at an

end and has a neck at its middle part (Fig. 3b). "Square" shaped particles (Fig.3a) possess a rosette structure similar to the rosettes of the acetylcholine receptor (9). The external diameter (∿50 Å) is very close to the size of the rod wide end. The region with low density is located at the central part of the rosette.
Rosettes type particles with the same structural parameters may be found by electron microscopy of the membranes reconstituted from a lipid (egg lecithine) and a purified protein (Fig. 2c). But in this case rosettes of two types are observed: the first type is identical to that discribed above, the second is characterized by the same structural features but a smaller external diameter (Fig. 3c). It is approximately equal to the size of the narrow "rod" end (∿40 Å).

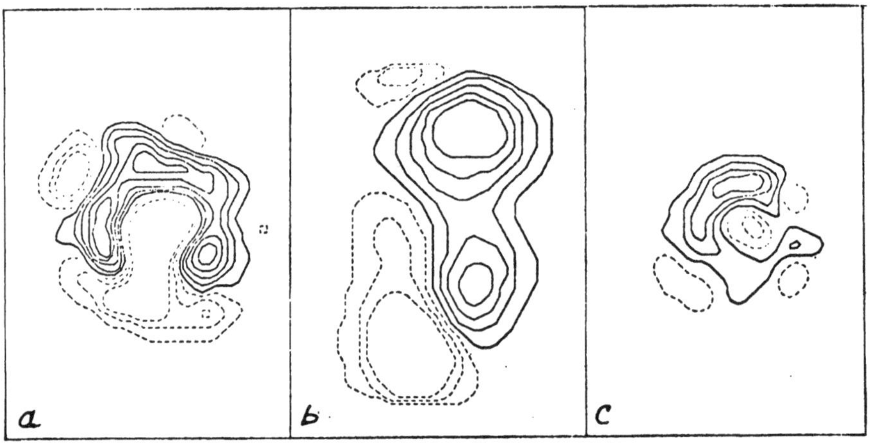

Fig. 3. Membrane (a,c) and side (b) projections of sodium channel structure.

Rosettes of two types are characteristic of membrane preparations and that allows us to conclude that they are projections of the sodium channel structure on extra- and intracellular membrane planes. "Rods", which are characteristic only of solubilized preparations, could be a side projection of the protein. These three main projections permit us to propose the putative three-dimensional structure of the sodium channel into the membrane (Fig. 4).

There are no direct evidences for the exact position of the lipid membrane. But the biggest dimension of the molecule (~85 Å) is much more than normal thickness of lipid bilayers and it means that a significant part of the protein should be exposed at the membrane surfaces. Membrane projections of the sodium channel are similar to those of the acetylcholine receptor (9). In both cases they have a rosette shape and low density regions in the central part of the rosettes. In the acetylcholine receptor this low density region is proposed to be like an ion channel forming part of the protein. This similarity suggests the existence of common principles in the structural organization of ion channels.

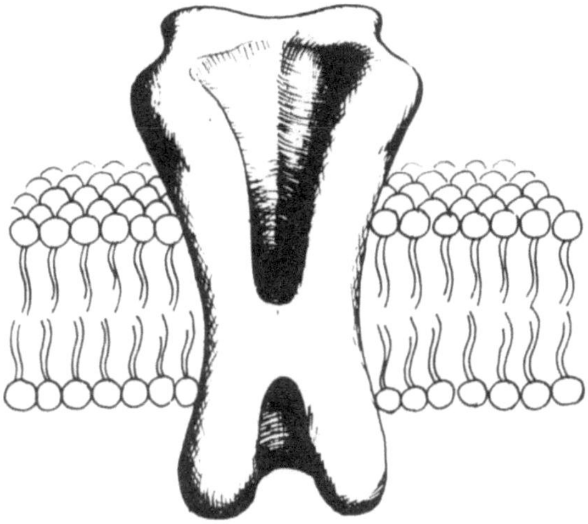

Fig. 4. Model of sodium channel structure in the membrane based on the projections presented on Fig. 2.

References

1. Miller, J.A., W.S. Agnew, S.R. Levinson. 1983. Biochemistry, 22, 462.

2. Hartshorne,R.P., D.J. Messner, J.C. Coppersmith, W.A. Catterall. 1982. J. Biol. Chem., 257, 13888.

3. Norman,R.J., A. Schmid, A. Lombet, J. Barhanin, M. Lazdunski. 1983. Proc. Nat. Acad. Sci. USA, 80, 4164.

4. Barchi,R.L., J.C. Tanaka, R.E. Furman. 1984. J. Cell. Biochem., 26, 135.

5. Grishin, E.V., V.A. Kovalenko, V.N. Pashkov, O.G. Shamotienko. 1984. Biol. Membranes (USSR), 1, 858.

6. Noda, M., S. Shimizu, T. Tanabe, T. Takai, T. Kagano, T. Ikeda, H. Takahashi, H. Nakayama, J. Kanaoka, N. Minamino, K. Kangawa, H. Matsuo, M.A. Raftery, T. Hirose, S. Inayama, H. Hayashida, T. Miyata., S. Numa. 1984. Nature, 312, 97.

7. Frank,J., A. Vershoor. 1981. Science, 214, 1353.

8. Orlova, E.V., M.B. Sherman. 1983. Izvestija Akad. Nauk USSR, 47, 1162.

9. Zingsheim, H.L., D.Ch. Neugebauer, J. Franck, W. Hanicke and E.J. Barrantes. 1982. EMBO J., 1, 541.

FURTHER EXPERIMENTAL EVIDENCE FOR THE FLIP-FLOP GATING MECHANISM OF ALAMETHICIN PORE FORMATION

G.Jung, K.-P.Voges, G.Becker
Institut für Organische Chemie, Universität Tübingen, D-7400 Tübingen, FRG

W.H.Sawyer
The Russel Grimwade School of Biochemistry, University of Melbourne, Australia

V.Rizzo, G.Schwarz
Biozentrum, Universität Basel, CH-4056 Basel, Schweiz

G.Menestrina, G.Boheim
Zellphysiologie, Ruhr-Universität Bochum, D-4630 Bochum, FRG

Introduction

A representative series of lipophilic and amphiphilic α-helices has been studied by spectroscopic methods and in electric bilayer experiments. The polypeptides constitute membrane active analogues of the eicosapeptide antibiotic alamethicin, a potential dependent ion conducting pore former in lipid bilayers. The formation of pores, which adopt various sizes depending on the number of monomers involved, has been explained by the flip-flop gating model (1).
Recent results of dielectric dispersion measurements on the lipophilic helices Boc-(Ala-Aib-Ala-Aib-Ala)$_n$-OMe (n = 10, 20) (2) and of biophysical studies on the amphiphilic helix $^+H_2$-Asn-Arg(H^+)-Arg(H^+)-Ala-Leu-Ile-Leu-Leu-Ala-Gln-(Ala-Aib-Ala-Aib-Ala)$_2$-OMe (3) an other α-helical analogues as well will be reported. Furthermore five heneikosapeptides were synthezised each carrying one tryptophane residue in the positions 1,6,11,16 and 21 along a conformationally stabilized (therefore ref. (4,5) α-helix consisting of four (Ala-Aib-Ala-Aib-Ala) units. A systematic study of the tryptophane fluorescence quenching by n-doxyl-stearic acids (6) revealed the relative orientation of the helices within lipid bilayers. This system allowed the determination of the partitioning, and via Stern-Vollmer plots one could verify the expected depths of the tryptophane residues within the bilayer membrane.

Chemistry of Peptides and Proteins, Vol. 3

It is obvious, that our results obtained from the studies with membrane modifying α-helical polypeptides may be also relevant for the design of new hormone and protein segments and of new carrier and membrane anchor systems for the presentation of antigens(7-9)and various biologically active agents.
With respect to earlier investigations on the structural and ion gating properties of alamethicin and analogues we refer to the proceedings of the 4th USSR-FRG symposium (10,11).

Aggregation of alamethicin in the non-conducting state

Recently CD measurements of alamethicin in liposomes (12) were interpreted in terms of a conformation different from the X-ray structure of alamethicin. Firstly it should be emphasized, that reversal of the ratios of the intensities of the Cotton effects at 207 and 221 nm from organic solvents (r = 0.8) to aqueous solutions (r ≥ 1.0) has been noted long before (13) and has been attributed to aggregation. This effect has been observed also in a nematic membrane simulating phase, in water and in aqueous ethanol (13) or in lecithine vesicles as well by our groups (Boheim, Methfessel, Schmitt, Jung, unpublished). Thus, the higher intensity of the long-wavelength Cotton effect is not an effect of the peculiar lipid environment as suggested by Cascio and Wallace (12), it is simply due to the presence of water. As found already by McMullen, Marlborough and Bayley (FEBS Lett. 16, 278-280 (1971)) even 1% sodium dodecylsulfate solutions show this CD band reversal. Secondly one should be aware of the fact, that bilayer experiments proof unequivocally that pores have to be formed out of aggregates, which exist even at extremely low concentrations (nanomolar level), Thirdly dielectric dispersion measurements show aggregation phenomena even in waterfree octanol/dioxane mixtures for alamethicin (14), trichotoxin (15) and the pore forming helical model peptides Boc-(Ala-Aib-Ala-Aib-Ala)$_n$-OMe (n = 2 and 4) (2). Aggregation increases with increasing dioxane (or water) content thereby the CD band ratio increases too (13).
There is no doubt, that an aqueous respectively strongly polar environment induces a partial decrease in helix content with simultaneous aggregation. However, it may also be speculative to draw conclusions on the α-helix content of alamethicin in vesicles respectively membranes based solely on CD data.
The quantitative and routine analysis of the CD spectra of Aib containing peptides is limited especially due to the non-occurrence of β-structures, the possible formation of 3_{10}-helices and due to their low molecular mass in comparison to the large proteins used for reference conformations. Therefore it may be appropriate to draw conclusions only from carefull systematic and comparative studies

of chain length, solvent and temperature dependencies (13,16 - 18) of peptides known also from X-ray structures (4,5,19 -22) and NMR studies.

New data supporting the α-conformation of alamethicin

The α-helix content of alamethicin derived from CD data even in methanol or octanol is definitely too low considering the X-ray structure of the fully extended α-helix (22). Since voltage-gated pore formation (1) depends directly on the length of the helix respectively the magnitude of the dipole moment (for experimental data see refs. (2,14,15)) the determination of the conformation of the membrane modifying peptides is of particular interest. In view of the difficulties experienced with quantitative CD analysis we used very early ^{13}C NMR (for refs. see (10) and (13)).

Recently convincing results for a complete α-helix of alamethicin in methanol and chloroform were obtained by ^{13}C NMR T_1 relaxation time measurements on highly pure synthetic alamethicin (23,24). These results were in full agreement with the data obtained from the helices Boc-(Ala-Aib-Ala-Aib-Ala)$_n$-OMe (n = 2 to 4) (10,25). Helical Aib residues exhibit a high magnetic non-equivalence (MNE) of the diastereotopic pro-_R_ and pro-_S_ Aib-C_β nuclei (3-5 ppm (10)). Due to different orientations toward the helix axis the two geminal Aib methyl groups differ also in relaxation time scales. Furthermore the extend of MNE seems to be directly correlated to the dihedral angles between one of the two methyl groups and the Aib-carbonyl group (25).

The partial β-structures proposed for alamethicin on the basis of IR reflection spectroscopy (ATR) in multilayers (26) or on the basis of ^1H NMR (27) and computer -aided design (28) seem to give only limited evidence for both the non-conducting and conducting state of the alamethicin pore, because β-structures are not allowed for steric reasons for Aib peptides as shown on an overwhelming number of X-ray structures.

The strongly conserved backbone conformation found for the synthetic C-terminal nonapeptide of alamethicin (5,20) in total agreement with the alamethicin structure (22) is indicative of the fact, that even partial segments - and even with protecting groups - adopt the α-conformation of the native antibiotic. This finding and the ^{13}C NMR results contradict a further recent note (12) suggesting that the long alamethicin helix found in the solid state (22) may result from crystal forces. The C-terminal nonapeptide adopts also in solution a mainly α-helical conformation according to CD and ^{13}C NMR (23).

In view of the lengths of the dipoles estimated from dielectric dispersion measurements of alamethicin (14), trichotoxin (15) and two model helices with 10 and 20 residues the existence of an equilibrium between monomers and antiparallely multimeric aggregated long α-helical rods must be accepted in organic solutions.

Recently we could confirm results obtained with alamethicin and analogues (13, 16 - 18) via conformational studies on the α-helical suzukacillin A and fragments (29), which is sequentially closely related to alamethicin. In contrast to alamethicin and suzukacillin the shorter trichotoxin A40 (30) shows different pore forming properties (31) and consequently, e.g. a different conformation for its C-terminus (for details see refs. cited in (20,23,29)).

On the orientation of the α-helices within lipid bilayers

With the exception of a binding study using a radioactive photolabeled lipid molecule (32) there are no reports on attempts to directly visualize the orientation of alamethicin molecules inserted in lipid membranes. Using five synthetic α-helical 21-peptide helices consisting of units of $(L\text{-Ala-Aib-Ala-Aib-Ala})_n$ with tryptophanyl residues in the positions 1,6,11,16 and 21 and fluorescence quenching techniques (6) with spin-labeled stearic acids the following information was obtained (32). The depth of each indolyl fluorophor within the lipid bilayer membrane of liposomes could be detected using a set of n-doxyl stearic acids with the quenching nitroxide label in positions 5,7,10,12 and 16. The results were illustrated by plotting the quenching efficiency QE = Io/I-1 versus the theoretical depths of the spin labels within a bilayer membrane (modified Stern-Vollmer plots) taking into account the different association parameters for each doxyl stearic acid (34).

The helices were investigated as N-Boc and N-protonated methyl esters. In any case the tryptophane bound to either N- or C-terminus of $(\text{Ala-Aib-Ala-Aib-Ala})_4$ is placed in the hydrophilic section of the membrane (strong quenching with 5- and 7- doxyl stearic acids, no quenching with 10-, 12- and 16-doxyl stearic acids). This was expected at least for the 21-peptide with N-protonated tryptophane at the N-terminus. However, all compounds showed the experimental behaviour of long helical rods spanning the lipid membrane. Thus, 21-peptides with tryptophane in position 11 were quenched by doxyl labels situated in the lipophilic core of the membrane. Consequently for the 21-peptides with Trp in positions 6 and 16 the fluorophor becomes localized within the steep gradient of the membrane potential as ver ified by the experiment, and in this case N-Boc and N-protonated species behave different!

New experiments on black lipid bilayer membranes

Simple helical oligomers such as Boc-(L-Ala-Aib-Ala-Aib-Ala)$_n$-OMe (n = 1 to 4) were shown to exhibit voltage-dependent conductance states (35). Although recorded at higher concentrations and voltage their behaviour was consistent with the flip-flop gating model (1). The shorter the helix the higher the concentrations needed to produce pore formation at a given voltage.
Whereas alamethicin enters the membrane via its more lipophilic N-terminus the following synthetic and α-helical eicosapeptide conjugate enters with its C-terminus: Boc(and H-)-Asn-Arg-Arg-Ala-Leu-Ile-Leu-Leu-Ala-Gln-Ala-Aib-Ala-Aib-Ala-Ala-Aib-Ala-Aib-Ala-OMe. This conjugate with the N-terminal interferon segment HuIFN-α(Ly)11-20 showing opposing asymmetry behaviour in current-voltage experiments constitutes one of the many new compounds supporting the flip-flop gating model (1).
Interestingly the Boc-alamethicin-2-20 benzyl ester (23) shows pore state properties comparable to those of the oligomers Boc-(Ala-Aib-Ala-Aib-Ala)$_n$-OMe with high levels and only one or two conductance states. On the other hand the oligomers with N-terminal Ac-Aib exhibited burst-like behaviour as registered for trichotoxin A40 (31).
The broad variety of different α-helical polypeptides including natural and modified melittin (31) led to the conclusion that the gating mechanism is based entirely on dipole-dipole interactions of adjacent helical segments depending on the applied membrane voltage (1). Groups inducing steric, hydrogen-bonding, and ionic forces may be introduced in simple α-helices for the systematic variation of pores in order to produce almost predictable properties. We also refer to the reports of Spach et al. (38), who showed pore formation with a number of non-Aib containing peptide oligomers.
Finally we should like to emphasize that we contradict in particular the many speculations of Fox and Richards (22) in their attempts to develop a model for the gating mechanism. Furthermore we cannot join the ideas reported recently by Hall et al. (28), and we do not support Balaram's extrusion mechanism (39). A detailed summary of the bilayer experiments carried out during the last years will appear elsewhere (40).

References

1. Boheim,G., Hanke,W., and Jung,G., Biophys. Struct. Mech. 9, 181-191 (1983).

2. Rizzo,V., Schwarz,G., Voges,K.-P., and Jung,G., Eur. Biophys. J. 12, 67-73 (1985).

3. Jung,G. Becker,G., Schmitt,H., Voges,K.-P., Boheim,G., and Griesbach,S., in Peptides, Structure and Function (ed. Hruby,V.J. and Rich,D.H.) p.491-494, Pierce, Rockford, 1983.

4. Jung,G., Bosch,R., Katz,E., Schmitt,H., Voges,K.-P., and Winter,W., Biopolymers 22, 241-246 (1983).

5. Bosch,R., Jung,G., Schmitt,H., Sheldrick,G.M., and Winter,W., Angew. Chem. 96, 440-443 (1984); Int. Ed. Engl. 23, 450-453 (1984).

6. Blatt,E., Chatelier,R.C., and Sawyer,W.H., Photochem. and Photobiol. 39, 477-483 (1984).

7. Jung,G. in Proc. First Int. Conf. on Synthetic Antigens, Ann. Sclavo 1984, 191-208.

8. Jung,G., Becker,G., Metzger,J., Wiesmüller,K.-H., Bessler,W.G., Biesert,L., Bühring,H.-J., and Muller,C., in these 5th Proc. USSR/FRG Symp., Odessa, 1985.

9. Jung,G., Wiesmüller,K.-H., Becker,G., Bühring,H.-J., and Bessler,W.G., Angew. Chem. Int. Ed. Engl. 24, 872-873 (1985).

10. Jung,G., Boheim,G., Bosch,R., Brückner,H., Hartter,P., Katz,E., König,W.A., Schmitt,H., Voges,K.-P., and Winter,W., in Chemistry of Peptides and Proteins, Vol.2 (eds. W.Voelter, E.Bayer, Y.A.Ovchinnikov, E.Wünsch), de Gruyter,Berlin, 1984, p. 215-226.

11. Boheim,G., Hanke,W., and Jung,G., in Chemistry of Peptides and Proteins, Vol. 2, (eds. W.Voelter, E.Bayer, Y.A.Ovchinnikov, E.Wünsch), de Gruyter, Berlin, 1984, p. 281-289.

12. Cascio,M. and Wallace,B.A., Ann. N.Y.Acad. Sci. 435, 527-529.

13. Jung,G., Dubischar,N., and Leibfritz,D., Eur. J. Biochem. 54, 395-409 (1985).

14. Schwarz,G. and Savko,P., Biophys. J. 39, 211-219 (1982).

15. Schwarz,G., Savko,P., and Jung,G., Biochim. Biophys. Acta 718, 419-428(1983).

16. Mayr,W., Oekonomopulos,R., and Jung,G., Biopolymers 18, 425-450 (1979).

17. Oekonomopulos,R. and Jung,G., Biopolymers 19, 203-214 (1980).

18. Mayr,W. and Jung,G., Liebigs Ann. Chem. 1980, 1489-1506.

19. Bosch,R., Jung,G., Schmitt,H., and Winter,W., Biopolymers 24, 961-978(1985).

20. Bosch,R., Jung,G., Schmitt,H., and Winter,W., Biopolymers 24, 979-999(1985).

21. Bosch,R., Jung,G., Schmitt,H., and Winter,W., Acta Cryst. C 41 (1985), in print.

22. Fox,R.O. and Richards,F.M., Nature 300, 325-330 (1982).

23. Schmitt,H. and Jung,G., Liebigs Ann. Chem. 1985, 321-344 and 345-364.

24. Jung,G. and Schmitt,H., in Peptides 1984, Proc. 18.Eur. Pept. Symp. (ed. U. Ragnarsson), Almquist and Wiksele, Stockholm, 1984, p. 569-572.

25. Jung,G., Voges,K.-P., Schmitt,H., Becker,G., Kolshorn,H., and Albert,K., in Proc. Forum Peptides, Cap d'Adge 1984, in print.

26. Fringeli,U.P. and Fringeli,M., Proc. Natl. Acad. Sci. USA 76, 3852 (1979).

27. Banerjee,O. and Chan,S.I., Biochemistry 22, 3709 (1983).

28. Hall,J.E., Balasubramanian,T.M., Hing,A., and Marshall,G., in Peptides, Structure and Function, Proc. 8th Amer. Pept. Symp. (eds. V.Hruby and Rich, D.H.), Pierce, Rockford, USA, 1983, p. 487-490.

29. Katz,E., Aydin,M., Lucht,N., König,W.A., Ooka,T., and Jung,G., Liebigs Ann. Chem. 1985, 1041-1062.

30. Brückner,H., König,W.A., Aydin,M., and Jung,G., Biochim.Biophys. Acta 827, 51-62 (1985).

31. Hanke,W., Methfessel,C., Wilmsen,H.-U., Katz,E., Jung,G., and Boheim,G., Biochim. Biophys. Acta 727, 108-114 (1983).

32. Jung,G., Brückner,H., and Schmitt,H., in Structure and Activity of Natural Products (eds. W.Voelter and G.Weitzel),de Gruyter,Berlin,1981, p. 75-114.

33. Voges,K.-P., Dissertation, Universität Tübingen, 1985.

34. Blatt,E., Chatelier,R.C., and Sawyer,W.H., Chem.Phys.Lett. 1984, 397-400.

35. Jung,G., Katz,E., Schmitt,H., Voges,K.-P., Menestrina,G., and Boheim,G., Stud. Phys. Theor. Chem. 24, 349-357 (1983).

36. Menestrina,G., Voges,K.-P., Jung,G., Boheim,G., submitted.

37. Boheim,G., Helbig,I., Becker,G., and Jung,G., submitted.

38. Spach,G., Heitz,F., and Trudelle,Y., Stud. Phys. Theor. Chem. 24, 375-382 (1983).

39. Balaram,P., in Peptides, Structure and Function, Proc. 8th Amer. Pept. Symp. (eds. V.J.Hruby and D.H.Rich), Pierce, Rockford, USA, 1983, pp. 477-486.

40. Boheim,G., Gelfert,S., Jung,G., Menestrina,G., in Proc. Int. Symp. on "Ion Transport through Membranes", Nov. 4-7, 1985, Nagoya, Japan (eds. K.Yagi and B.Pullman), Academic Press, in print.

AUTHOR INDEX

SUBJECT INDEX

Peptide Separations Using Cation-exchange Polymers I.

High efficiency column useful for separating acidic peptides as well as general peptide mapping.

Figure 1.

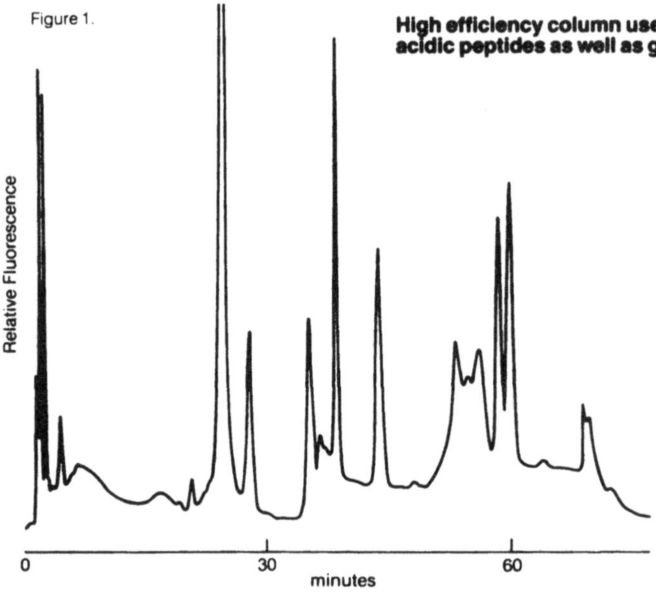

Figure 1.
Map of Human Globin Tryptic Peptides.
The α and β chains of human globin were digested with trypsin and the subsequent peptide mixture chromatographed on the BIO-10 column. Approximately 27 peptides should be obtained. Acidic peptides elute early, not strongly retained under the conditions used. Neutral and basic peptides are more strongly retained and are eluted by using higher pH eluents. Detection was accomplished by reacting effluent peptides with *o*-phthalaldehyde and monitoring fluorescence of resulting fluorophores.
Analysis conditions: Eluents, step gradient using aqueous buffers each containing 0.2 *N* Na^+ at pH values 3.8, 4.3, 4.6, 6.2 and 9.9. Each eluent pumped for 15 minutes. Flow rate, 0.3 mL/min; Column temperature, 60°C; Pressure, 50 atm; Detection, Waters 420 fluorometer.

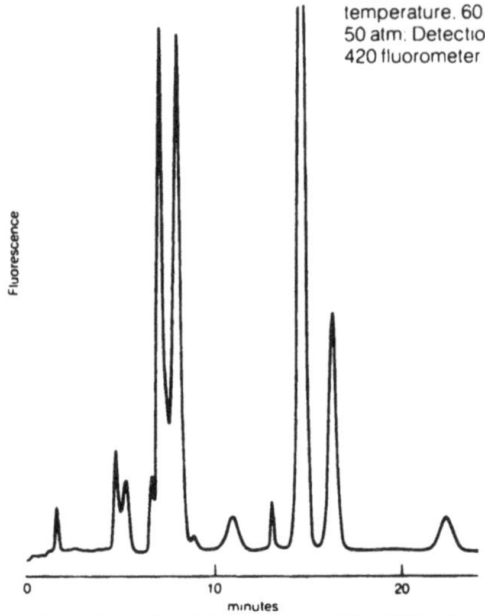

Figure 2.
Separation of Acidic Peptides from Tp Globin.
Acidic peptides are usually not retained on cation-exchange columns, but are easily separated on the BIO-10 using low pH eluents (see text for explanation). In this example, a citrate buffer was used isocratically. Peptides often can be separated isocratically by groups (acidic, neutral, basic) simply by taking advantage of their ion-exchange characteristics and using the appropriate eluent.
Analysis conditions: Eluent, 0.05 *N* Na^+ (citrate), pH 3.2; Flow rate, 0.3 mL/min; Temperature, 60°C; Pressure, 50 atm; Detection, Waters 420 fluorometer.

ict Handelsgesellschaft m. b. H.
Antoniterstrasse 27
6230 Frankfurt 80
Telefon 069/316003
Telex 411 320 ict d

ict AG
Türkheimerstrasse 5 · Postfach
4009 Basel
Telefon 061/390025
Telex 64005 ict ch

ict Handelsgesellschaft m. b. H.
Bösendorferstrasse 2
1010 Wien
Telefon 0222/657189
Teletex (61) 3222289 ict a

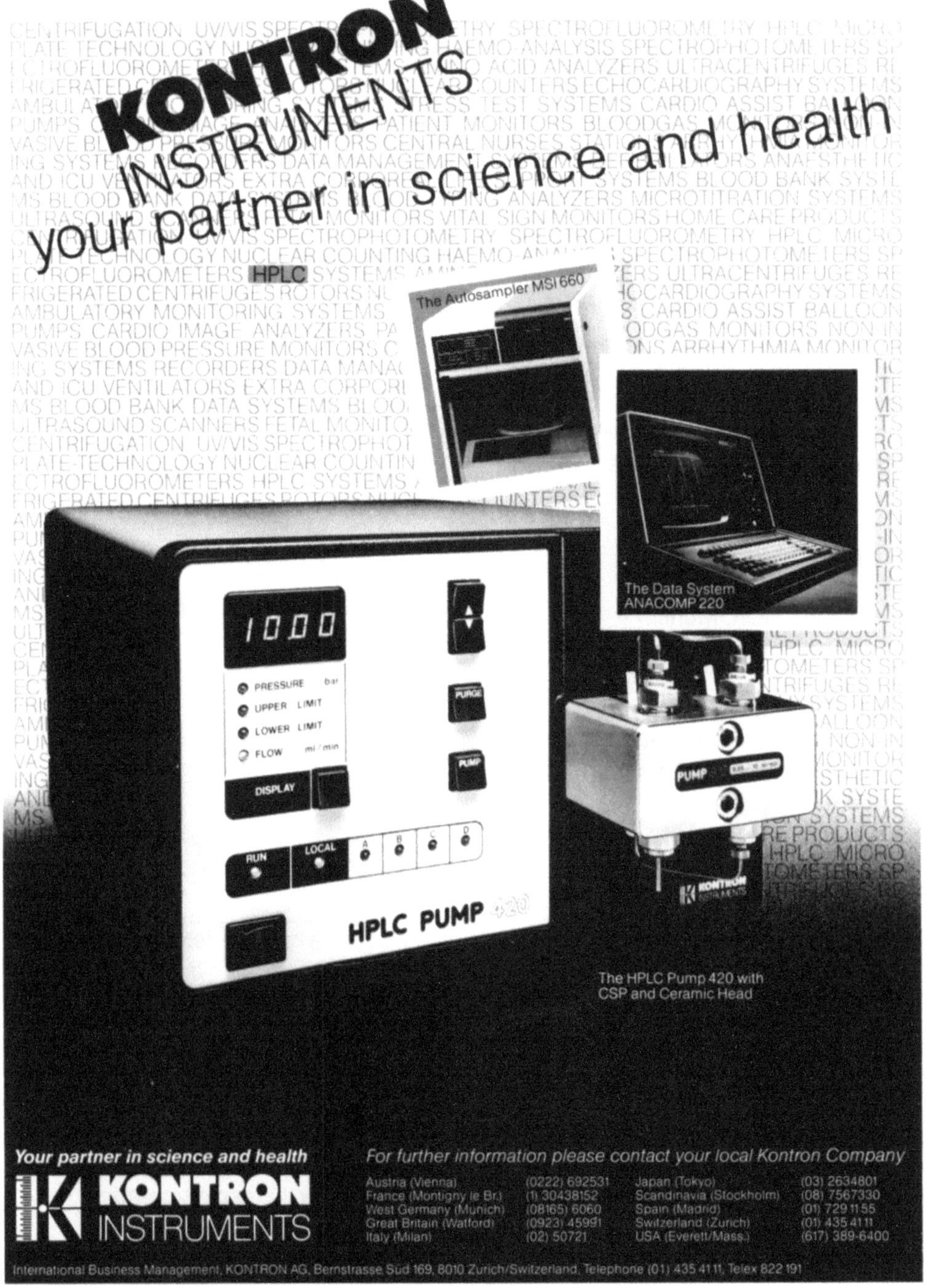
KONTRON
INSTRUMENTS
your partner in science and health
HPLC
The Autosampler MSI 660
The Data System ANACOMP 220
PRESSURE bar
UPPER LIMIT
LOWER LIMIT
FLOW ml/min
DISPLAY
PURGE
PUMP
RUN
LOCAL
HPLC PUMP
The HPLC Pump 420 with CSP and Ceramic Head
Your partner in science and health
KONTRON INSTRUMENTS
For further information please contact your local Kontron Company
Austria (Vienna) (0222) 692531
France (Montigny le Br.) (1) 30438152
West Germany (Munich) (08165) 6060
Great Britain (Watford) (0923) 45991
Italy (Milan) (02) 50721
Japan (Tokyo) (03) 2634801
Scandinavia (Stockholm) (08) 7567330
Spain (Madrid) (01) 729 11 55
Switzerland (Zurich) (01) 435 41 11
USA (Everett/Mass.) (617) 389-6400
International Business Management, KONTRON AG, Bernstrasse Süd 169, 8010 Zurich/Switzerland, Telephone (01) 435 41 11, Telex 822 191

Bei Fragen zur Produktsicherheit wenden Sie sich bitte an:
If you have any questions regarding product safety,
please contact:

Walter de Gruyter GmbH
Genthiner Straße 13
10785 Berlin
productsafety@degruyterbrill.com